THE PRESENCE OF THE PAST

Rupert Sheldrake studied natural sciences at Cambridge and philosophy at Harvard. He took a Ph.D. in biochemistry at Cambridge and in 1967 became a Fellow of Clare College, where he was Director of Studies in biochemistry and cell biology. As a Research Fellow of the Royal Society, he carried out research on the development of plants and the ageing of cells. From 1974 to 1978 he worked on the physiology of tropical legume crops at the International Crops Research Institute for the Semi-Arid Tropics, in Hyderabad, India, where he has continued to act as Consultant Physiologist. He is married and lives in London.

By the same author

The Presence of the Past

MORPHIC RESONANCE
AND THE HABITS OF NATURE

RUPERT SHELDRAKE

HarperCollins*Publishers*

HarperCollins*Publishers*
77–85 Fulham Palace Road,
Hammersmith, London W6 8JB

This paperback edition 1994
3 5 7 9 8 6 4 2

Previously published in paperback by Fontana 1989
Reprinted once

First published in Great Britain by
HarperCollins*Publishers* 1988

Copyright © A. Rupert Sheldrake 1988

Rupert Sheldrake asserts the moral right to
be identified as the author of this work

ISBN 0 00 637466 2

Set in Bembo

Printed in Great Britain by
Hartnolls Ltd., Bodmin, Cornwall

To all my teachers,
past and present

PREFACE

This book carries further the ideas I expressed in *A New Science of Life*, first published in 1981, in which I put forward the far-ranging hypothesis of formative causation and discussed some of its many implications, especially in the chemical and biological realms.

In this book, which is less technical in style, I place the hypothesis of formative causation in its broad historical, philosophical, and scientific contexts, summarize its main chemical and biological implications, and explore its consequences in the realms of psychology, society, and culture. I show how it points towards a new and radically evolutionary understanding of ourselves and the world we live in, an understanding which I believe is in harmony with the modern idea that all nature is evolutionary.

The hypothesis of formative causation proposes that memory is inherent in nature. In doing so, it conflicts with a number of orthodox scientific theories. These theories grew up in the context of the pre-evolutionary cosmology, predominant until the 1960s, in which both nature and the laws of nature were believed to be eternal. Throughout this book, I contrast the interpretations provided by the hypothesis of formative causation with the conventional scientific interpretations, and show how these approaches can be tested against each other by a wide variety of experiments.

In 1982, the Tarrytown Group of New York announced an international competition, with substantial monetary prizes, for the experimental testing of the hypothesis of formative causation—tests which could, of course, either support the hypothesis or go against it. At the same time, the British magazine *New Scientist* mounted a competition for new *designs* for such experimental tests. The winning designs, selected by a panel of British scientists, were published in *New Scientist* in April 1983, and indeed proved to be a fruitful stimulus to research, by both myself and others. The results of the Tarrytown competition were announced and the prizes were awarded in June 1986. The winning experiments, selected by an international panel of scientists, are summarized in chapter 10.

These competitions have done much to put the testing of the hypothesis of formative causation into the public domain, and for this I am very grateful to Robert L. Schwartz for his imaginative conception of the Tarrytown competition, for organizing it, and for providing the first prize; and to the Tiger Trust, of Holland, and to Meyster Verlag, of Munich, the publishers of the German translation of *A New Science of Life,* for providing the second and third prizes. I am likewise grateful to the *New Scientist,* and in particular to Michael Kenward, the editor, and Colin Tudge, then the features editor, for running their competition; and to the scientists (named in chapter 10) who served as judges in these competitions.

I have been privileged to discuss the idea of formative causation at a variety of seminars and conferences in universities and other institutions in Austria, Britain, Canada, Finland, France, Germany, Holland, India, Sweden, Switzerland, and the United States. I would like to express my gratitude to all those who made these meetings possible, and to more people than I can possibly name for their criticisms, comments, suggestions, questioning, information, enthusiasm, and encouragement. I am especially grateful to four groups in the United States which have provided contexts for stimulating discussions, renewed and continued over a period of years: in New York, the International Center for Integrative Studies; and in California, the Esalen Institute, the Institute of Noetic Sciences, and the Ojai Foundation.

This book has greatly benefited directly and indirectly from these various discussions, from correspondence with people all over the world, and from many conversations and arguments with friends and colleagues. I would like to thank all those who have so generously shared their knowledge, experience, and insight. I am particularly grateful to Ralph Abraham, David Abram, Andra Akers, Patrick Bateson, John Beloff, Anthony Blond, David Bohm, Richard Braithwaite, John Brockman, Keith Campbell, Fritjof Capra, Jennifer Chambers, Jeremy Cherfas, Christopher Clarke, Isabel Clarke, Virginia Coyle, Tom Creighton, Paul Davies, Larry Dossey, Lindy Dufferin and Ava, Dorothy Emmet, Susan Fassberg, Marilyn Ferguson, Jim Garrison, Alan Gauld, Adele Getty, Elmar Gruber, Brian Goodwin, George Greer, David Griffin, Bede Griffiths, Stanislav Grof, Lola Hardwick, David Hart, Nick Herbert, Rainer Hertel, Joan Halifax, Jean Houston, Caroline Humphrey, Nicholas Humphrey, Tim Hunt, Francis Huxley, Brian Inglis, Brother John, Colleen Kelley, Arnold Keyserling, Stanley Krippner, the late J. Krishnamurti, Peter Lawrence, David Lorimer, the late Margeret Masterman, Terence McKenna, Noel McInnis, Ralph Metzner, John Michell, Joan Miller, Michael Murphy, Tom Myers, Claudio Naranjo, Jim Nollman, the late Frank O'Meara, Brendan O'Reagan, Robert Ott, the late Michael Ovenden, Alan Pickering, Nigel Pennick, Jeremy Prynne, Anthony Ramsay, Martin Rees, Micky Remann, Keith Roberts, Steven Rose, Janis Roze, Peter Russell, Gary Schwartz, Robert L. Schwartz, Irene Seeland, John Steele, Ian Stevenson, Dennis Stillings, Harley Swiftdeer, Jeremy Tarcher, George Tracy, Patrice van Eersel, Francisco Varela, Melanie Ward, Lyall Watson, Renée Weber, Christopher Whitmont, George Wickman, Ion Will, Roger Williams, Arthur Young, and Connie Zweig.

Over twenty people have been kind enough to read drafts of this book, either in whole, or parts relevant to their own special field. The rewriting of this book has benefited greatly from their comments, criticisms, and advice. In particular, I thank Christopher Clarke, Paul Davies, Peter Fry, Brian Goodwin, Bede Griffiths, David Hart, Anthony Laude, my wife Jill Purce, Anthony Ramsay, Steven Rose—and especially Nicholas Humphrey, who has been a continual source of good sense and illuminating intelligence throughout the three-year-long process of writing and rewriting. The final version has also been considerably improved through many helpful suggestions from my editors: Helen Fraser of Collins, in London; and Hugh O'Neill of Times Books, in New York.

The writing of this book has involved an exciting journey of exploration. My companion throughout this entire journey has been my wife, Jill, and I am deeply thankful to her for her unfailing inspiration and encouragement, for countless conversations in which ideas in this book took shape, and for the help she has given me in so many ways.

I thank Keith Roberts, Jeni Fox, and Craig Robson for doing the drawings and diagrams.

I am grateful to Melanie Ward for typing the various drafts of this book, and for her secretarial help.

Finally, I would like to acknowledge all the plants and animals from which I have learned, and in particular the animal I know best, our cat, Remedy.

CONTENTS

Contents

INTRODUCTION

The Habits of Nature

"They say that habit is second nature. Who knows but nature is only first habit?"
—Blaise Pascal, *Pensées*

This book explores the possibility that memory is inherent in nature. It suggests that natural systems, such as termite colonies, or pigeons, or orchid plants, or insulin molecules, inherit a collective memory from all previous things of their kind, however far away they were and however long ago they existed. Because of this cumulative memory, through repetition the nature of things becomes increasingly habitual. Things are as they are because they were as they were.

Thus habits may be inherent in the nature of all living organisms, in the nature of crystals, molecules, and atoms, and indeed in the entire cosmos.

A beech seedling, for example, as it grows into a tree takes up the characteristic shape, structure, and habits of a beech. It is able to do so because it inherits its nature from previous beeches; but this inheritance is not just a matter of chemical genes. It depends also on the transmission of habits of growth and development from countless beech trees that existed in the past.

Likewise, as a swallow grows up, it flies, feeds, preens, migrates, mates, and nests as swallows habitually do. It inherits the instincts of its species through invisible influences, acting at a distance, that make the behaviour of past swallows in some sense present within it. It draws on and is shaped by the collective memory of its species.

All humans too draw upon a collective memory, to which all in turn contribute.

If this view of nature is even approximately correct, it should be possible to observe the progressive establishment of new habits as they spread within a species.

For example, when birds such as blue tits learn a new habit, such as stealing milk from milk bottles by tearing off the bottle caps, then blue tits elsewhere, even beyond the range of all normal means of communication, should show an increasing tendency to learn the same thing.

When people learn something new, such as wind-surfing, then as more people learn to do it, it should tend to become progressively easier to learn, just because so many other people have learned to do it already.

When crystals of a newly synthesized chemical substance, for example a new kind of drug, arise for the first time they have no exact precedent; but as the same compound is crystallized again and again, the crystals should tend to form more readily all over the world, just because they have already formed somewhere else.

In the same way that this inheritance of habits may depend on direct influences from previous similar things in the past, so the memory of individual organisms may depend on direct influences from their *own* past. If memory is inherent in the nature of things, then the inheritance of collective habits and the development of individual habits, the development of the individual's "second nature," can be seen as different aspects of the same fundamental process, the process whereby the past in some sense becomes present on the basis of similarity.

Thus, for example, our own personal habits may depend on cumulative influences from our past behaviour to which we "tune in." If so, there is no need for them to be stored in a material form within our nervous systems. The same applies to our conscious memories—of a song we know, or of something that happened last year. The past may in some sense become present to us directly. Our memories may not be stored inside our brains, as we usually assume they must be.

All these possibilities can be conceived of in the framework of a scientific hypothesis, which I call the hypothesis of formative causation. According to this hypothesis, the nature of things depends on fields, called morphic fields. Each kind of natural system has its own kind of field: there is an insulin field, a beech field, a swallow field, and so on. Such fields shape all the different kinds of atoms, molecules, crystals, living organisms, societies, customs, and habits of mind.

Morphic fields, like the known fields of physics, are non-material regions of influence extending in space and continuing in time. They are localized within and around the systems they organize. When any particular organized system ceases to exist, as when an atom splits, a snowflake melts, an animal dies, its organizing field disappears from that place. But in another sense, morphic fields do not disappear: they are potential organizing patterns of influence, and can appear again physically in other times and places, wherever and whenever the physical conditions are appropriate. When they do so they contain within themselves a memory of their previous physical existences.

The process by which the past becomes present within morphic fields is called morphic resonance. Morphic resonance involves the transmission of formative causal influences through both space and time. The memory within the morphic fields is cumulative, and that is why all sorts of things become increasingly habitual through repetition. When such repetition has occurred on an astronomical scale over billions of years, as it has in the case of many kinds of atoms, molecules, and crystals, the nature of these things has become so deeply habitual that it is effectively changeless, or seemingly eternal.

All this obviously contrasts with currently orthodox theories. There is no such thing in contemporary physics, chemistry, or biology as morphic resonance; and the known fields of physics are generally assumed to be governed by eternal laws of nature. By contrast, morphic fields arise and evolve in time and space, and are influenced by what has actually happened in the world. Morphic fields are conceived of in an evolutionary spirit, but the known fields of physics are not. Or at least, until quite recently they were not.

Until the 1960s, the universe was generally believed by physicists to be eternal; so were the properties of matter and of fields; so were the laws of nature. They always had been and always would be the same. But the universe is now thought to have been born in a primordial explosion some fifteen billion years ago and to have been growing and evolving ever since.

Now, in the 1980s, theoretical physics is in ferment. Theories are reaching back into the first moments of creation. Entirely new, evolutionary conceptions of matter and of fields are coming into being.

The cosmos now seems more like a growing and developing organism than like an eternal machine. In this context, habits may be more natural than immutable laws.

This is the possibility that this book explores. But before beginning this exploration, it is helpful to consider in more detail the habitual assumptions we make about the nature of things. The hypothesis of formative causation conflicts with a number of scientific theories which have been orthodox for decades, or even for centuries, so it is important to be aware of what these theories are and how they have developed and to take account of their successes and limitations.

At various stages throughout this book, the interpretations of phenomena in terms of the orthodox theories are compared with those in terms of the hypothesis of formative causation. This comparison enables the alternative approaches to be understood more clearly, and it also enables us to see where they make different predictions that can be tested by experiment. By means of such tests, it should be possible to find out which approach is in better accordance with the world we live in.

The Plan of This Book

Any new way of thinking has to come into being in the context of existing habits of thought. The realm of science is no exception. At any given time, the generally accepted models of reality, often called paradigms, embody assumptions that are more or less taken for granted and which easily become habitual.

In the first three chapters, we examine the two predominant models of reality in contemporary science: on the one hand the idea that physical reality is constant and entirely governed by eternal laws, and on the other hand the idea that nature is evolutionary.

In chapter 1, we consider the way these two models of reality have coexisted for over a century, and how they are now in conflict as a result of the recent revolution in cosmology. All nature is now thought to be evolutionary, and consequently the assumption of eternal laws of nature is thrown into question. Rather than being governed by eternal laws, the nature of things may be habitual. This possibility was already being considered by philosophers and biologists towards the end of the last century, but it was ruled out by the orthodox assumption of an eternal physical reality that was essentially constant.

In chapter 2, we examine the history of the idea of the eternity of nature. It is rooted in mystical intuition, and came down to modern science through traditions of thought inherited from classical Greece. The theoretical eternities of physics have evolved from ancient, pre-evolutionary conceptions of reality, and are now at variance with the new evolutionary cosmology.

In chapter 3 we look at the evolution of the idea of evolution. Its historical roots can be found in Christians' faith in the progressive movement of human history towards the fulfilment of God's purposes. From this belief, in seventeenth-century Europe a new vision of human progress began to develop: a faith in the transformation of the world for the benefit of humanity through progress in science and technology. This conviction was continually reinforced by the advances of science, industry, medicine, and agriculture and has by now become predominant on a global scale. In the course of the nineteenth century, the progress of humanity came to be seen in a much wider context: it became one aspect of a great evolutionary process which had given rise to all forms of life on earth. Finally, in the new cosmology the idea of evolution has been taken to its ultimate limits: the whole universe is evolutionary.

As a result, we can no longer take the eternal laws of nature for granted. But if we think of them as habitual, we find ourselves in conflict with the conventional assumptions of physics, chemistry, and biology, which were formulated in the context of an eternal mechanistic universe. In chapter 4, we consider the nature of atoms, molecules, crystals, plants, and animals. They are all complex structures of activity which come into being spontaneously. Why do they have the structures they do? How are they organized? How do complex living organisms such as trees develop from much simpler structures such as seeds? We look at the orthodox answers to these questions and at the assumptions they embody, and in chapter 5 we see that the coming into being of living organisms—the growth of a fly, for example, from a fertilized egg—still remains mysterious, despite the many impressive discoveries of twentieth-century biology. In contemporary biology, one of the most promising ways of thinking about the development of living organisms is in terms of organizing fields, called morphogenetic fields. However, the nature of these fields has itself remained mysterious.

In chapter 6 we discuss the nature of these fields, and the interpretation of them provided by the hypothesis of formative causation; and in chapter 7 we see how this hypothesis applies to the development of molecules and crystals as well as living organisms. The morphic fields of all these systems can be thought of as containing an inherent memory, due to morphic resonance from all previous similar systems.

In chapter 8 we consider the new interpretation of biological heredity that this hypothesis provides and look at ways in which it could be tested experimentally.

The next four chapters are concerned with memory, learning, and habit in animals and human beings. The idea of morphic resonance enables memory to be understood in terms of direct causal influences from an organism's own past. This therefore provides a radical alternative to the conventional theory that habits and memories are somehow stored as material "traces" within the nervous system. This way of looking at the phenomena is unfamiliar, but it seems to be more consistent with the available evidence than the conventional theory. It leads to a range of empirically testable predictions, and I describe some experiments that have already been done to test it.

In chapter 13 the concept of morphic fields is extended to the organized societies of social animals, such as termite colonies and flocks of birds, and chapter 14 considers the structures of human societies and cultures in the light of this idea. In chapter 15, I suggest that the concept of morphic resonance could provide a new interpretation of rituals, customs, and traditions, including the traditions of science.

The evolution of morphic fields by natural selection and the role of morphic resonance in the evolutionary process is discussed in chapter 16, and in chapter 17 the nature of morphic fields is considered in relation to the new evolutionary theories of physics. Chapter 18 addresses the question of evolutionary creativity: What are the possible sources of new patterns of organization? How do new morphic fields arise in the first place?

I have tried to keep technical terminology to a minimum, but the use of some specialized scientific and philosophical terms is unavoidable. These terms are explained as the book goes along, and I hope their meanings will become clear even if they are unfamiliar to start with. There is also a glossary at the end of the book which summarizes what these words and phrases are taken to mean.

CHAPTER 1

Eternity and Evolution

Evolution in an Eternal World

We inherited a dual vision of the world from nineteenth-century science: on the one hand a great evolutionary process on earth, and on the other, the physical eternity of a mechanistic universe. All the matter and the energy in the cosmos were believed to be eternal, everything governed by eternal laws of nature.

From this dual perspective, life evolved on earth within a physical eternity. The evolution of life made no difference to the fundamental realities of the physical universe. Nor would the extinction of life on earth. The total amount of matter and energy and electric charge would remain exactly the same, and so would all the laws of nature. Life evolves, but fundamental physical reality does not.

This double world view has become deeply habitual, and in many ways it continues to shape scientific thinking. In this chapter we examine this conventional world view in more detail, and see where it has already begun to be transcended. What is emerging in its place is an evolutionary vision of reality at every level: subatomic, atomic, chemical, biological, social, ecological, cultural, mental, economic, astronomical, and cosmic.

Physical Eternity

The mechanical universe that we inherited from nineteenth-century physics was eternal. It was a vast machine governed by eternal laws.

The world machine of physics started life in the seventeenth century. To begin with, it was assumed to have been made by God, set in motion by his will, and to run inexorably in accordance with his immutable laws. Nevertheless, for the first century of its existence, the Newtonian world machine had a persistent tendency to run down. From time to time the celestial clockwork had to be wound up again by God.

But by the beginning of the nineteenth century, the theoretical machinery was perfected and the world became a perpetual motion machine. The machinery was eternal, and it would always go on, as it always had done, in an entirely deterministic and predictable way: or at least in a way that would in principle be entirely predictable by a superhuman all-knowing intelligence, if such an intelligence existed.

For the great French physicist Pierre Laplace and for many subsequent scientists, God was no longer needed to wind things up or start things off. He became an unnecessary hypothesis. His universal laws remained, but no longer as ideas in his eternal mind. They had no ultimate reason for existing; they were purposeless. Everything, even physicists, became inanimate matter moving in accordance with these blind laws.

Towards the end of the nineteenth century, the world machine started to run down again. It could not be a perpetual motion machine because, according to the laws of thermodynamics, perpetual motion machines are impossible. The universe must be running down towards a final heat death, a state of thermodynamic equilibrium in which the machinery would stop working, never to start again. The machine would run out of steam, and a God who had become an unnecessary hypothesis could hardly be expected to stoke it up again. Nevertheless, all the matter and energy of the world would endure forever and ever; the remnants of the exhausted machinery would never decay.

The revolutions in twentieth-century physics have in a variety of ways transcended the old mechanistic metaphors.[1] The indestructible billiard-ball atoms have become complex systems of vibrating and orbiting particles, which are themselves complex structures of activity. The rigorous determinism of classical mechanistic theory has softened into a science of probabilities. And spontaneity has emerged in everything. Even the vacuum has ceased to be an empty void; it has become a seething ocean of energy, producing countless vibrating particles all the time and taking them back again. "A vacuum is not inert and featureless, but alive with throbbing energy and vitality."[2]

The world machine of matter in motion has been transformed by relativity and quantum physics into a cosmic system of fields and energy. As Einstein conceived of it, the universe exists eternally within the universal field of gravitation. He did not conclude that the universe was essentially constant because of his equations. Rather, he adjusted his equations to endow the universe with an eternal stability:

> When Einstein first applied his field equations of general relativity to the cosmological problem he discovered that static solutions were impossible. Since there was at that time no observational evidence to suggest that the Universe was in a non-static state and the philosophical prejudices of centuries underpinned the notion of a changeless background universe, Einstein altered his field equations to include the cosmological constant, Λ. The Einstein equations with cosmological constant have a static cosmological solution: the Einstein static universe.[3]

Static models of the universe remained orthodox until the the 1960s, and many of the habits of thought engendered by the idea of a physical eternity still persist with great power.

Evolution

We also inherited from nineteenth-century science a great evolutionary vision, very different in spirit from the eternal universe of physics. All the many kinds of living organism, centipedes, dolphins, bamboos, sparrows, and millions of other forms of life, have come into being through a vast creative process. The evolutionary tree has been growing and branching spontaneously for well over three billion years. We ourselves are products of evolution, and evolution continues at an ever-accelerating pace in the realm of humanity. Societies and cultures evolve, civilizations evolve, economies evolve, and science and technology evolve.

We experience the evolutionary process directly in our own lives: the world around us is changing as it has never changed before. Stretching back behind the changes that we ourselves have seen is the evolution of modern civilization, itself rooted in earlier civilizations and more primitive forms of society. Beyond these is a long, mysterious period of prehistoric humanity; further back still, our apelike ancestors; beyond them, more primitive mammals, then reptiles, then fish, then primitive vertebrates, then perhaps some sort of worm, right back to single cells, to microbes, and ultimately to the first living cells on earth. Beyond these we go back into a chemical realm of molecules and crystals, and finally to atoms and subatomic particles. This is our evolutionary lineage.

In the course of our growing up and education, most of us as modern people have implicitly or explicitly accepted both models of reality: a physical eternity and an evolutionary process. Within the sciences, both models coexisted peacefully until quite recently. They were kept safely apart. Evolution was kept down to earth, whereas the heavens were eternal. Terrestrial evolution is the province of geology, biology, psychology, and the social sciences. The celestial realm is the province of physics, as are energy, fields, and the fundamental particles of matter.

Charles Darwin and biologists who followed him had to try to fit the evolutionary tree of life into a mechanical universe which did not itself evolve and which, if anything, was running down. The world machine had no ultimate purpose, and no such things as purposes could be admitted within it. From the mechanistic point of view, living organisms are complex machines, inanimate and purposeless. The Darwinian doctrine is that the evolution of living organisms in no sense involves a process of purposive striving, nor is it divinely designed or guided; rather, organisms vary by chance, their offspring tend to inherit their variations, and through the blind workings of natural selection, the various forms of life evolve with no design or purpose, either conscious or unconscious. Eyes and wings, mango trees and weaver birds, ant and termite colonies, the echo-location system of bats, and indeed all aspects of life have come into being by chance, through the mechanistic operation of inanimate forces and by the power of natural selection.

The Darwinian theory of evolution has always been controversial, and remains so today. Some people still deny that evolution has happened at all; but others who have accepted the reality of the evolutionary process have gone much further than Darwinism does; they have seen the evolutionary process not just as a local and temporary affair on earth within an eternal world machine, but as part of a universal evolutionary process.

Philosophies of universal evolution, such as the theories of general progress so popular in Victorian England, conflicted with the universe according to physics. So did evolutionary visions, such as that of Teilhard de Chardin,[4] who saw the evolutionary process being drawn towards an end or goal, an inconceivable state of final unity. From the point of view of mechanistic science, such philosophies and visions have generally been regarded as illusory: the evolution of life on earth is not part of a cosmic evolutionary process that is leading somewhere; it is a kind of local fluctuation within a mechanistic universe that has no purpose at all.

We are all familiar with this point of view, which has had a deep and pervasive influence on twentieth-century thought. This is how Bertrand Russell expressed it in the context of the devolving world machine:

> That man is the product of causes which had no prevision of the end they were achieving; that his origin, his growth, his hopes and fears, his loves and beliefs, are but the outcome of accidental collisions of atoms; that no fire, no heroism, no intensity of thought and feeling,

can preserve an individual life beyond the grave; that all the labours of the ages, all the devotion, all the inspiration, all the noonday brightness of human genius, are destined to extinction in the vast death of the solar system; and that the whole temple of Man's achievement must inevitably be buried beneath the debris of a universe in ruins—all these things, if not quite beyond dispute, are yet so nearly certain, that no philosophy which rejects them can hope to stand. Only within the scaffolding of these truths, only on the firm foundation of unyielding despair, can the soul's habitation henceforth be built.[5]

This cheerless prospect has indeed seemed inevitable to many modern people, and the replacement of the devolving world machine by an Einsteinian static universe made little difference to this pessimistic outlook. The mechanistic theory is more than just a scientific theory: it has been taken to be a dreadful truth that no rational person can deny, whatever existential anguish it may cause. In this austere faith the molecular biologist Jacques Monod proclaimed:

> Man must at last wake out of his millenary dream and discover his total solitude, his fundamental isolation. He must realize that, like a gypsy, he lives on the boundary of an alien world; a world that is deaf to his music, and as indifferent to his hopes as it is to his sufferings and his crimes.[6]

But scientific theories are subject to change, and in the 1960s the theoretical universe of physics broke out of its eternity. It looks no longer like an eternal machine, but more like a developing organism. Everything is evolutionary in nature. The evolution of life on earth and the development of humanity are no longer a local fluctuation in an eternal physical reality; they are aspects of a cosmic evolutionary process. A variety of philosophers and visionaries have been saying this for years, but now this is orthodox physics as well.[7]

The Evolutionary Universe

Most cosmologists now believe that the universe began in a primordial explosion some fifteen billion years ago and that it has been growing ever since. This expansion is thought to be caused not by a sort of cosmic repulsion, but rather by the Big Bang itself. The speed at which the galaxies are rushing apart is gradually declining under the influence of gravitation. If the matter density of the universe is sufficiently low, the expansion will continue forever. But if there is more than a certain amount of matter in the universe, the expansion will stop, and the universe will begin to contract, ultimately resulting in a reversal of the Big Bang in a terminal implosion called the Big Crunch. Most physicists seem to favour continued expansion; but some prefer the Big Crunch, and find in it a way to return to a repetitive eternity: for the Big Crunch could be the Big Bang of the next universe, and so on forever.

However, even if we assume for the purpose of argument that our universe is one in an endless series, we could never know whether they all develop in exactly the same way or evolve differently each time. All we *can* know about is the evolution of the universe we live in.

Opinions differ as to what happened in the first 10^{-30} second, but according to the currently popular "inflationary" model, the universe had a very brief period of extraordinarily rapid expansion during which all the matter and energy in the universe were created from virtually nothing.[8] After this, the inflationary model coincides with what is now called the "standard" Big Bang model.

About a hundredth of a second after the beginning, when the universe had cooled to a mere hundred billion degrees, it consisted of an undifferentiated soup of matter and radiation. Within three minutes, the neutrons and protons began to combine into helium nuclei. Within thirty minutes, most of them were combined in this way, or remained as free protons, hydrogen nuclei.[9]

After a further 700,000 years of expansion and cooling, the temperature dropped low enough that electrons and nuclei could form stable atoms. The lack of free electrons then made the universe transparent to radiation, and the "decoupling" of matter and radiation allowed galaxies and stars to begin to form.

The evolution of matter continued within the stars, where nuclear reactions produced the many chemical elements that are found in dust clouds between the stars, in comets, meteors, and planets. Such elements are thought to be formed with particular intensity when stars explode as supernovae. In the cold conditions of interstellar space the formation of molecules becomes possible; and in cool aggregates of matter, for instance planets, a great variety of crystals come into being, such as those that make up the rocks of the earth.

In this sequence, the one, the primordial "singularity," becomes many, as ever more complex forms differentiate within the universe as it grows.

This image is far removed from the constant mechanical universe of classical physics. The evolutionary conception is now being extended to everything, even to the fundamental particles and fields of physics. Here is a recent description by a theoretical physicist, Paul Davies:

In the beginning the universe was a featureless ferment of quantum energy, a state of exceptionally high symmetry. Indeed, the initial state of the universe could well have been the simplest possible. It was only as the universe rapidly expanded and cooled that the familiar structures in the world "froze out" of the primeval furnace. One by one the four fundamental forces separated out from the superforce. Step by step the particles which go to build all the matter in the world acquired their present identities. . . . One might say that the highly ordered and intricate cosmos we see today "congealed" from the structureless uniformity of the big bang. All the fundamental structure around us is a relic or fossil from that initial phase. The more primitive the object, the earlier the epoch at which it was forged in the primeval furnace.[10]

The universe would have developed very differently if the laws and constants of physics had been even slightly different. There is no a priori reason known to physics why they should be as they are. Yet they are as they are, and so life on earth and we ourselves have been able to evolve. The laws of physics have to take into account the fact that physicists exist. This consideration is essential to modern cosmology, and is expressed in the Anthropic Cosmological Principle. The "weak form" of this principle is now widely accepted:[11] "The observed values of all physical and cosmological quantities are not equally probable, but they take on values restricted by the requirement that there exist sites where carbon-based life can evolve and by the requirement that the Universe be old enough for it to have already done so."[12]

Some physicists go further in advocating a "strong form" of the Anthropic Principle: "The Universe must have those properties which allow life to develop within it at some stage in its history."[13]

At first sight this seems tautological, a rather ponderous restatement of an obvious truth. Nevertheless it is intensely controversial because of its implication that the universe may, after all, have a grand purpose and design. Some cosmologists go even further:

Suppose that for some unknown reason the Strong Anthropic Principle is true and that intelligent life must come into existence at some stage in the Universe's history. But if it dies out at our stage of development, long before it has had any measurable non-quantum influence on the Universe in the large, it is hard to see why it *must* have come into existence in the first place. This motivates the following generalization of the Strong Anthropic Principle: *Final Anthropic Principle: Intelligent information-processing must come into existence in the Universe, and, once it comes into existence, it will never die out.*[14]

This is clearly a matter of opinion. But the very existence of such debates among contemporary physicists shows how far modern cosmology has already moved beyond the double world view that has been orthodox for so many years. For generations of scientists, a purposeless physical eternity seemed to be the basis of all reality. But this was not an absolute scientific truth, even though it was often regarded as such; it was just a theory—a theory that has now been superseded by physics itself. Whether the cosmic evolutionary process has any purpose or not, according to the new cosmology life on earth and we ourselves have evolved within an evolving universe.

Do the Laws of Nature Evolve?

Do the laws of nature evolve? Or does physical reality evolve while the laws of nature stay the same? In any case, what do we mean by the "laws of nature"?

Water boils in the same way in Scotland, Thailand, and New Guinea, and everywhere else too. Under given conditions it boils at predictable temperatures—for example at 100°C at standard atmospheric pressure. Sugar crystals form in much the same way under similar conditions all over the world. Chick embryos develop in much the same way everywhere when fertilized hens' eggs are incubated under appropriate conditions. We usually assume that all these things happen because the appropriate materials, under the appropriate physical and chemical conditions, are under the influence of natural laws—laws that are invisible and intangible, but are nevertheless present everywhere and always. There is order in nature; and the order depends on law.

These hypothetical laws of nature are somehow independent of the things they govern. For example, the laws governing the formation of sugar crystals do not just operate only inside and around the growing crystals, but exist outside them. They have an existence that somehow transcends particular times and places. Thus the sugar crystals that are forming today in sugar factories in Cuba are not following local Cuban laws, but rather laws of nature which apply everywhere on earth, and indeed everywhere in the universe. These laws of nature cannot be altered by any laws the government of Cuba may pass, and they are not affected by what people think—not even by what scientists think. Sugar crystals formed perfectly well (as far as we know) before the structure of sugar molecules was worked out by organic chemists and before the structure of their crystals was worked out by crystallographers; indeed these crystals were forming perfectly well before there were any scientists at all. Scientists may have discovered and more or less precisely described the laws governing the formation of these crystals, but the laws have an objective existence quite independent of human beings, and even independent of the actual existence of the crystals themselves. They are eternal. They existed before the first sugar molecules arose anywhere in the universe. Indeed they existed before there was a universe at all—they are eternal realities which transcend time and space altogether.

But wait a minute. How could we possibly *know* that the laws of nature existed before the universe came into being? We could not ever hope to prove it by experiment. This is surely no more than a metaphysical *assumption*. Nevertheless, this assumption is still taken for granted by most scientists, including evolutionary cosmologists, and has been incorporated into the common sense of the modern world. Probably all of us can recognize it in the background of our own thinking.

This assumption became habitual when physical reality was still thought to be eternal, and has persisted in spite of the revolution in cosmology. But then where or what were the laws of nature before the Big Bang?

> The nothingness "before" the creation of the universe is the most complete void that we can imagine—no space, time, or matter existed. It is a world without place, without duration or eternity, without number—it is what mathematicians call "the empty set." Yet this unthinkable void converts itself into a plenum of existence—a necessary consequence of physical laws. Where are these laws written into that void? What "tells" the void that it is pregnant with a possible universe? It would seem that even the void is subject to law, a logic that exists prior to time and space.[15]

This assumption that the laws of nature are eternal is the last great surviving legacy of the old cosmology. We are rarely even conscious of making it. But when we do bring this assumption into awareness, we can see that it is only one of several possibilities. Perhaps all the laws of nature came into being at the very moment of the Big Bang. Or perhaps they arose in stages, and then, having arisen, persisted changelessly thereafter. For example, the laws governing the crystallization of sugar may have come into being when sugar molecules first crystallized somewhere in the universe; they may have been universal and changeless ever since. Or perhaps the laws of nature have actually evolved along with nature itself, and perhaps they are still evolving. Or perhaps they are not laws at all, but more like habits. Maybe the very idea of "laws" is inappropriate.

The concept of laws of nature is metaphorical. It is based on an analogy with human laws, which are binding rules of conduct prescribed by authority and extending throughout the realm of the sovereign power. In the seventeenth century, the metaphor was quite explicit: the laws of nature were framed by God, the Lord of all Creation. His laws were immutable; his writ ran everywhere and always.

Although many people no longer believe in such a God, his universal laws have survived him to this day. But when we pause to consider the nature of these laws, they rapidly become mysterious. They govern matter and motion, but they are not themselves material nor do they move. They cannot be seen or weighed or touched; they lie beyond the realm of sense experience. They are potentially present everywhere and always. They have no physical source or origin. Indeed, even in the absence of God, they still share many of his traditional attributes. They are omnipresent, immutable, universal, and self-subsistent. Nothing can be hidden from them, nor lie beyond their power.

Eternal laws made sense when they were ideas within the mind of God, as they were for the founding fathers of modern science. They still seemed to make sense when they governed an eternal universe from which God's mind had been dissolved. But do they any longer make sense in the context of the Big Bang and an evolving universe?

When we look again at the source of the legal metaphor, to human legal systems, we see at once that *real* laws do indeed develop and evolve. In the English tradition, the common law that governs so much of our lives has grown up over many centuries, rooted in ancestral customs and judicial precedents, continually developing as circumstances change and as new situations arise. And in all countries, new laws are enacted and old ones modified or repealed by the powers that be. Constitutional governments are themselves subject to legal constitutions, which likewise change and evolve. And from time to time, old constitutions are overturned by revolutions and replaced by new ones, drawn up by constitution-makers. We apply this idea to science itself in the metaphor of scientific revolutions. They establish new scientific constitutions, within which scientific laws are framed.

If we are to persist with the legal metaphor, it might be appropriate to suppose that the evolving natural world is governed by a system of natural common law, rather than by a preformed legal system established at the outset, like a universal Napoleonic code.

But then who or what corresponds to the judicial system that establishes the precedents? And who or what framed the constitution of the Big Bang in the first place? And by what power or authority are they maintained? These questions arise inevitably, because they are implicit in the metaphor of law. Laws imply lawgivers, and they are maintained by the power of authority. If we drop the idea that the laws of nature are framed and maintained by God, then we must ask: what makes them up and how are they sustained?

Many philosophers would deny that these questions have any meaning. From the point of view of the empiricist tradition, what we call the laws of nature are in fact human concepts that merely refer to regularities which scientists observe, describe, and model. They have no real, objective existence. They are theories and hypotheses in human minds.[16] So there is no point in asking how they arose as objective realities or by what power they are maintained.

But then what about the observable regularities that these laws refer to? What is the basis of the regularities of nature? They cannot depend on natural laws if these laws are only in human minds. And there is no basis for assuming that these regularities are eternal. The regularities within an evolving universe evolve: this is what evolution means.

The Growth of Habits

If the evolving regularities of nature are not governed by transcendent laws, then could they not be more like habits? Habits develop over time; they depend on what has happened before and on how often it has happened. They are not all given in advance by eternal laws which are quite independent of anything that actually happens—and even independent of the existence of the universe. Habits develop *within* nature; they are not imposed on the world ready-made. Sugar crystals, for example, may form in the way they do now because countless sugar crystals have already formed that way before.

This general possibility—the possibility that the regularities of nature are more like habits than products of transcendent laws—is what this book explores. This exploration takes place in the context of a specific, scientifically testable hypothesis, the hypothesis of formative causation. This hypothesis is described in chapter 6 and subsequent chapters. But the general idea that nature is habitual is by no means new: it has been tried out before, and was indeed widely discussed towards the end of the last century and at the beginning of this one. But the wave of interest in this idea ebbed after World War I. It went out of fashion and sank into obscurity. Why?

The idea of the habits of nature was conceived of in an evolutionary spirit, rather than under the aspect of a theoretical eternity. For example, about a century ago the American philosopher C. S. Peirce pointed out that the idea of fixed and changeless laws imposed upon the universe from the start is inconsistent with a thoroughgoing evolutionary philosophy. Rather, he thought, the "laws of nature" are more like *habits*. The tendency to form habits grows spontaneously, as follows: "Its first germs arose from pure chance. There were slight tendencies to obey rules that had been followed, and these tendencies were rules which were more and more obeyed by their own action."[17]

Peirce considered that "the law of habit is the law of mind" and concluded that the growing cosmos is alive. "Matter is merely mind deadened by the development of habit to the point where the breaking up of these habits is very difficult."[18]

The German philosopher Friedrich Nietzsche, writing around the same time, went so far as to suggest that the "laws of nature" not only evolved, but underwent some sort of natural selection:

> At the beginning of things we may have to assume, as the most general form of existence, a world which was not yet mechanical, which was outside all mechanical laws, although having access to them. Thus the origin of the mechanical world would be a lawless game which would ultimately acquire such consistency as the organic laws seem to have now. . . . All our mechanical laws would not be eternal, but evolved, and would have survived innumerable alternative mechanical laws.[19]

And somewhat later William James wrote in a vein similar to Peirce:

> If . . . one takes the theory of evolution radically, one ought to apply it not only to the rock-strata, the animals and plants, but to the stars, to the chemical elements, and to the laws of nature. There must have been a far-off antiquity, one is then tempted to suppose, when things were really chaotic. Little by little, out of all the haphazard possibilities of that time, a few connected things and habits arose, and the rudiments of regular performance began.[20]

Other philosophers towards the end of the last century and at the beginning of this advocated similar ideas,[21] but then this entire line of thought more or less fizzled out. Physicists held firm to the idea of an eternal universe governed by eternal laws; and indeed this idea gained a new lease of life through Einstein's general theory of relativity. Einstein postulated not a relative but an absolute, eternal universe. Events within this universe were relative to each other; but the background reality was changeless. We should remind ourselves again that it was not until the 1960s that an evolutionary cosmology became predominant in physics.

The idea of habit was also explored in the realm of biology. Living organisms seem to have within themselves a kind of memory. Embryos develop in ways that repeat the development of their ancestors. Animals have instincts that seem to embody ancestral experience. And all animals can learn;

they build up habits of their own. All this was pointed out with admirable clarity over a hundred years ago by Samuel Butler. Memory, he concluded in *Life and Habit*, is the fundamental characteristic of life: "Life is that property of matter whereby it can remember—matter which can remember is living. Matter which cannot remember is dead." Two years later, in *Unconscious Memory*, he went even further: "I can conceive of no matter which is not able to remember a little, and which is not living in respect of what it can remember. I do not see how action of any kind is conceivable without the supposition that every atom retains a memory of certain antecedents."[22]

As embryos develop they pass through stages that recall the embryonic forms of remote ancestral types; in some way the development of an individual organism seems to be related to the entire evolutionary process that gave rise to it. Human embryos, for example, pass through a fishlike stage with gill slits (Fig. 1.1). Butler saw in this a manifestation of the organism's memory of its own past history. "The small, structureless, impregnate ovum from which we have each of us sprung, has a potential recollection of all that has happened to each one of its ancestors."[23]

Such ideas were widely discussed by biologists until about the 1920s,[24] and the theory that "heredity is a form of unconscious organic memory"[25] was worked out in considerable detail.[26] But by then the development of genetics seemed to have shown that heredity could be explained in terms of genes, made up of complex molecules. The genetic material is now known to consist of DNA. The memory of which Butler and others spoke appeared to be embodied in inanimate matter after all and to be produced mechanistically. The notion of inherited habits of form and behaviour dropped out of biology.

However, as we will see in chapters 4 to 8, in spite of all the successes of genetics, molecular biology, neurophysiology, and so on, biologists have still not managed to explain the development of embryos or the inheritance of instincts in mechanistic terms. Chemical genes and the synthesis of specific proteins certainly have something to do with it; but how does the inheritance of a certain set of chemical genes and the synthesis of certain proteins make swallows, for example, migrate from a certain part of England to southern Africa before the English winter begins, and then make the birds migrate back to the same place in England in the spring? No one knows. No one knows how embryos progressively take up their forms or how instincts are inherited or how habits develop or how memories work. And, of course, the nature of minds is obscure.

In short, all these aspects of life are still mysterious. Many biologists believe that in due course they will cease to be so because they will all be

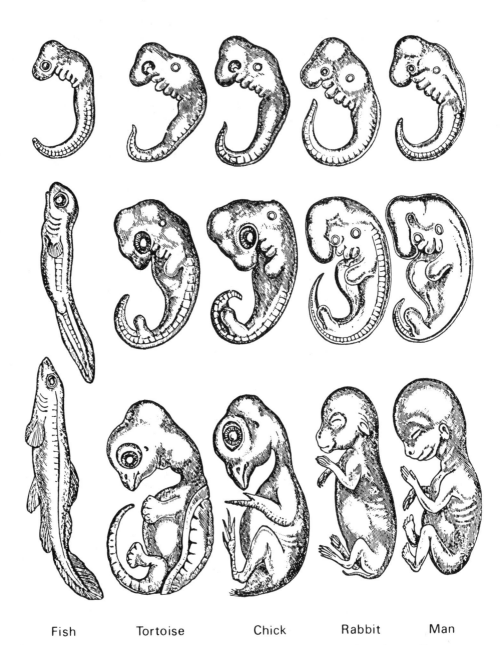

Fish   Tortoise   Chick   Rabbit   Man

Figure 1.1 The embryonic development of five species of vertebrates, illustrating the striking similarities at the early stages of development. Note the embryonic gill slits between the eye and forelimb. (After Haeckel, 1892)

explained mechanistically. That is to say, they will be interpreted in terms of physical and chemical models, and hence will ultimately be accounted for in terms of the eternal properties of matter, fields, and energy. As conventionally conceived of, this process would not involve invoking mysterious non-material memories or fields which evolve in time, but rather would rely on the assumption of eternal laws of nature transcending time and space.

The vision of eternity which has inspired the theories of physics for many centuries remains a powerful force, and to understand why we have to consider its history. We do so in the next chapter; and then in chapter 3 we turn again to the evolutionary vision of reality, a vision which is still growing and extending its scope, and which is proving itself to be more powerful than the vision of a physical eternity—even in the heart of theoretical physics.

CHAPTER 2

Changeless Laws, Permanent Energy

Intuitions of a Timeless Reality

In the context of the new cosmology, all physical reality is evolutionary. But the old idea of eternity lives on in the conception of eternal laws that transcend the physical universe.

If we question this assumption we find that it is very deeply held. But is there any persuasive reason, other than the power of tradition, why we should accept the idea of eternal physical laws? In an evolutionary universe, how can we rule out the possibility that the laws of nature evolve, or that there is memory in nature and that the regularities of nature are habitual?

Even to entertain such notions involves a radical break with tradition. It means contemplating the possibility of a new understanding of the nature of nature. It would involve carrying forward towards completion the change of paradigm which has already gone so far; namely, the change from the idea of physical eternity to an evolutionary conception of the cosmos.

But the power of tradition is strong, often stronger than we are aware of, because so much of its influence is unconscious. If we are to question the assumption of a theoretical eternity, we should be aware of the long traditions that lie behind it. In this chapter we examine its historical development.

The idea of a physical eternity—an eternity of matter in motion governed by eternal laws—has come down to us through mechanistic science, but it is rooted in far older traditions, with origins more mystical than scientific.

The intuition of a timeless state of being, a reality where nothing alters, has been described, in so far as it can be described, by mystics throughout the centuries. For many who have experienced it, this vision of a changeless reality has been so powerful and so self-evidently true that they have concluded that the changing world of everyday experience is somehow less real. The impermanence of things in this world is an appearance or reflection or illusion. Underlying everything is the true reality which neither comes into being nor passes away.

The Pythagoreans

One of the main currents of scientific thought can be traced back to the Greek religious community founded by Pythagoras in the sixth century before Christ. The Pythagoreans were influenced by ideas from the ancient civilizations of Egypt, Persia, and Babylon. They worshipped the god Apollo and followed a variety of mystical practices.

In common with other Greek seekers, they looked beyond the changing world of experience for the divine, which they thought of as that which was without beginning or end. They found this principle in numbers. Numbers were divine and were the changeless principles underlying the changing world of experience. They were at the same time the symbols of ordering, the designators of position, the determiners of spatial extent, and also, through ratio and proportion, the principles of natural law.[1]

Pythagoras himself is said to have made the seminal discovery that musical tones could be understood in terms of mathematical ratios. The properties of stretched strings are such that if the ratio of lengths is 1:2, the tones are an octave apart; the ratio of 3:2 gives the fifth, and 4:3 the fourth. He found that such relationships are not confined to stretched strings, but apply equally to pieces of metal and to flutes. Here were harmonic proportions that could be expressed exactly, understood by reason, and at the same time be *heard*. This discovery provided an astonishing synthesis of quality and quantity—tone and number—which was complemented by the synthesis of arithmetic and geometry, where numerical ratios and proportions could be *seen* and comprehended in geometrical figures. Thus ratio and proportion could be directly experienced through the senses and at the same time be understood as timeless, fundamental principles. The cosmos itself was understood to be a vast harmonic system of ratios. Pythagoras is said to have claimed that he actually heard this cosmic music, the harmony of the spheres, although "not with normal hearing."[2]

Pythagorean mystical experience was not in conflict with reason, but rather in harmony with it: for reason itself was considered above all the ability to experience proportions and ratios. Indeed this insight helped to shape the Greek understanding of the rational—that which is concerned with ratio. Reason came to be regarded as the highest aspect of the soul, that which is not only closest to the divine, but actually participates in the divine nature.

According to Pythagorean cosmology, there were two primordial first principles, *peras* and *apeirion*, which can be roughly translated as Limit and the Unlimited. These primary opposites produced the One through the imposition of limits on the Unlimited. But some of the Unlimited remained outside the cosmos as a void, which the One breathed in to fill up the space between things.[3] From the One, which is both odd and even, proceed numbers. These are the substance of the cosmos, both cause and substrate, modifications and states in the things that exist.

Although the Pythagoreans are often regarded as prototypic natural scientists, they were in fact steeped in a mystical and prescientific experience of the world. In non-literate cultures, numbers are not mere abstract concepts, but mysterious beings with a life of their own. "Each number has its own peculiar character, a kind of mystic atmosphere and 'field of action' peculiar to itself." Pythagoreanism took to an extreme such number mysticism, which is found in one form or another in traditional cultures all over the world.[4]

The Pythagorean vision continues to exert its fascination, neither just because of the rational methods of mathematics nor just because of the successes of mathematical physics: "More important is the feeling that there is a kind of knowing which penetrates to the very core of the universe, which offers truth as something at once beatific and comforting, and presents the human being as cradled in a universal harmony."[5]

This vision has been caught again and again by mathematicians and scientists over the centuries, and has motivated and inspired most of the leading physicists, among them Albert Einstein.[6]

Platonism, Aristotelianism, and the Rise of Western Science

The insights of the Pythagoreans had a major influence on Plato and on the Platonic tradition that followed him. Impressed by the certainty of mathematics, Plato assumed that knowledge must be real, unitary, and unchanging. However, the world is full of a multitude of changing things. Hence these must be in some sense reflections of eternal Forms, Ideas, or essences, which exist outside space and time and independently of any particular manifestations of them in the world of sense experience. The eternal Forms cannot be perceived with the senses, but grasped only by intellectual intuition. This intuition is not reached by mere thinking, but by mystical insight.

Particular things, for example a horse, were said to imitate, participate in, or be made by their Form, in this case the Horse-Idea. This is the essence of what it means to be a horse; it is, in other words, the eternal "horseness." This conception of eternal Ideas remained the central tenet of the Platonic and neo-Platonic tradition; and in Christian neo-Platonism, which developed within the Roman Empire in the first few centuries of the Christian era, the Platonic Forms were taken to be Ideas in the Mind of God.

The other great philosophical tradition inherited by Christendom from the classical world was Aristotelian. Aristotle, a student of Plato, denied the existence of the transcendent Forms; he saw instead the forms of particular kinds of things as inherent in the things themselves. The form of the horse species, for example, exists in particular animals, known as horses, but not in a transcendent Horse-Idea.

Aristotle's philosophy was animistic. He believed that nature was animate and that all living beings had psyches, or souls. These souls were not transcendent, like Plato's Ideas, but immanent in actual living beings. For example, the soul of a beech tree draws the developing seedling towards the mature form of its species, and towards flowering, fruiting, and the setting of seed. The soul of the beech gives the matter of the tree its form and guides its progressive development. Souls contain within themselves the goals of the development and the behaviour of living organisms; they give them their forms and purposes, and are the source of their purposive activity.[7]

In the Aristotelian system, natural processes of change were drawn towards ends or goals which were immanent in nature; nature was alive and was permeated with natural purposes. Even stones had a purpose in falling: they were going home to earth, their proper place.

However, the forms and purposes of things—the ends in which their souls were actualized, to use the Aristotelian terminology—were changeless. Souls did not evolve. Their natures were fixed.

In medieval Europe, a great synthesis of Aristotelian philosophy and Christian theology took place—a synthesis systematically expounded by Thomas Aquinas in the thirteenth century and developed in the medieval Schools. According to this scholastic philosophy, nature was alive, and all the many kinds of living beings had souls. These souls had been created in the first place by God, and had remained the same ever after. Their nature was changeless. By contrast, in the human realm there had been a process of progressive development, revealed in the divinely guided history of the Jews, and above all by the incarnation of God in human form in Jesus Christ. The journey of humanity after the Fall and expulsion from the Garden of Eden towards a new knowledge and experience of God was proclaimed by the prophets of Israel, made evident by God's revealing of himself in human history, and drawn onwards by faith in God's purposes. But only human beings could develop in this way; the souls of plants and animals and other living beings could not. They remained as they were when God first created them, and so they would remain until the end of this world.

This Christianized animistic philosophy became the dominant orthodoxy of the medieval universities, and continued to be taught in the universities of Europe into the seventeenth century and beyond; and indeed it is still taught in a modernized form in many Roman Catholic seminaries.

But at the time of the Renaissance there was a great revival of the Pythagorean and Platonic traditions. The founders of modern science drew their inspiration from these intertwined philosophies, carrying over from them assumptions about eternal Ideas which were built into the foundations of the science they created. They rejected the Aristotelian philosophy.

From Nicholas of Cusa to Galileo

The fifteenth-century mathematician Nicholas of Cusa formed a Pythagorean conception of the world, which had an enduring influence on sixteenth- and seventeenth-century natural philosophy. He saw in the world an infinite harmony in which all things had their mathematical proportions. He considered that "knowledge is always measurement" and that cognition consists in the determination of ratios and therefore cannot be attained without the aid of numbers. He thought that "number is the first model of things in the mind of the creator,"[8] and that all certain knowledge that is possible for man must be mathematical knowledge.[9]

Copernicus shared these opinions, and became convinced that the whole universe is made of number. Hence what is mathematically true is also "really or astronomically true."[10] He made a detailed study of the ancient writings of astronomers of the Pythagorean school and adopted an old idea that had been taught in their tradition: the earth is not at the centre of the cosmos, but rather circles around the sun. According to the theory then orthodox, the earth was a sphere around which the moon, the sun, the planets, and the stars moved in a concentric series of spheres. Copernicus's reasons for preferring everything to move around the sun came from the strong intellectual appeal of this idea, and also from his reverence for the sun:

> Who, in our most beautiful temple, could set this light in another or better place, than that from which it can at once illuminate the world? Not to speak of the fact that not unfittingly do some call it the light of the world, others the soul, still others the governor.[11]

On this assumption he calculated the orbits of the earth and planets, and found that a "more rational" and harmonious geometry of the heavens could be constructed. The intellectual appeal of this theory drew the interest and support of mathematicians, but over sixty years elapsed before Copernicus's theory began to be supported in a more empirical manner.

Kepler was one of those who enthusiastically adopted this mathematical vision. He also had a strong sense of the centrality of the sun, "whose essence is nothing else than the purest light," and regarded it as the first principle and prime mover of the universe. The sun "alone appears, by virtue of his dignity and power, suited for this motive duty and worthy to become the home of God Himself."[12] He found to his delight that the orbits of the planets bore a rough resemblance to the hypothetical spheres which could be inscribed within and circumscribed around the five regular Platonic solids (tetrahedron, octahedron, cube, icosahedron, and dodecahedron—Fig. 2.1).

His third law (that the squares of the periodic times of the planets are proportional to the cube of their mean distance from the sun), published in his *Harmonices Mundi* (1619), was embedded in a lengthy attempt to determine the music of the spheres according to precise laws and to express it in musical notation. But he went further than discovering such mathematical relationships: he believed that the mathematical harmony discovered in the observed facts was the *cause* of these facts, the reason why they are as they are. God created the world in accordance with the principle of perfect numbers; hence the mathematical harmonies in the mind of the creator provide the cause "why the number, the size and the motives of the orbits are as they are and not otherwise."[13]

Kepler believed that the knowledge of things we have through the senses is obscure, confused, and untrustworthy; the only features of the world that can give certain knowledge are its quantitative characteristics: the real world is the mathematical harmony discoverable in things. The changeable qualities that we actually experience are at a lower level of reality; they do not so truly exist. God created the world in accordance with numerical harmonies and that is why he made the human mind in such a way that it could truly know only by means of quantity.

To Galileo also, nature appeared as a simple, orderly system in which everything happened with inexorable necessity; she "acts only through immutable laws which she never transgresses." This necessity followed from her essentially mathematical character:

> Philosophy is written in that great book which ever lies before our eyes—I mean the universe—but we cannot understand it if we do not first learn the language and grasp the symbols in which it is written.

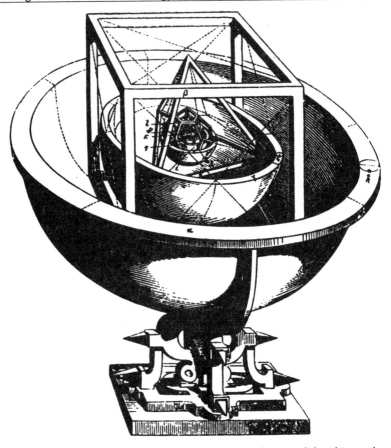

Figure 2.1 Kepler's version of the solar system as one Platonic solid within another, the radii of the intervening concentric spheres corresponding to the orbits of the planets.

> This book is written in the mathematical language, and the symbols are triangles, circles, and other geometrical figures, without whose help it is impossible to comprehend a single word of it; without which one wanders in vain through a dark labyrinth.[14]

This mathematical order is owing to God, who thinks into the world its rigorous mathematical necessity, and who also permits by the mathematical method an absolute certainty of scientific knowledge.

In accordance with these assumptions, Galileo made a clear distinction between that which is absolute, objective, immutable, and mathematical and that which is relative, subjective, and fluctuating. The former is the realm of knowledge, human and divine; the latter the realm of opinion and illusion. The objects we know by means of our senses are not the real or mathematical objects; nevertheless they have certain qualities which, handled by mathematical rules, lead to a true knowledge. These are the real or primary qualities, such as number, magnitude, position, and motion. All other qualities, which are so prominent to the senses, are secondary, subordinate effects of the primary qualities; moreover, they are *subjective*. "These tastes, odours, colours, etc., on the side of the object in which they seem to exist, are nothing else than mere names, but hold their residence solely in the sensitive body; so that if the animal were removed, every such quality would be abolished and annihilated."[15]

This distinction was of great importance in the subsequent development of science and was a major step towards banishing direct human experience from the realm of nature. Until the time of Galileo it had been taken for granted that humanity and nature were both part of a larger whole. But now all those aspects of experience that could not be reduced to mathematical principles were excluded from the objective, external world. Practically the only thing that was left in common between human beings and the mathematical universe was the ability of human beings to comprehend the mathematical order of things.

Descartes and the Mechanical Philosophy

Descartes took this mathematical theory of reality to an extreme which has dominated Western science ever since. On the one hand there was the material universe, extended in mathematical space and entirely governed by mathematical laws. On the other hand there were rational human minds which, like the mind of God, were non-material in nature. They were spiritual substances that were not extended in space.

All plants and animals became inanimate machines, and so did human bodies. Only rational minds were non-mechanical—they were spiritual— and human minds had the Godlike capacity to comprehend the mathematical order of the world. Mathematical knowledge was certain and true.

Descartes had already developed a deep interest in mathematics in his youth, but his faith was established in a visionary experience which was a turning point in his life. When he was living in Neuberg on the Danube, on the eve of St. Martin's Day in 1619 the Angel of Truth appeared to him in a dream and revealed to him that mathematics was the sole key needed to unlock the secrets of nature. He "was filled with enthusiasm, and discovered the foundations of a marvellous science."[16]

In this mathematical science, geometry was the science of resting bodies, and physics the science of moving bodies in mathematical space. The geometric properties of bodies, their form and size, could not account for the fact that they moved; and so Descartes accounted for motion by supposing that God had set the material universe in motion in the beginning, and maintained the same quantity of motion by his "general concourse." Since the creation, the world had therefore been nothing but a vast machine, with no freedom or spontaneity at any point. Everything continued to move mechanically in accordance with the eternal mathematical principles of extended space and the eternal mathematical laws of motion.

This new philosophy of nature was called the mechanical philosophy. Here in a youthful form was the mechanistic world view.[17]

Descartes's mechanical philosophy of nature involved a conscious rejection of the old scholastic orthodoxy still taught in universities. In this Aristotelian tradition, the world was alive; nature was animate and contained within herself her own principle of life and her own ends; all living beings had souls. But in his philosophy Descartes expelled all souls and purposes from nature; only human beings had conscious minds and conscious purposes, because their rational minds, like God, were spiritual and therefore not part of the material world. The human spirit was supposed by Descartes to interact with the human brain in the pineal gland, in a manner that remained unexplained by him or anyone else. The pineal gland has now been replaced by the cerebral cortex as the supposed seat of consciousness, but the problem of "the ghost in the machine" is still with us today.[18]

Everything in nature worked entirely mechanically; in other words, everything was inanimate—except for human minds. Thus Descartes eliminated from the world all such disturbances as life, will, and intentions. Nothing had its own principle of life or its own source of movement: these came from God. And the mathematical laws of nature were God-given metaphysical truths: "The metaphysical truths styled eternal have been established by God, and, like the rest of his creation, depend entirely on him."[19]

The orthodox Christian conception of nature was very different from Descartes's. The world was alive, and the living God had created living beings with souls; he had not created inanimate machines. For Descartes, however, with the world and all living beings within it inanimate, God became the sole living principle of everything, including rational human minds. Descartes was proposing a far more extreme form of monotheism than the orthodox doctrine of the Church. He thought his was a more elevated conception of God, and had a low opinion of conventional ideas. As he said himself: "The majority of men do not think of God as an infinite and incomprehensible being, and as the sole author from whom all things flow; they go no further than the letters of his name. . . . The vulgar almost imagine him as a finite thing."[20]

In the twentieth century, it is easy for us to forget that the mechanistic world view started off with an elevated intellectual conception of God; it involved a new kind of theology as well as a new kind of science. God the all-powerful designer, maker, and motive force of an inanimate world machine is not the God of traditional theology; nor is this idea of God taken seriously by many modern scientists. But the modern conception of eternal physical laws is rooted in this kind of theology, a theology taken further by Newton in his new interpretation of the world machine and its corresponding God.

Atomism and Materialism

So far we have confined our attention to the influence of the Pythagorean-Platonic tradition on the development of science. But seventeenth-century science was heir to another ancient Greek tradition: the philosophy of atomism. The marriage of these two traditions in Newtonian physics was extremely fruitful, and continued harmoniously for over two centuries; today it survives in a modernized form in which the invisible atoms have been replaced by elusive "fundamental particles."

The philosophy of atomism was first propounded in the fifth century before Christ by Leucippus and Democritus. The atomists, like the Pythagoreans and like Plato, were seeking a changeless reality which underlay the changing world. Their starting point was the philosophy of Parmenides, who tried to form an intellectual conception of ultimate changeless being. He concluded that being must be a changeless, undifferentiated sphere. There could only be one changeless thing, not many different things which change. But in fact the world we experience contains many different things that change. Parmenides could only regard this as the result of illusion.

This conclusion was unacceptable to philosophers who followed him, for obvious reasons. They looked for more plausible theories of Absolute Being; the Pythagoreans found it in numbers, and Plato in eternal Ideas. But the atomists found another answer: Absolute Being is not a vast, undifferentiated, changeless sphere, but rather consists of many tiny, undifferentiated, changeless things—material atoms moving in the void. These atoms are permanent: the very word *atom* means that which cannot be split up. Changes are due to the movement and combination and rearrangement of these real but invisible particles. Thus the permanent atoms are the changeless basis of the changing phenomena of the world: matter is Absolute Being.[21]

This is the essence of the philosophy of materialism, which in various forms remains so influential in the modern world. For materialists, unlike

Platonists, there is no such thing as a universal mind, or spirit, or God. Human thoughts are merely an aspect of material changes in the body, and there is no reality other than matter in motion in which they can participate or to which they can refer.

This ancient philosophy was revived in the seventeenth century, and in his great synthesis Isaac Newton brought atomism together with the concept of eternal mathematical laws, producing a dual vision of changelessness—permanent matter in motion governed by permanent non-material laws. A cosmic dualism of physical reality and mathematical laws has been implicit within the scientific world view ever since.

The tradition we have inherited is both materialist and Platonic in spirit. Some scientists (for example many biologists) have emphasized its materialist aspect; others (including many physicists) have emphasized its Platonic aspect; and mechanistic science does indeed have both these aspects. It was born of a marriage between the eternal laws and the mathematical time and space of the Heavenly Father, and the ever-changing physical reality of Mother Nature. The great Mother became the forces of nature and matter in motion;[22] and indeed the word *matter* still carries a dim memory of her, for *mother* and *matter* come from a common Indo-European root. In Latin, these words are *mater* and *materia,* and from *materia* came the English words *material* and *materialism.*

The Newtonian Synthesis

The world machine of Descartes was not made up of atoms in a void; there was no void in his theoretical universe. Seemingly empty space was full of vortices of subtle matter. Each star was at the centre of a huge vortical system, and planets such as the earth were lesser vortical systems swept along by the greater vortex of the solar system. Indeed the entire universe was a vast system of whirlpools of varying size and velocity.

By contrast, Newton's universe was made up of permanent atomic matter moving in the void. Massive bodies such as the earth did not move around the sun because of vortices of subtle matter, but rather because of immaterial forces. The earth and the sun were linked by the attractive force of gravitation, which acted across empty space.

Gravitation was like a magical force in that it involved unseen connections that acted at a distance. Newton spent many years in alchemical research and in the study of ancient doctrines of cosmic intelligences, angelic powers, and the soul of the world. What influence these interests had on his scientific theories is a matter of debate.[23] Nevertheless, his law of universal gravitation involves what would now be called a holistic vision: every particle of matter attracts every other particle; everything is interconnected. But in Newton's opinion, such a force could not arise from the particles of matter themselves; they had no such attractive power. Rather, gravitational force depended on the being of God; it was an expression of his will. Likewise, the absolute mathematical space and time in which all matter existed was none other than an aspect of God, "containing in himself all things as their principle and place."

> He is eternal and infinite, omnipotent and omniscient; that is, his duration reaches from eternity to eternity; his presence from infinity to infinity: he governs all things, and knows all things that are or can be done. . . . He endures forever and is everywhere present; and by existing always and everywhere, he constitutes duration and space. . . . He is all similar, all eye, all ear, all brain, all arm, all power to perceive, to understand, and to act; but in a manner not at all human, in a manner not at all corporeal, in a manner utterly unknown to us.[24]

This aspect of Newton's thought was soon forgotten. The hidden forces permeating the space of the universe were soon attributed to matter itself: they arose from material reality rather than from God. And when God was finally dissolved away from Newton's vision, what was left was a world machine in absolute mathematical space and time, containing inanimate forces and matter, and entirely governed by eternal mathematical laws.

This mechanistic paradigm, supported and enlarged through the experimental methods of science, was successful. It enabled many physical phenomena to be understood in terms of mathematical models; it enabled predictions to be made; and above all it proved to be extremely useful in the control and exploitation of the material world. The growing understanding of nature in mechanistic terms stimulated the development of new technologies, through which material reality could be manipulated ever more effectively for human ends. We see evidence of the power of this paradigm all around us today in the technology that surrounds and sustains our lives.

The Theory of Relativity

Maxwell's unified theory of electromagnetism, developed in the 1860s, enabled electricity, magnetism, and light to be brought within a broad mathematical framework. Physics was expanded, but it was also radically changed, for Maxwell's theory placed in the heart of physics the concept of fields. What exactly *are* fields? Maxwell thought of them as modifications of a subtle medium, the aether. But the failure of experimental attempts to detect the aether led Einstein in his special theory of relativity (1905) to account for electro-magnetic phenomena in terms of fields alone: fields which are non-material in nature.

Einstein revolutionized the Newtonian world view by abandoning the idea that mass, space, and time are absolute quantities; rather, he took the speed of light as absolute. He unified the previously separate conceptions of mass and energy, and showed that both are aspects of the same reality, related through his famous equation $E = mc^2$, where c is the velocity of light. Light itself is non-material; it consists of energetic vibrations moving in the electro-magnetic field.

In his general theory of relativity, Einstein extended the field concept to gravitation, treating gravity as a property of a space-time continuum curved in the vicinity of matter. His equations are based on a four-dimensional geometry that treats time as if it were a spatial dimension: time is therefore essentially spatialized or geometricized.

Far from undermining the mathematical vision of classical physics, this theory can be regarded as its culmination. In it the timeless mathematical principles are primary, and enable all relative movements to be seen within the framework of a universal geometry. In a manner reminiscent of Kepler, Einstein spoke of gravitation as having a "geometrical cause." Also like Kepler, he was strongly imbued with a sense of the mathematical rationality of the universe:

> The individual feels the futility of human desires and aims and the sublimity and marvellous order which reveal themselves both in nature and in the world of thought. Individual existence impresses him as a kind of prison and he wants to experience the universe as a single magnificent whole. . . . What a deep conviction of the rationality of the universe and what a yearning to understand, were it but a feeble reflection of the mind revealed in this world, Kepler and Newton must have had to enable them to spend years of solitary labour in disentangling the principles of celestial mechanics! Those whose acquaintance with scientific research is derived chiefly from its practical results easily develop a completely false notion of the mentality of the men who, surrounded by a sceptical world, have shown the way to kindred spirits scattered wide through the world and the centuries. Only one who has devoted his life to similar ends can have a vivid realization of what has inspired these men and given them the strength to remain true to their purpose in spite of countless failures.[25]

One of the first physicists to grasp Einstein's theory of relativity fully was Arthur Eddington, who led the expedition to photograph the solar eclipse of 1919, which provided the first evidence to favour the theory. He wrote widely about the implications of this theory, and concluded that it pointed to the idea that "the stuff of the world is mind stuff." But "the mind stuff is not spread out in space and time; these are part of the cyclic scheme ultimately derived out of it."[26]

James Jeans, Eddington's contemporary, concluded in a similarly Platonic vein that "the universe can be best pictured, although still very imperfectly and inadequately, as consisting of pure thought, the thought of what, for want of a wider word, we must describe as a mathematical thinker."[27]

Quantum Theory

Quantum mechanics represents a far more radical break with classical physics than does the theory of relativity. One of its most important consequences has been the abandonment of strict determinism: its equations permit predictions only in terms of probabilities. However, in spite of its radical features, it remains a major development of the Pythagorean-Platonic tradition, for it enables the properties of atoms to be understood in terms of numbers, and moreover, harmonic series of numbers: it represents a further step towards the traditional goal of science. Louis de Broglie, one of the founders of quantum mechanics, described this goal as "to succeed in penetrating further into the realm of natural harmonies, to come to have a glimpse of a reflection of the order which rules in the universe, some portions of the deep and hidden realities which constitute it."[28] Quantum theory extends the Platonic approach into the very heart of matter, which Democritus and succeeding atomists had regarded as solid and homogeneous. As Werner Heisenberg stated it:

> On this point modern physics has definitely decided for Plato. For the smallest units of matter are, in fact, not physical objects in the ordinary sense of the word; they are forms, structures, or—in Plato's sense— Ideas, which can be unambiguously spoken of only in the language of mathematics.[29]

Nevertheless, quantum physicists have still proceeded in the spirit of atomism to try to find the ultimate particles of matter. As they have penetrated further into the atom, into its nucleus, and into its nuclear particles, one of the surprises has been that there are so many kinds of quantum particles—over two hundred have been identified so far. Attempts are still being made to fit them into numerical schemes, such as eight- and ten-membered families, which are thought to reflect different permutations and combinations of yet more fundamental components such as quarks (Fig. 2.2). This is the area in which the Pythagorean quest is being pursued at present with most vigour: the attempt to find behind the changing world of experi-

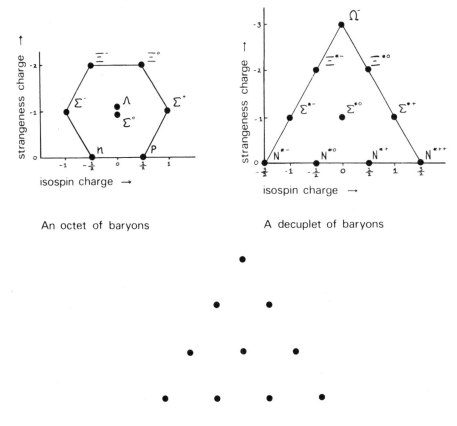

An octet of baryons

A decuplet of baryons

The Pythagorean tetrachtys

Figure 2.2 Two "family groups" of baryons. (After Pagels, 1983) Baryons are elementary particles which have a half-integral spin and take part in strong interactions. Each one contains three quarks, which come in three "flavours": *up, down,* and *strange.* The different kinds of baryons contain characteristic combinations of quarks; for example the proton has two *up* and one *down,* and the neutron has one *up* and two *down.* The octet of baryons is often called the "eightfold way." The decuplet of baryons is arranged in the same way as the tetrachtys, the ancient symbol which lay at the heart of the Pythagorean number wisdom.

ence an eternal mathematical reality, a reality that does not evolve through time and is unaffected by anything that actually happens.

Eternal Energy

As well as eternal laws, both Newtonian and modern physics presuppose other theoretical eternities in the form of physical quantities that are assumed to be conserved in the same total amounts forever.

In Newtonian physics, the atoms of matter were regarded as indestructible; hence the total number of atoms in the universe always remained the same. This concept was expressed in a general form in the law of conservation of matter: matter is neither created nor destroyed.

Historically, the law of conservation of energy was introduced as an expression of the constancy of motion in the universe. The universe keeps going on its own; it does not need to be rewound like a mechanical clock. This law was therefore complementary to the law of conservation of matter: both the substance of the universe and its activity are eternal.

At first, the concept of mass was linked with matter, and both were considered to be conserved together: the mass of every atom is constant, and all atoms are conserved. This straightforward view was thrown into turmoil in the twentieth century when it was found that not only could atoms be split into particles, but some particles could themselves split or fuse; therefore, the total number of particles is *not* conserved. Moreover, the mass of a particle can vary. But order was restored again when it was realized that the mass of a particle or system is simply another manifestation of its energy, or motion. The formula $E = mc^2$ expresses the conversion between these two alternative ways of measuring the same thing. Thus the law of conservation of mass has now been subsumed within an expanded version of the law of conservation of energy.

Thus the total amount of energy in the universe is constant. Neither the coming into being of our galaxy nor the advent of life on Earth has made any difference to the universal energy, which neither increases nor diminishes in its total amount: it is unaffected by anything that actually happens.[30]

The conservation laws mean that physical changes in isolated systems can be represented by means of equations: in spite of all the changes, the total amount of energy, electric charge, and so on, is the same before and afterwards.

A conservation law means that there is a number which you can calculate at one moment, then as nature undergoes its multitude of changes, if you calculate this quantity again at a later time it will be the same as it was before, the number does not change. . . . It comes out the same answer always, no matter what happens.[31]

The equivalence of "before" and "after" in such equations means that changes can occur in either direction: they are, in principle, reversible. Things could go either way; in the world they describe there is no real and irreversible change, in other words, no *becoming*. The fundamental realities of physics, ever conserved, do not evolve; nor are they affected by anything that does in fact develop in time, for example the birth of a star or a new species of insect, or the extinction of either. As Ilya Prigogine has expressed it:

Everything is given in classical physics: change is nothing but a denial of becoming and time is only a parameter, unaffected by the transformation that it describes. The image of a stable world, a world that escapes the process of becoming, has remained until now the very ideal of theoretical physics. . . . Today we know that Newtonian dynamics describes only part of our physical experience. . . . As the scales of very small objects (atoms, "elementary" particles) or of hyperdense objects (such as neutron stars or black holes) are approached, new phenomena occur. To deal with such phenomena, Newtonian dynamics is replaced by quantum mechanics and by relativistic dynamics. However, these new forms of dynamics—by themselves quite revolutionary—have inherited the idea of Newtonian physics: a static universe, a universe of *being* without *becoming*.[32]

The only major physical principle that deals with irreversible change is the second law of thermodynamics, which used to be interpreted to mean that the universe is running down. However, thermodynamics does not challenge the eternity of energy: on the contrary it affirms it. The first law of thermodynamics is in fact none other than a statement of the law of conservation of energy.

The Survival of Eternal Laws

The laws of nature in the form found in scientific textbooks are, of course, man-made. They are continually modified and updated as science progresses. Nevertheless, as this brief history of theoretical physics shows, scientists have generally assumed that they somehow point towards, or reflect, eternal mathematical principles of order. This is, of course, a metaphysical assumption, and has been the subject of debate among philosophers ever since David Hume challenged it in the eighteenth century. However, the enduring prevalence of this assumption has been little affected by such philosophical discussions. It is an integral part of the mechanistic paradigm, and the power of this paradigm has been sustained by the spectacular successes of physics and of the new technologies that have grown out of them.

But over and above the successes of science and technology, the assumption of eternal mathematical realities is sustained by the enduring fascination of the realm of mathematics itself. Mathematical relationships seem to express strangely timeless truths, valid everywhere and forever. These truths are objective, and yet clearly part of the world of thought rather than the world of things. They do indeed appear to be like ideas in a universal mind.

Mathematicians and physicists are naturally far more aware of this mysterious, even mystical, aspect of mathematics than those who have never penetrated into these subjects. Heinrich Hertz, the nineteenth-century physicist who gave his name to our unit of frequency, put it as follows:

> One cannot escape the feeling that these mathematical formulae have an independent existence and an intelligence of their own, that they are wiser than we, wiser even than their discoverers, that we get more out of them than was originally put into them.[33]

Over the present century, the prevailing influence of empiricism and positivism in academic philosophy has made Platonism unfashionable and favours instead a philosophy of mathematics called formalism, according to which much, if not all, of mathematics is merely an intellectual game, without any ultimate meaning. However, the allegiance of mathematicians themselves to formalism is less than wholehearted:

> The majority of writers on the subject seem to agree that most mathematicians, when doing mathematics, are convinced that they are dealing with an objective reality, but then if challenged to give a philosophical account of this reality find it easiest to pretend that they do not believe in it after all. . . . The typical mathematician is both a Platonist and a formalist—a secret Platonist with a formalist mask that he puts on when the occasion calls for it.[34]

Even though energy, fields, and matter are currently thought to have arisen in time as the universe was born and grew, the mathematical laws of nature are still generally assumed to be eternal and to have existed in some sense "before" the cosmos began. Perhaps few scientists make this assumption explicit, but the idea of universal changeless laws is implicit in the very method of science as we know it, and is present in the background of all conventional scientific thinking. This assumption underlies the ideal of scientific repeatability.

Repeatable Experiments

An essential aspect of the scientific method is that observations should be reproducible. Science deals with the regularities of nature, with those aspects of the world that are objective and repetitive. Under the same conditions, the same experiments should always give the same results for any competent experimenter, anywhere in the world, and at any time. Why? Because the laws of nature are the same everywhere and always. Whether we are aware of it or not, this metaphysical assumption underlies the ideal of reproducibility on which the traditional method of science is founded. In the words of Heinz Pagels:

> The universality of physical laws is perhaps their deepest feature—all events, not just some, are subject to the same universal grammar of material creation. This fact is rather surprising, for nothing is less evident in the variety of nature than the existence of universal laws. Only with the development of the experimental method and its interpretive system of thought could the remarkable idea that the variety of nature was a consequence of universal laws be in fact verified.[35]

Karl Popper, a leading philosopher of science, argues that the metaphysical assumption of universal laws is actually *necessary* for science: "Only if we require that explanations shall use universal laws of nature (supplemented by initial conditions) can we make progress to realizing the idea of independent, or non-*ad hoc,* explanations."[36]

Without this requirement, there would be no basis for the principle of objective reproducibility which is so essential for the scientific method. Popper is here simply making explicit what most scientists take for granted.

Then what are these universal laws of nature? Popper proposes that they state "structural properties of the world." In doing so, he fully recognizes an inherent ambiguity: for on the one hand the structures explain the laws, and on the other the laws explain the structures. But he conceives that "at some level, structure and law may become indistinguishable—that the laws *impose* a certain kind of structure on the world, and that they may be interpreted alternatively, as *descriptions* of that structure. This seems to be aimed at, if not yet actually achieved, by the field theories of matter."[37]

However, the fundamental field theories of matter are now in a state of flux, and in contemporary theoretical physics evolutionary conceptions of fields are coming into being. In an evolutionary universe, the "structural properties of the world" evolve. How can we any longer take it for granted that these structural properties are entirely governed by pre-existing laws?

What if they are more like universal habits that have grown up within the growing universe?

To consider the possibility that nature is habitual involves more than just challenging the assumption that everything is governed by transcendent laws that are unaffected by anything that happens: it seems to challenge the basis of the scientific method itself. For if the structural properties of the world change, then how could experiments be reproducible? And how could the idea of repeatability have been so impressively verified by the successes of the scientific method?

A moment's reflection shows that in physics it would probably make little difference in practice if nature were habitual. Entities such as electrons, atoms, stars, fundamental fields, and indeed most of the things studied by physicists are thought to have been around for billions of years. The nature of these kinds of things may long since have been so deeply habitual that they are in effect changeless. They can be modelled by timeless mathematical laws. The idea that their nature is fixed eternally would be an idealization which is for most intents and purposes appropriate. Experiments on them would in general be more or less exactly reproducible. The same would be true of repeatable experiments on most of the systems studied by chemists, geologists, crystallographers, biologists, and other scientists: systems that have existed countless times, over many thousands or millions of years. If nature is habitual, then well-established phenomena will indeed appear to behave as if they are governed by transcendent, changeless laws.

The difference between the two approaches becomes apparent in the case of new phenomena that are not yet well established. An essential feature of the evolutionary process is that *new* organized systems come into being, with patterns of organization that have never existed before: for example the appearance of a new kind of molecule, or crystal, or plant, or instinct, or piece of music. In so far as these things are truly new, they cannot be explained in terms of a simple repetition of what has gone before. They cannot already be habitual, although through repetition they will become so. But from the conventional point of view, everything new is determined by pre-existing laws which have always been in existence. These laws are not altered by anything that happens, and remain the same whether or not the phenomena they govern actually occur in the world.

Thus, from the orthodox standpoint, new kinds of molecules, crystals, organisms, instincts, and ideas are governed by the same unalterable laws on the first occasion they appear or the thousandth or the trillionth.

By contrast, if memory is inherent in the nature of things, the phenomena will not arise in exactly the same way on the first or thousandth or trillionth occasion. Their subsequent appearances will be affected by the very fact that they have existed before. They will be influenced by a cumulative memory of these previous occurrences and will tend to become increasingly habitual. Other things being equal, they will tend to arise everywhere more readily, or with a higher probability, the more often they are repeated.

For example, the crystals of a newly synthesized kind of molecule should tend to form more readily all over the world the more frequently the substance is crystallized. Or when animals such as rats are trained to learn a new trick in one laboratory, other rats of the same breed everywhere else should subsequently tend to learn the same trick more quickly.

There is already evidence that such effects actually occur, and we will return to it in chapter 7 and subsequent chapters. For the purpose of the present discussion, we need consider only the *possibility* that nature is habitual. This possibility means that we can no longer take for granted the conventional assumption that all scientific experiments should in principle be exactly repeatable. For new phenomena would tend to become more probable through repetition, and experimental observations of them would therefore give quantitatively different results as time went on. By the same token, it should be possible to detect the growth of habits by measuring the rate or frequency with which they appear under standardized conditions. If a phenomenon is becoming more habitual, it should tend to arise with a higher probability as it is repeated again and again.

But how could the idea that nature is habitual ever be established scientifically if it undermines the scientific ideal of exact repeatability? At first sight, this appears to involve a paradox: for if nature is habitual, it will not be possible to study the growth of any particular habit over and over again, because the habit will already have grown. However, the growth of habits could be studied again and again with *fresh* kinds of molecule, crystal, behavioural pattern, and so on. The same *kinds* of experiments could be repeated. And through such repeated experimentation, it could be established whether or not there is a general tendency for new natural phenomena to become more deeply habitual the more often they arise.

CHAPTER 3

From Human Progress to Universal Evolution

The double vision we inherited from nineteenth-century science, of evolution on earth within a physical eternity, is rooted in a far older cultural duality. It reflects the double cultural heritage of Europe: on the one hand the intellectual traditions of the civilizations of Greece and Rome, and on the other hand the Christian faith. The eternities of physics are rooted in our Greek heritage, and our faith in progressive development in the religion of the Jews.

In the medieval synthesis of these two traditions, humanity was believed to undergo a progressive historical development through God's revelation of himself in historical events and through faith in God's purposes. But the rest of the world did not progress: the nature of nature was constant.

By the end of the eighteenth century, faith in human progress through the increase of human understanding was widespread; the advance of science itself strongly reinforced this faith, as did the growing industrial revolution. But still the old division held: humanity progressed, but the natural world did not.

As the nineteenth century wore on, a great new vision of progressive development opened up: not only human beings but all living things were seen to have evolved. But still the theory of evolution was kept down to earth.

Now, finally, the entire cosmos is thought to have grown and developed in time: all nature is evolutionary. We can no longer think of nature under the aspect of eternity.

In this chapter, we look at the religious roots of the faith in human progress; at the way the concept of progress led to an evolutionary conception of all life on earth; and at the Darwinian attempt to fit evolution into a mechanistic world. We end by considering the possibility of a new evolutionary synthesis in which the evolution of life is seen as one aspect of the cosmic evolutionary process.

Faith in God's Purposes

The Greek philosophers, like the philosophers of other ancient civilizations, generally thought of time in terms of endlessly repeated cycles: cycles of breathing, of day and night, of the moon, of the year, great astronomical cycles of years, and great cycles of cycles. In some Hindu systems, for example, a great cycle or *mahayuga* lasts for 12,000 years; and beyond this are further cycles, up to the great cycle of Brahma comprising 2,560,000 *mahayugas*.[1]

Almost all the ancient theories of cycles of great time were found in combination with myths of a golden age. The cycle starts with a golden age, followed by successive ages of decadence and degeneration in all things. At the end of the last age of the cycle the world undergoes a general dissolution and is then regenerated. There is a new golden age, and so on in an eternal recurrence.[2]

Consistent with this eternal cyclical vision of things, Hindu and Buddhist philosophies see life itself in terms of repeated cycles of birth, growth, and death, with human lives passing through successive cycles of rebirth. With a similar consistency, the Pythagoreans believed in reincarnation, and so did Plato.

By contrast, in the Judaeo-Christian tradition there is just one process of development in time. The Bible begins with the story of the creation, when "God created the heavens and the earth," and ends with the vision of the new creation in the Book of Revelation: "I saw a new heaven and a new earth, for the first heaven and the first earth had passed away."[3] The whole story of the Bible is thus set within a cosmic vision of creation, destruction, and recreation. But this is not part of a system of eternal recurrences: the new creation of the Book of Revelation is not followed by another stage of dissolution, but is the consummation of all things, in which the whole creation is taken up into the divine life, passing beyond its present state of existence in space and time into its final state of fulfilment.[4] The six days of creation in the Book of Genesis represent the week of time and of earthly activity, while the seventh day is the day of eternity when all labour ceases.

This is the Judaeo-Christian "myth of history."[5] It starts, like many other myths, with a golden age—our first parents in the Garden of Eden in harmony with each other, with the world, and with God. Then comes the Fall, through the eating of the fruit of the tree of knowledge of good and evil, and the expulsion from Eden into a world of toil, suffering, and death. But then a great journey begins towards a new Eden, towards the new country promised by God.

The prototype of this historical process was the journey of the people of Israel out of captivity in Egypt, through the sufferings and the covenant with God in the wilderness, and to the promised land. This metaphor of the journey underlies the concept of progress, of going forward. There can be no going forward unless there is a direction to advance in; and journeys have a direction because they have a destination or purpose or goal.

A belief in progressive development was not absent from the ancient civilizations. Indeed, cities themselves were seen to be an advance over the primitive or barbarous state of man. The evidence was there for all to see in the splendour of the buildings, in the advances in the arts and the skills of artisans, and in the organization of empires.[6] But the development of civilization was set against the myth of decline from the golden age. The future could only hold even further decadence and destruction.

By contrast, in the Judaeo-Christian tradition there was an intense religious faith in the future. As the Epistle to the Hebrews expressed it:

> Faith is the substance of things hoped for, the evidence of things not seen. . . . By faith Noah, being warned of God of things not seen as yet, moved with fear, prepared an ark to the saving of his house. . . . By faith Abraham, when he was called to go out into a place which he should after receive for an inheritance, obeyed; and he went out, not knowing whither he went. . . . These all died in faith, not having received the promises, but having seen them afar off, and were per-suaded of them, and embraced them, and confessed that they were strangers and pilgrims on the earth. For they that say such things declare plainly that they seek a country. And truly, if they had been mindful of that country from whence they came out, they might have had opportunity to have returned. But now they desire a better country, that is, an heavenly: wherefore God is not ashamed to be called their God, for he hath prepared for them a city.[7]

According to one current of Christian faith, based on the authority of the Book of Revelation, after his second coming Christ will establish a messianic kingdom here on Earth and will reign over it for a thousand years, until the Last Judgement. This is the millennium. At intervals throughout Christian history, millenarian groups of believers have sprung up over and over again. Characteristic of millenarian faith is a belief in the imminent coming of the new age here on Earth, not just in some other-worldly heaven, and not just for individual souls. The salvation of the faithful will be collective, and life on Earth will be totally transformed.[8]

A strong millenarian faith in the imminent coming of God's Kingdom filled many of the Puritans in seventeenth-century England. In this faith the Pilgrim Fathers left the old country for the new—a New England in the New World. In England itself, the king was beheaded and the old order overthrown; and it was in this revolutionary atmosphere that an entirely new vision of the coming of the new age on earth began to develop: a transformation of the world through human progress, with science in the vanguard.

Faith in Human Progress

The prophet of this new vision was Francis Bacon. In *New Atlantis,* written in 1624, shortly before his death, the new age of the millenarian faith became a kind of scientific utopia. The advancement of "the whole of mankind" would be achieved through man's dominion over nature through mechanical means. Only scientific knowledge, founded on the empirical method, could further the ambition "to endeavour to establish and extend the power and dominion of the human race itself over the universe," as Bacon put it. In this way, the human race could "recover that right over nature which belongs to it by divine behest."[9]

In Bacon's New Atlantis, progress was placed in the hands of a group of scientists and technicians who studied nature by the experimental method. Nature was to be forced to give up her secrets, so that they might be used to benefit mankind.[10] These scientists and technicians worked in a prototypic scientific research institute called Salomon's House, wore special robes, and were in effect a kind of scientific priesthood.

In England, during the revolutionary regime of the Puritans, such a visionary group of scientists and philosophers began to meet informally. This group, known as the Invisible College, formed the nucleus of the Royal Society, founded in 1660, soon after the restoration of the monarchy. Here, in the "Royal Society of London for Improving Natural Knowledge," was a deliberate realization of Bacon's vision. The Royal Society was Salomon's House. Similar bodies of scientists were likewise officially established in state academies of science throughout the Western world.

The successes of science and the growth of new industries increasingly confirmed the faith in scientific progress, and this faith continually grew and spread—in the eighteenth century throughout Europe and America; in the nineteenth century throughout the empires of the European powers; and in the present century to the remotest corners of the earth. Where Christian missionaries failed, the missionaries of technological progress have succeeded.

This faith has been carried from its homelands in the West in Marxist forms throughout the Soviet Union and China; in capitalist forms to Japan and the Far East; and in various forms to all the nations of the world, which have consequently become "developing countries." And the process of conversion is now being extended to the remotest villages and tribes through education and economic development.

The aspiration for progress helps make development happen. And even for those without education, there is compelling evidence of industrial progress everywhere. Where in the modern world are there no transistor radios, for example, or digital wristwatches? And where in the world had anything like them ever been seen before? They are not just repetitions of things that have always been known; they are truly new. Through science and technology new things are happening in the world that have never happened before.

We may of course wonder whether all such changes are truly progressive. Nevertheless, whether we like it or not, the processes of accelerating change all around us are born out of a *faith* in progress, a faith that is still very strong. But the ideal of the transformation of the world through scientific progress is only one version of millenarianism. We live also within the fields of others.

New England was founded in the seventeenth century by the Pilgrim Fathers in a millenarian spirit. The revolutionary political movements of the late eighteenth century were millenarian: the old order would be overthrown and a new era established—an era of Liberty, Equality, and Fraternity, in the slogan of the French Revolution. And the vision of an entirely new age was built into the foundations of the newly formed United States. It is proclaimed on the Great Seal of the nation: *Novus ordo seclorum,* a new order of ages. It can be seen on every dollar bill.

Communism is another form of messianic faith; and now, as we approach the turn of a millennium, the great millenarian powers of the Soviet Union and the United States confront each other in continuous preparation for an apocalyptic war. In the last days of this age, according to the Book of Revelation, there will be plagues, the casting down of fire, darkness over the earth, a great war in heaven—and much more. This apocalyptic aspect of the Judaeo-Christian vision of history has not gone away: on the contrary, it has taken on a new and dreadful plausibility.

Progressive Evolution

The progress of science has taken place within a larger vision of human progress, which itself developed within the context of a religious faith in God's guidance of history towards the new creation. In the course of the nineteenth century, this vision of progressive development was extended to encompass the whole of life on Earth. The evolution of science paved the way for the science of evolution.

By the end of the eighteenth century, it had become obvious to many Europeans and Americans that human progress and the increasing power of man over nature were taking place through the growth of human understanding and above all through the progress of science. But was this progressive development in accordance with God's purposes, and was it guided by God's will? Many people believed that it was, and many still do. But for the atheists of the Enlightenment, human progress was the achievement of human reason itself. Human reason was the supreme form of consciousness in a mechanistic universe; and human purposes were the only purposes there were. In the course of the French Revolution, the churches of Paris were closed, and the cathedral of Notre-Dame was transformed into a Temple of Reason.

But if human reason was developing, how and why was it doing so? In the early nineteenth century, the philosopher Hegel found an answer in terms of an evolutionary system that describes the dynamic processes of progressive development. Hegel saw the evolution of human thought as an aspect of the process of the Absolute, or in religious language the manifestation of God. It was a rhythmic process of developing wholeness, in which thought progresses dialectically, through contradiction and argument. Each such process starts with an initial proposition, the thesis; this proves to be inadequate, and generates its opposite, the antithesis. This in turn proves inadequate, and the opposites are taken up into a higher synthesis. The synthesis leads to a new thesis, to which a new antithesis springs up, and so on.

Hegel's system proved to be self-fulfilling; to his thesis, Karl Marx opposed the antithesis: not spirit but matter develops dialectically. Dialectical materialism in the tradition of Marx and Engels is a progressive, evolutionary philosophy that regards historical progress as governed by objective, scientific laws. Human progress is just one aspect of the general progressive development of matter, from which mind itself arises.

In the evolutionary philosophy of Herbert Spencer, progress was taken to be not just an objective scientific reality, but the supreme law of the entire universe. Spencer, like Marx, was interested primarily in human progress; his philosophy of universal evolution was a grand generalization that enabled human evolution to be seen as an aspect of a universal process. Spencer and other nineteenth-century philosophers of evolution, such as C. S. Peirce (pp. 13–14), were proposing a sweeping vision of evolution as a universal process long before an evolutionary cosmology became orthodox in physics. The idea of evolution started life in such evolutionary philosophies; only later did it become the dominant idea in biology, and much later still in physics.

It was Spencer, rather than Darwin, who popularized the word *evolution,* even before the publication of Darwin's *Origin of Species* in 1859. In the first edition of this book, Darwin scarcely used the word *evolution*; only in the sixth edition did he begin to use it in reference to his theory, and then only sparingly. He employed instead such phrases as "descent with modification" or simply "progress."[11]

The word *evolution* literally means "unrolling." It was originally used to refer to the progressive unfolding of embryonic structures such as buds. The "evolutionist" school of eighteenth-century biology maintained that the development of embryos took place by the evolution of a microscopic preformed structure that was present in the fertilized egg in the first place.

Thus the word *evolution* implied a pre-existing plan or structure that progressively unrolled in time. This is probably the reason why Darwin did not choose to use this word when he first put forward his theory.[12] For the "evolution" of life would imply the existence of a pre-existing structure or plan—presumably a divine plan—and this is just what Darwin wanted to rule out. But if they were not divinely planned, how could the forms of life on Earth have evolved by spontaneous natural processes?

Darwin found an answer in terms that mirrored processes at work in the progress of commerce and industry: innovation, competition, and the elimination of the inefficient. And, of course, the inheritance of wealth.

In the realm of life, Darwin pointed out, organisms vary spontaneously, offspring tend to inherit the characteristics of their parents, and in the competition that inevitably results from the prodigious fertility of plants and animals, the unfit are eliminated by natural selection. Thus natural selection could, he thought, account for both the wonderful adaptations to their environment shown by plants and animals and the progressive development of new forms of life.[13] This conception was summarized in the title of his most famous book, *The Origin of Species by Means of Natural Selection, or the Preservation of Favoured Races in the Struggle for Life.*

Darwin's theory was cast within the context of a mechanistic universe; his evolutionary tree of life grew up within a world of physical eternities. We now consider in more detail how this pre-evolutionary framework of thought has shaped the Darwinian theory of evolution. We then go on to consider the possibility of a new evolutionary synthesis—a synthesis in which the evolution of life can be understood as one aspect of a cosmic evolutionary process in which not just nature but the "laws of nature" evolve.

Time for Very Slow Change

An essential precondition for Darwin's theory of progressive evolution by gradual change was the expansion of terrestrial time. The biblical account of creation was generally supposed to refer to events that occurred only thousands of years ago: according to one well-known chronology the world was created in 4004 B.C. But the mechanistic cosmology provided a very different context for the origin of the earth: the universe of astronomy and of celestial mechanics, a universe that went on forever.

Descartes, for instance, supposed that the planets were carried around the sun in a vortex of transparent aether (p. 28), and saw no reason why one vortex should not run down while another appeared elsewhere. In this way a sun and planetary system, such as ours, could be formed within the ceaseless motions of the physical universe.

Or, according to other theories, the earth had been a comet, formed by the condensation of dust particles in space under the influence of gravity into a solid body, which had then been trapped in orbit around the sun. Or the earth had been formed by the cooling of hot matter thrown out by the sun when a comet dived into it.[14]

The most successful theory was proposed by the philosopher Kant in 1755. According to his "nebular hypothesis" the whole solar system began as a cloud of dust particles that condensed under the force of its own gravity and gradually acquired a tendency to rotate. Small amounts condensed into solid bodies circling around the main concentration, which ignited to form the sun. In Laplace's *System of the World* (1796) it was assumed that all stars condensed in this way: hence the majority had planets circling them. The gradual formation of a planetary system such as ours therefore became a perfectly natural, mechanistic phenomenon. There was no need for God to have created the earth, or the sun, or indeed anything at all.

Such theories provided one background for speculations about the history of the earth. The Book of Genesis provided the other: the earth and the living creatures on it were created in stages, represented by the days of creation. After the creation, there had been a series of catastrophes on Earth, most notably the Flood.

Throughout the long history of evolutionary debate, these two models have continued to conflict and to interact with each other. Mechanists have generally preferred slow, gradual change; those who have believed in the divine guidance of evolution have generally seen in it a series of stages and sudden jumps. Of course, abrupt changes do not necessarily imply divine intervention, but with the Bible in the background, they have often been thought so to do.

As the science of geology developed in the late eighteenth and early nineteenth centuries, some geologists saw evidence in the rock strata for processes not unlike those described in the Book of Genesis: clear evidence for a flood or a series of floods; evidence for sudden discontinuities; and in the strata above the primary rocks, the occurrence of fossils roughly in the order of Genesis—fish, animals on land, and finally man.[15]

Others, in the light of the physical eternity of the world machine, tried to find a conception of the earth that was as gradual and as non-progressive as possible. At the end of the eighteenth century, James Hutton insisted that the scientific geologist should do his utmost to explain the structure of the earth through causes he can observe in action now. "We find no vestige of a beginning, no prospect of an end." He dismissed as unscientific the idea of catastrophes on a scale we no longer observe. What we can observe is that land masses are continually eroded by wind and water; the debris is carried out to sea and deposited on the ocean floor, where it can harden into rock strata; and these new rocks can subsequently be elevated by earthquakes to form dry land. Earthquakes are driven by heat and pressure from the earth's core, and volcanoes occur when molten rock from the interior finds its way to the surface.[16]

Because the changes we observe today are very slow, Hutton's scheme demanded a great antiquity for the earth—an innovation of the greatest importance.[17]

This system was taken further by Charles Lyell, whose *Principles of Geology* (1830–33) so strongly influenced Darwin. Like Hutton, Lyell adopted a steady-state view of the earth and emphasized the role of gradual changes in accordance with universal physical laws. He denied any directional trend in the development of life. He tried to account for the growing fossil evidence in terms of fluctuating climates, and supposed that all life forms had always been present in every geological period; there had been no sequential development of higher forms from lower—except for the appearance of man.[18]

However, the investigations of rock strata by geologists increasingly supported the idea of directional changes in the earth's development. Sudden breaks between rock formations indicated sudden changes in conditions. Even more striking were the different kinds of fossils found in successive rock formations. Most spectacular of all were the remains of giant reptiles such as dinosaurs. The sequence of fossil remains convinced many naturalists that the history of animal life began with an age of invertebrates, followed by fishes, reptiles, mammals, and finally man.

Some theologians saw in this process the creative guidance of God. New species did not appear gradually through the operation of everyday laws of nature; they appeared suddenly through divine interventions in the history of life. Periodic extinctions occurred as a result of catastrophes, and then new forms of life were created.[19]

Darwin, by contrast, rejected such ideas of divine intervention. He insisted that evolution took place gradually through the smooth operation of the ordinary laws of nature: there were no sudden changes. This aspect of his theory was controversial from the outset, but Darwin stuck to the principle of gradual evolution in the face of all criticism. To admit the existence of any abrupt and inexplicable changes would, he believed, be "to enter into the realm of miracle, and to leave those of science."[20]

In the sixth edition of *The Origin of Species,* Darwin did make one concession to his critics:

> One class of facts, however, namely the sudden appearance of man and distinct forms of life in our geological formations, supports at first sight this belief in abrupt development. But the value of this evidence depends entirely on the perfection of the geological record, in relation to periods remote in the history of the world. If the record is as fragmentary as many geologists strenuously assert, there is nothing strange in new forms appearing as if suddenly developed.[21]

This argument has a familiar ring, and is still in widespread use today. For Darwinians have generally followed Darwin in emphasizing the role of gradual change, and the absence of evidence for missing links has always been explained in terms of imperfections in the fossil record. But the evidence for catastrophes and for the abrupt appearance of new forms of life has not gone away. On the contrary, it has been strengthened by increasingly detailed studies of the fossil record. Evolution that occurs by fits and starts seems to fit the facts better than a process of slow and steady change, and the former idea has been advanced again and again. Its most recent form is the hypothesis of "punctuated equilibria."[22]

Meanwhile, the notion of great global catastrophes has undergone a recent revival in scientifically respectable form. In 1980, abnormal quantities of iridium and other metals were found in clay layers at the boundary between Cretaceous and Tertiary rock strata—in other words layers which were formed some 65 million years ago, at the time that the dinosaurs, as well as many other animals and plants, became extinct. The following explanation was suggested: an asteroid collided with the earth and caused so much dust to be thrown into the atmosphere that sunlight was blotted out for weeks, causing the extinction of dinosaurs and many other forms of life.[23] This hypothesis has gained in plausibility as a result of calculations of the effects of a nuclear war, and in particular the prospect of a "nuclear winter" caused by the blotting out of sunlight by the smoke and debris in the atmosphere.[24]

Subsequently, a variety of calculations have suggested that mass extinctions have occurred over the last 250 million years with a periodicity of about 26 million years. The regularity of this cycle suggests the need for an astronomical explanation, and several have been proposed. We find ourselves back in the realm of great cycles of astronomical time.

One proposed explanation is that the sun has a dark companion star, Nemesis, in a highly eccentric orbit. When Nemesis comes close to the comet cloud at the outer limits of the solar system, it disturbs it, triggering an intense shower of comets. The ensuing series of impacts on the earth lasts up to a million years. Another model proposes a cycle due to the sun's oscillation about the plane of the galaxy, resulting in changes in cosmic radiation sufficient to cause major climatic changes. Yet another proposes that the earth may have passed periodically through interstellar clouds of dust or gas.[25] But some scientists argue that the great extinctions have followed no such regular cycle after all.[26] The debate continues.

The Tree of Life

In the beginning, according to the Book of Genesis,

The Lord God planted a garden eastward in Eden. . . . And out of the ground made the Lord God to grow every tree that is pleasant to the sight, and good for food; the tree of life also in the midst of the garden.[27]

In Darwin's great evolutionary vision, the whole of life had developed in time like a great tree: the evolutionary tree of life (Fig. 3.1). Ever since the first seed of life appeared on Earth, this tree had been growing by itself, entirely naturally and in accordance with the laws of the natural world. Evolution, like the growth of a tree, was an organic, spontaneous process of continual growth and adaptation to the prevailing conditions of life. It all happened naturally.

For Darwin, it was not God who planted the tree of life, nor God who tended it: Darwin conceived of God very differently. God was the great

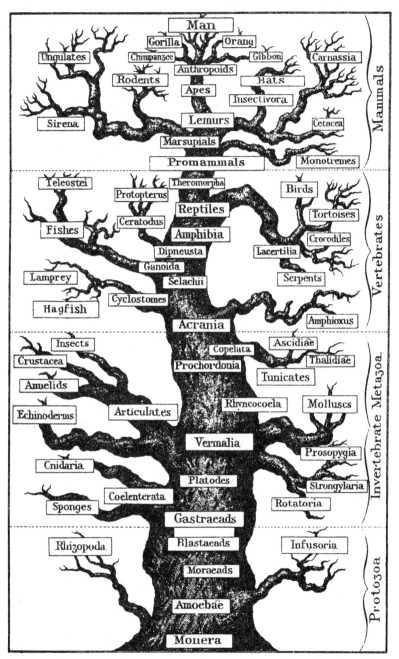

Figure 3.1 The evolutionary tree of life, according to Ernst Haeckel. (From Haeckel, 1910)

designer and creator of the world machine, who had designed all living things within it in the most wonderful and intricate of ways. All creatures but man were inanimate; they were machines whose designing intelligence was outside themselves, in the mind of God, just as the designing intelligence of man-made machines is not inside the matter of the machines, but outside them, in their human makers.

One of the exponents of this kind of theology was William Paley. His *Natural Theology* (which deeply influenced Darwin in his youth) takes the beautiful and appropriate designs of living organisms as proof of a designing intelligence, and hence as proof for the existence of God. This book begins with his famous example of the watch. Suppose, he wrote, when walking on a heath I find a watch. Without knowing how it came to be there, its precision and intricacy of design would force us to conclude

> that the watch must have had a maker: that there must have existed, at some time, and at some place or other, an artificer or artificers, who formed it for the purpose which we find it actually to answer; who comprehended its construction and designed its use.

He then extended this argument by analogy to the works of nature:

> Every indication of contrivance, every manifestation of design, which existed in the watch, exists in the works of nature; with the difference, on the side of nature, of being greater or more, and that in a degree which exceeds all computation.

Paley compared the human eye to a man-made instrument such as a telescope, and concluded that "there is precisely the same proof that the eye was made for vision, as there is that the telescope was made for assisting it."[28]

In a mechanistic universe designed by such a God, there was no freedom or spontaneity anywhere in nature. Everything had already been perfectly designed. For Darwin's tree of life to grow of its own accord, Darwin had to get rid of this all-designing God. But he could do so only by finding some other way of explaining the intricate designs and purposive adaptations of flowers, wings, eyes—indeed of everything alive. He, like Paley, found this designing agency outside living organisms—but it was not in God, it was in nature. Natural selection chose the best designs from those which Nature herself threw up spontaneously. And, working gradually over many genera-tions, natural selection has shaped all the forms of life there are, and all the forms that have ever existed.

Darwin started from the analogy of human selection whose powers can be seen so clearly in the great range of varieties of dogs, pigeons, and other domesticated animals and plants. All these had been produced through spontaneous variation and by selective breeding, through conscious or unconscious human selection. Natural selection was like this, except that no consciousness and no purposes were involved. The term "natural selection" could, he admitted, be taken to imply some conscious choice, but this was not what he meant. Nor was natural selection an active power:

> It has been said that I speak of natural selection as an active power or Deity: but who objects to an author speaking of the attraction of gravity as ruling the movements of the planets? Everyone knows what is meant by such metaphorical expressions; and they are almost necessary for brevity.... With a little familiarity such superficial objections will be forgotten.[29]

And thus Darwin replaced the designing intelligence of Paley's machine-making God with the blind workings of natural selection. Darwinians have continued to do so ever since.

The Blind Watchmaker

Richard Dawkins, one of the most forceful modern defenders of Darwinism, has recently replied to Paley all over again. His book *The Blind Watchmaker* (1986) opens with a statement of faith:

> This book is written in the conviction that our own existence once presented the greatest of all mysteries, but that it is a mystery no longer because it is solved. Darwin and Wallace solved it, though we shall continue to add footnotes to their solution for a while yet.... I want to persuade the reader, not just that the Darwinian world view *happens* to be true, but that it is the only known theory that *could,* in principle, solve the mystery of existence.[30]

Dawkins's argument, like Darwin's, stands in antithesis to Paley's. But notice that both sides of this debate share an assumption which is not questioned by either: the mechanistic world view. Plants and animals are like machines; either they are intelligently designed by the God of the world machine, or they are the product of the blind workings of evolution by natural selection. But what if we change our way of thinking about the external designing intelligence, or about the nature of life? Different possibilities then open up which do not fit into either of these standard positions. Several such possibilities have already been explored; here we consider only two examples. The first involves a modified conception of external designing intelligence, and the second the idea of creative organizing principles within life itself.

Alfred Russel Wallace, like Darwin, understood the power of natural selection; he, like Darwin, discovered it. But he became convinced that Darwinian mechanisms alone could not explain the evolution of life. His last book was called *The World of Life: A Manifestation of Creative Power, Directive Mind and Ultimate Purpose* (1911). In it he proposed that "higher intelligences" had directed the main lines of evolutionary development in accordance with conscious purposes.

> We are led, therefore, to postulate a body of what we may term organizing spirits, who would be charged with the duty of so influencing the myriads of cell-souls as to carry out *their* part of the work with accuracy and certainty. . . . At successive stages of development of the life-world, more and perhaps higher intelligences might be required to direct the main lines of variation in definite directions in accordance with the general design to be worked out. . . . Some such conception as this—of delegated powers to beings of a very high, and to others of a very low grade of life and intellect—seems to me less grossly improbable than that the infinite Deity not only designed the whole of the cosmos, but that He Himself alone is the consciously acting power in every cell of every living thing that is or ever has been upon the earth.[31]

By contrast, Henri Bergson saw the purposive organizing principles of the evolutionary process as internal to the evolving forms of life. He compared the evolutionary process to the development of mind through the onward movement of the current of life, the *élan vital*.

> This current of life, traversing the bodies it has organized one after another, passing from generation to generation, has become divided among species and distributed amongst individuals without losing any of its force, rather intensifying in proportion to its advance. . . . Now, the more we fix our attention on this continuity of life, the more we see that organic evolution resembles the evolution of a consciousness, in which the past presses against the present and causes the upspringing of a new form of consciousness, incommensurable with its antecedents.[32]

Bergson did not, however, believe that this process of creative evolution had any ultimate, external purpose. If there was a God of the evolutionary process, he was not an external God, but a god who created himself in the very process of evolution.

The evolutionary theories of Wallace and Bergson just go to show what sorts of things can happen if we step outside the Paley-Darwin antithesis. But when we step back into the mechanistic world view, the choice narrows again to that between the designing intelligence of the Great Artificer, or the blind inanimate mechanisms of Darwinian evolution.

But why should we keep forcing living organisms into mechanistic metaphors? Why should we not think of them as they really are: living organisms?

Evolving Organisms

For over sixty years, an alternative to the mechanistic philosophy of nature has gradually been developing: the philosophy of organism. This philosophy, sometimes called the holistic or organismic philosophy, or the "systems" approach, is in one sense a new form of animism: nature is once again seen as alive, and all organisms within it contain their own organizing principles within themselves. These are no longer thought of as souls, as they are in the Aristotelian philosophy, but are given a variety of other names such as "systems properties" or "emergent principles of organization" or "patterns which connect" or "organizing fields." But the modern philosophy of organism differs in two essential respects from pre-mechanistic animism: first, it is post-mechanistic, and is developing in the light of the insights and discoveries of mechanistic science; and second, it is evolutionary.

As the philosopher Alfred North Whitehead pointed out over sixty years ago:

> A thoroughgoing evolutionary philosophy is inconsistent with materialism. The aboriginal stuff, or material, from which a materialistic philosophy starts is incapable of evolution. This material is itself the ultimate substance. Evolution, on the materialistic theory, is reduced to the role of being another word for the description of changes of the external relations between portions of matter. There is nothing to evolve, because one set of external relations is as good as any other set of external relations. There can merely be change, purposeless and unprogressive. But the whole point of the modern doctrine is the evolution of the complex organisms from antecedent states of less complex organisms. The doctrine thus cries aloud for a conception of organism as fundamental for nature.[33]

In Whitehead's phrase, organisms are "structures of activity" at all levels of complexity. Even subatomic particles, atoms, molecules, and crystals are organisms, and hence in some sense alive.

From the organismic point of view, life is not something that has emerged from dead matter, and that needs to be explained in terms of the added vital factors of vitalism. *All* nature is alive. The organizing principles of living organisms are different in degree but not different in kind from the organizing principles of molecules or of societies or of galaxies. "Biology is the study of the larger organisms, whereas physics is the study of the smaller organisms," as Whitehead put it.[34] And in the light of the new cosmology, physics is also the study of the all-embracing cosmic organism, and of the galactic, stellar and planetary organisms which have come into being within it.

> The universe confronts us with this obvious but far-reaching fact. It is not a mere confusion, but is arranged in units which attract our attention, larger and smaller units in a series of discrete "levels," which for precision we call a hierarchy of wholes and parts. The first fact about the natural universe is its organization as a system of systems from larger to smaller, and so also is every individual organism![35]

Think, for example of a termite colony, which is an organism made up of individual insects, which are organisms made up of organs, made up of tissues, made up of cells, made up of organized subcellular systems, made up of molecules, made up of atoms, made up of electrons and nuclei, made up of nuclear particles. At each level are organized wholes, which are made up of parts which are themselves organized wholes. And at each level the whole is more than the sum of its parts; it has an irreducible integrity.

What are these elusive principles of organization which are manifested in organisms or systems at all levels of complexity? In the words of L. L. Whyte:

> A neglected principle of order, or better, a process of ordering runs through all levels; the universe displays a tendency towards order, which I have called morphic; in the viable organism this morphic tendency becomes the tendency to organic co-ordination (not yet understood), and in the healthy human mind it becomes the search for unity which gives rise to religion, art, philosophy and the sciences.[36]

In an evolutionary universe, the organizing principles of all systems at all levels of complexity must have evolved—the organizing principles of gold atoms, for example, or of bacterial cells, or of flocks of geese, have all come into being in time. None of them was there in the first place, at the time of the Big Bang.

But were they already present as transcendent Platonic archetypes in an immaterial form, as it were awaiting the moment when they would first be manifested in the physical universe? Or are they more like habits which have developed in time?

These are the questions we explore in the following chapters of this book. We begin by considering the structures of molecules, crystals, plants, and animals and the ways in which they come into being.

The whole of this book represents an attempt to develop a new conception of the evolutionary nature of things. In the final three chapters we return to a discussion of the evolution of life and of the physical universe, and conclude with a discussion of the nature of evolutionary creativity.

The perennial question of whether or not the evolutionary process has any ultimate purpose is left open.

CHAPTER 4

The Nature of Material Forms

The Elusiveness of Form

We see many different forms every day—trees, people, cars, spoons, written words, cats—and we recognize them with no trouble at all. We take them for granted. But forms are surprisingly elusive when we try to think about them or pin them down. We can represent them in pictures and diagrams, photograph them, imagine them, see them in our dreams; but we cannot weigh them or capture them as readings on dials. They are not the same as energy, mass, momentum, electric charge, temperature, or any of the other quantities of physics. Any particular thing we can directly see and experience has certain quantitative characteristics, but it somehow remains more than these: it also has a form, or shape and structure. Consider a particular foxglove plant. It has a definite position, mass, energy, and temperature; it is composed of a variety of chemicals in particular proportions; it has measurable electrical phenomena going on within it; it absorbs a certain percentage of the light falling on it; it transpires a certain amount of water per hour; and so on. But it is more than all these measurable quantities and rates; it remains irreducibly a foxglove.

As the plant grows, it incorporates into itself matter and energy taken from its surroundings; when it dies, all this matter and energy are released, and the form of the plant breaks down and disappears. As the material form of the foxglove comes into being and as it vanishes, there is no change in the total amount of matter and energy in the world, but there is a change in the way in which the matter and energy are organized.

In the case of human artifacts, this elusive quality of form or organization is somewhat easier to think about. For example, as a house is built it takes up a particular structure. Its form is symbolically represented in advance in architects' diagrams, and to start with it originated in someone's mind. But this form cannot be understood by weighing or chemically analysing the house nor the blueprints nor the architects' brains. Nor can it be captured by demolishing the house and examining its constituent parts. With the same building materials and the same amount of labour, houses of different shape and structure could have been built. None of these houses could exist without the building materials or without the energy expended by the builders; but neither the materials nor the amount of work done in building fully explain the form of the house. So what is it? It exists materially in the house, but it is not itself material. It is a pattern, or arrangement, or structure of information that can be repeated more or less exactly in many individual houses, as in housing developments. It is more like an idea than a thing, but nevertheless it is essential to these actual houses and cannot be separated from them: it is not only or merely an abstract idea.

This is the paradox of all material forms. The form is in one sense united with matter, but the form aspect and the material aspect are also separable. Every spoon, for example, has the form of a spoon, and this is what makes it a spoon. Spoons can be made of different kinds of matter, such as silver, steel, wood, or plastic; and conversely the same matter that is in the spoons could equally well have been made into different forms, for instance into forks. Spoons come and go, but when they are broken up or melted down or burnt, all the matter and energy within them persist: the existence or non-existence of spoons makes no difference to these fundamental physical realities.

As a plastic spoon is burnt, for example, the carbon atoms within it are incorporated into carbon dioxide molecules which are dispersed in the air. Let us think of the possible fate of one of these molecules. It may be absorbed by a nettle leaf, and the carbon atom may then be assimilated by photosynthesis into a sugar molecule, and thence through a series of biochemical transformations into a protein molecule within one of the leaf cells. This part of the leaf may be eaten and digested by a caterpillar of the peacock butterfly, and the carbon atom may end up in a DNA molecule in the butterfly's body. The butterfly may be eaten and digested by a bird, and so on through endless food chains and carbon cycles.

The matter of any given carbon atom has the potential to be part of any one of countless millions of different forms, natural or artificial; it could be in a diamond crystal or an aspirin molecule, a gene or a protein, a mushroom or a giraffe, a telephone or an aeroplane, a Russian or an American.

It is generally the case that the matter and energy of which things are composed have the potential to be present in many different forms, and so these forms cannot be fully explained just in terms of their material constituents and the energy within them. The form seems to be something over and above the material components that make it up, but at the same time it can be expressed only through the organization of matter and energy. So what is it?

Philosophies of Form

The question of form has been discussed by Western philosophers for well over two thousand years, and the same kinds of arguments have reappeared century after century and are still alive and well today. If we are to arrive at an evolutionary conception of form we need to pass beyond the traditional, non-evolutionary theories. But they still exert a deep and habitual influence on our thinking.

There are three main traditional ways of thinking about the forms of things: the Platonic, the Aristotelian, and the nominalist. As we saw in chapter 2, according to the Platonic philosophy, the forms of actual material things are reflections of eternal Forms, or Ideas in the mind of God, or transcendent mathematical laws: the source of the form is outside the material object, and indeed outside time and space. By contrast, in the Aristotelian understanding, the sources of material forms are immanent in nature, rather than transcendent. The forms of all kinds of organisms arise from non-material organizing principles inherent in the organisms themselves.

The nominalist tradition grew up in the Middle Ages in reaction against both Platonism and Aristotelianism; and nominalists and empiricists ever since have been like an opposition party to the ruling Aristotelians or Platonists. Nominalists keep reminding us that human words, categories, concepts, and theories are all produced by human minds, but have a perpetual tendency to take on a life of their own, as if they exist outside our minds as well. We give things names (*nomen,* as in nominalism, is the Latin for *name*), which depend on human convention or convenience; but this does not mean that they refer to things that have an independent, objective existence. The things we characterize as horses, for instance, may indeed resemble each other in relevant respects; but if we say a horse form exists outside our minds as well as within them, we are making an unnecessary duplication. We are violating the principle of economy of thought. This principle is the famous Occam's razor, invented by the fourteenth-century English nominalist, William of Occam. By means of this mental razor, the Platonic Ideas and Aristotelian species-forms are simply cut away.

If all forms and concepts exist only in our own minds, we cannot know what is really there in the world, underlying the phenomena of our experience: in fact, in a nominalist world we cannot know any objective reality that is independent of our minds and languages, because all knowledge depends on minds and languages.

This philosophical tradition has been especially strong in Britain, and in its positivist and empiricist forms continues to dominate academic philosophy in the English-speaking countries. Within science, its main influence has come about through its long-established alliance with materialism. For example, in the seventeenth century Thomas Hobbes, as a nominalist, rejected the idea that forms exist objectively outside our own minds, as both Platonists and Aristotelians thought they did. These philosophical concepts were just words: "Words are wise men's counters, they do but reckon with them, but they are the money of fools."[1] On the other hand, as a materialist, Hobbes believed in the reality of material atoms in motion. The invisible realities of the other philosophies of nature were merely words and empty concepts; but the invisible atoms of materialism were not just words and concepts: they were real.

This alliance of nominalism and materialism leads to the familiar doctrine that concepts, names, and ideas exist only in our minds, but at the same time our minds are only aspects of the material processes in our bodies and are in principle ultimately explicable in terms of matter in motion. Thus in some mysterious way the material processes in terms of which the mind is to be explained are more real than the mind that does the explaining. The matter is real in a way that the mind that conceives of it is not.

The combination of materialism and nominalism is inevitably paradoxical; there is always an internal tension, because at any time the nominalist critique can be brought to bear on the particles of matter themselves. These are also words and concepts in human minds, so why should they have any more reality or objective existence than any other categories or concepts? All that can be scientifically known of nature are observations and measurements. And even these are not independent facts; they depend on the conscious activity of observers and measurers, which in turn is guided by human interests, concepts, and theories. Indeed, in the context of quantum mechanics, we have been reminded repeatedly that observations necessarily involve the minds of the observers; they cannot be regarded as objective facts that are independent of human activity.[2]

From here it is but a short step to solipsism or idealism: everything is in the mind. For the solipsist, everything is in his own mind; for the idealist, everything is in a universal or Absolute mind. And since human minds, especially the minds of physicists, find within themselves mathematical principles of order that have a curious objective and timeless quality, this line of thought easily leads back to the familiar territory of Platonism.[3]

We now consider briefly the ways in which these traditional philosophies of form have shaped the contemporary scientific understanding of chemical and biological forms.

Platonic Physics and Chemistry

What is the nature of atomic, molecular, and crystalline forms?

As we saw in chapter 2, physics had been inspired again and again by the Platonic vision of an eternal, rational order that transcends the physical universe. And to a large extent, atomic, chemical, and crystalline forms are still conceived of in a Platonic spirit.

First, the atoms of the elements, of which over a hundred kinds have been identified, each have a characteristic and unalterable number. Hydrogen is atomic type number 1, for example, sodium is type 11, lead type 82, and so on. When the symbols of the atoms are arranged in a series in accordance with their characteristic atomic numbers, they show a periodic pattern, with 2, 8, 8, 18, 18, and 32 elements in successive periods. This mathematical pattern is represented in the periodic system of the elements (Fig. 4.1). The atomic numbers are now understood in terms of the internal structures of the various kinds of atoms; they represent the numbers of protons in the atomic nucleus: lead, for example, has 82. The 82 positive charges of these protons are balanced by 82 negatively charged electrons, all of which are continuously orbiting the nucleus. It is precisely this number of protons and electrons that characterizes the lead atom in its electrically neutral form—if it had 83 it would be not lead but bismuth; if it had 81 it would be thallium.

Atomic forms are currently explained in terms of quantum physics: the nature of the different kinds of atoms is believed to be fully determined by quantum theoretical laws which in principle specify all the details of the structure of the nuclei and the orbitals of the electrons around them. In practice, the detailed calculations are too complex to carry out for any but the simplest of all atoms, hydrogen, with a single proton and electron. Nevertheless, it is taken for granted that if such calculations could be performed for all other atomic types, they would give the right answer and thus vindicate the adequacy of existing theories. But this is a matter of faith.

When the principles of atomic structure were worked out in the first few decades of this century, the universe was still believed to be eternal, and so were atoms and the laws that governed them. But atoms are now thought to have evolved over time. Once there were no lead atoms, or sodium atoms,

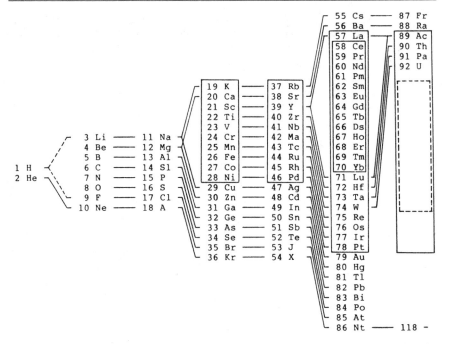

Figure 4.1 The Periodic System of the Elements, in the version of Niels Bohr. The atomic numbers correspond to the numbers of protons and electrons in each kind of atom. (After van Spronsen, Elsevier Science Publishers B.V., Biomedical Division, Amsterdam, 1969)

or atoms of any kind at all. In so far as atomic forms are still conceived of in a Platonic spirit, the periodic system of the elements already existed before the Big Bang; as the universe evolved, one by one the possible kinds of atomic form took on material existence. It is as if the eternal Forms of the atoms were awaiting their opportunities to be actualized in time and space.

The forms of molecules, like those of atoms, are usually conceived of as if they were Platonic Ideas. Chemists represent them symbolically as formulae. One kind, the *rational* formula, expresses the numerical ratios of atoms in a molecule; glucose, for example, is made up of 6 carbon, 12 hydrogen, and 6 oxygen atoms: $C_6H_{12}O_6$. But this rational formula is not unique to glucose; several other kinds of sugar molecule have the same ratios of atoms, but they are arranged in different spatial patterns, which can be represented in *structural* formulae (Fig. 4.2), and more effectively in three-dimensional models.

It is conventionally taken for granted that the structures and properties of molecules have an eternal reality independent of the actual material

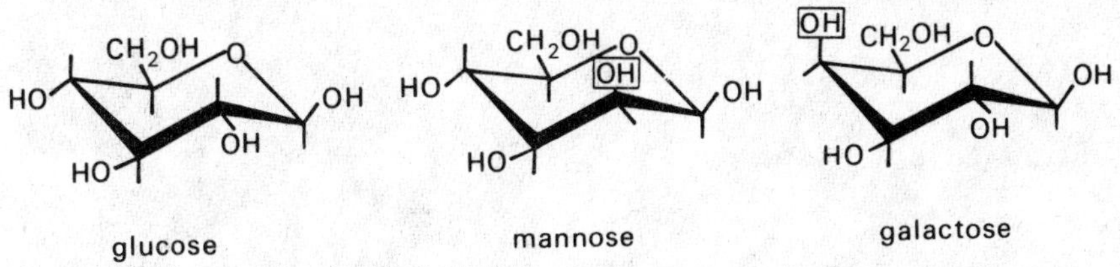

Figure 4.2 Structural formulae of three kinds of sugar molecule. The lines represent chemical bonds. Carbon atoms are present wherever four lines meet, and hydrogen atoms where bonds end where no hydroxyl (OH or HO) group is indicated. Mannose and galactose differ from glucose in the position of one of their hydroxyl groups, shown with a box around it.

existence of these compounds. Thus, for example, the orthodox assumption is that everything about a new kind of molecule could in principle be calculated in advance before the molecule is ever synthesized by research chemists for the first time; the structure and properties of the molecule are determined by transcendent principles of order that exist prior to its material existence.

As we shall see in chapter 7, it is not in fact possible to predict in any detail the structures and properties of molecules—for example the three-dimensional structure of proteins—on the basis of quantum mechanics and the other theories of present-day physics. It is a pure assumption that they are predetermined by timeless mathematical laws; and it is an even greater assumption that they are completely explicable in terms of the current theories of physics. But in so far as this assumption is still taken for granted, chemistry, biochemistry, and molecular biology continue to operate within a Platonic paradigm.

Just as chemists study the forms and properties of molecules, crystallographers study the forms and properties of crystals. Each kind of crystal has a characteristic kind of symmetrical structure, and within the crystals the molecules and atoms are arranged in repetitive three-dimensional patterns, the smallest unit of which is called the "unit cell" of the crystal.

The diagrams and models made by crystallographers (Fig. 4.3) are in one sense idealizations of the actual physical structure of the crystals; but in the context of Platonism, they are more than mere man-made models. They are symbolic representations of the eternal archetypal Form of the crystal. This transcendent pattern exists prior to the crystals that crystallographers study. Hence it is assumed that as new kinds of crystals come into being for the first time, they are only materializing or reifying archetypal patterns that have always been present in a non-physical form.

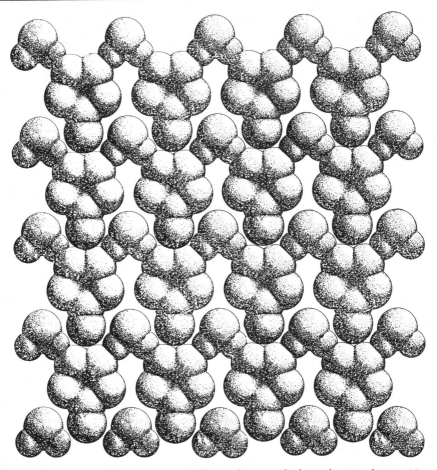

Figure 4.3 A layer in a crystal of tetrazolate monohydrate, showing the repetitive arrangement of the tetrazolate and water molecules. (After Franke, 1966)

At first sight, this conventional assumption seems to be entirely metaphysical. Nevertheless, if we think of it as a scientific hypothesis, we can see that it leads to empirically testable predictions. It would lead us to expect that, other things being equal, the crystals of a newly synthesized chemical compound should form in the same way and at the same average rate the first time this kind of crystallization happens, the thousandth time, or the billionth. This prediction has never been systematically tested, and there is an open possibility that when it is tested it will turn out to be wrong. We return to this question in chapter 7.

Platonic Biology

In the eighteenth century, a great systematic framework for biology was constructed by Linnaeus. His system of classification of animals and plants, in an expanded and modified form, remains fundamental to biology today. Linnaeus grouped together species within a hierarchy of taxonomic categories: species, genus, class, order, and so forth. At each higher level there were ever more basic similarities of form. For example the common English oak *Quercus robur* belongs to the genus *Quercus,* which includes other oak species such as the evergreen holm-oak, *Quercus ilex.* This genus is in the family Fagaceae, which includes beeches and sweet chestnuts as well. This family is within the class of dicotyledons, flowering plants with two cotyledons or seedling leaves in their embryos. Together with the monocotyledons (which include the families of grasses, orchids, and palms) they are Angiospermae, plants with flowers, as opposed to plants that bear seeds without flowers (the Gymnospermae), such as conifers.

Linnaeus believed that he had been privileged to see the outline of the divine plan of creation and that the rational Creator had formed plants and animals according to a meaningful order that man himself could recognize by means of his God-given reason.[4]

In the pre-Darwinian period, the comparative study of form, the science of morphology, revealed deep similarities between the body plans, skeletons, and other structures of broad groups of organisms. The "rational morphologists" of this period believed, like Linnaeus, that the biological realm is rationally intelligible and that in it are reflected eternal laws of form and organization. They developed the concept of the typical form or archetype of each group of organisms and saw the species in the group as variations on this archetypal theme. Structures that were variations of the same archetypal pattern were called homologous structures (Fig. 4.4).

Richard Owen, for example, in his book *On the Archetype and Homologies of the Vertebrate Skeleton* (1848) described the form of the idealized vertebrate, an imaginary creature that represented the essence of the type, without the specialized details of any actual animal. Just as Goethe before him had tried to visualize the form of the archetypal plant, Owen wanted to discover a principle of unity within the type, a unity at a level of reality deeper than the material world. His understanding of homologies enabled him to think of evolutionary branching in terms of modifications of the same basic pattern; for example the archetypal five-fingered forelimb of vertebrates had been transformed into a whale's flipper, a bat's wing, or a human hand. On the basis of the fossil record, he argued that the earliest members of any

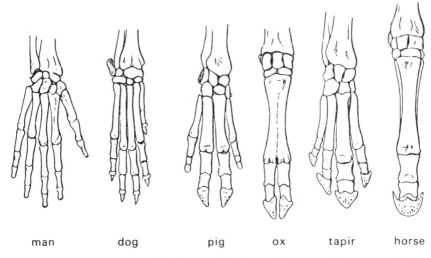

| man | dog | pig | ox | tapir | horse |

Figure 4.4 Skeletons of the hand or fore-foot of six mammalian species. All are modifications of an ancestral five-toed foot. (After Haeckel, 1910)

class were generally of unspecialized structure; the subsequent history of the class involved the development of specialized variations on this basic structural theme.

Owen did not believe that the evolution of forms was propelled by natural selection; rather, it was the unfolding of a rational plan, and worked through "causes" or "laws" that governed the appearance of new forms of life. Likewise, the great Swiss-American naturalist Louis Agassiz thought of the sequential development of living forms as the working out of variations on basic plans. Each basic type, and the ideal form of each specific variation on it, was fixed in accordance with the Creator's will.[5]

Darwin and his followers rejected such ideas. They attempted to account for the archetypal forms and homologies entirely historically, through descent from common ancestors. The Darwinian and neo-Darwinian interpretation of evolution through the interplay of chance and natural selection differs radically from a rational process of unfolding and transformation. There is no longer any attempt to understand evolution "from a higher and more rational standpoint."[6]

But the spirit of the rational morphologists has never been completely displaced from biology. A major twentieth-century contribution to this tradition was made by D'Arcy Thompson in his classic study *On Growth and Form*. He shed much light on the form of organisms through both geometrical considerations and physical analogies (Fig. 4.5); and he showed that

Figure 4.5 D'Arcy Thompson's comparisons of drops falling in fluid with the forms of jellyfish. (a) drops of ink falling in water, (b) a drop of fusel oil falling in paraffin, (c) *Cordylophora* (d) *Cladonema*. (After D'Arcy Thompson, *On Growth and Form*; Cambridge University Press, 1942)

within broad groups, organisms could be understood as permutations or deformations of one another (Fig. 4.6). These transformations were orderly and appeared to be governed by mathematical laws. For example, in the case of the Foramanifera: "We can trace in the most complete and beautiful manner the passage of one form into another among these little shells." But, he added,

> The question stares us in the face whether this be an "evolution" which we have any right to correlate with historic *time*. The mathematician can trace one conic section into another, and "evolve," for example, through innumerable graded ellipses, the circle from the straight line: which tracing of continuous steps is a true "evolution" though time has no part therein. . . . Such a conception of evolution is not easy for the modern biologist to grasp, and still harder to appreciate.[7]

An approach to biological form in this mathematical spirit is currently advocated by Brian Goodwin and Gerry Webster, among others.[8] They hope that a mathematical understanding of the generation of form in developing embryos will lead to a "knowledge of the world of natural forms and their relationships in terms of a theory of generative transformations."[9] They

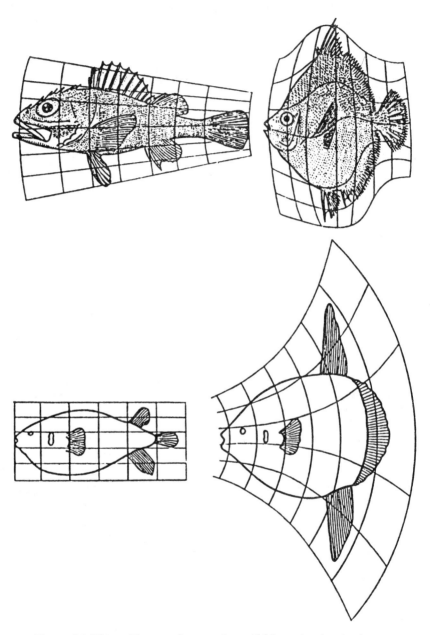

Figure 4.6 D'Arcy Thompson's comparisons of fish species, showing how one can be "deformed" into another. Top left, *Scorpaena* sp.; right, *Antigonia capros;* bottom left, the porcupine fish *Diodora;* right, the sunfish *Orthagoriscus mola.* (After D'Arcy Thompson, *On Growth and Form*; Cambridge University Press, 1942)

explicitly acknowledge that this approach recalls the spirit of the rational morphologists:

[It] attempts to shift the focus from a preoccupation with the contingent and the historical, which leaves biology without intelligible macroscopic structure, to concern with general principles of organization and transformation which could give biology a rational taxonomy and a theory of directed evolutionary change.[10]

In so far as living organisms can be understood mathematically, the historical aspect of biology recedes into the background, as it does in physics and chemistry. Chemists do not usually ask themselves about the evolutionary origins of atoms and molecules; they generally take the Platonic paradigm for granted. A Platonic biology would resemble physics and chemistry in this respect, as Goodwin has made very clear. A rational taxonomy "would be quite independent of the actual historical sequence of appearance of species, genera, and phyla, just as the periodic table of the elements is independent of their historical appearance and is compatible with a great variety of possible sequences."[11]

Aristotelian Biology

The Aristotelian tradition lived on in biology in the form of vitalism. Whereas mechanists maintained that living organisms are inanimate machines, vitalists argued that they are truly alive. The inherent organizing principles of plants and animals, which Aristotle called souls, were referred to by a variety of terms such as vital factors, the *nisus formativus* (the formative impulse), or entelechy. Vitalists thought that these non-material vital factors organized the bodies and behaviour of living organisms in a holistic and purposive manner, drawing organisms towards a realization of their potential forms and ways of behaving, and that when organisms die, the vital factors disappear from them.

Although vitalism is rarely advocated nowadays in an explicit form, it continues to exert a strong, though often unconscious, influence on the thinking of biologists. In contemporary biology, theoretical entities such as genetic programs and "selfish genes" play similar roles to vital factors, as we shall see in the next chapter.

The organismic philosophy of nature has much in common with the Aristotelian tradition (pp. 21–22). It is more radical than vitalism in that it sees organisms at all levels of complexity, from subatomic particles to galaxies, and even the entire cosmos, as alive. The organizing roles that used to be attributed to souls and vital factors are now thought of in terms of systems properties, patterns of information, emergent organizing principles, or organizing fields.

The concept of morphic fields developed in this book represents an attempt to understand such organizing fields in an evolutionary spirit.

Materialistic Biology

The orthodox approach to biological form is provided by the mechanistic theory of life.

As we saw in chapter 2, the mechanistic world view grew out of a synthesis of the Platonic and the materialist philosophies of nature: on the one hand, all nature was governed by eternal, non-material laws; and on the other hand, at the basis of all physical reality were the permanent atoms of matter. An emphasis on the materialist aspect of this synthesis leads to a reductionist approach, the attempt to reduce more complex systems to less complex ones. From the atomistic point of view, the lower something is in the hierarchy of order, the more real it is; atomism emphasizes the supreme material reality of the smallest and most fundamental particles of matter.

In practice, there is no attempt in mechanistic biology to reduce the phenomena of life to the level of the fundamental particles of modern physics; reduction to the molecular level is generally considered quite sufficient. From molecules downward, reduction is assumed to be plain sailing; it is simply taken for granted that the structures and properties of molecules can be reduced to the properties of atoms and subatomic particles, and in principle be understood in terms of the current theories of physics. This is the job of physicists and chemists.

Morphogenesis

So far we have been considering the main theoretical approaches to biological form. Platonists try to understand it in terms of transcendent archetypes or eternal mathematical laws, Aristotelians in terms of non-material organizing principles immanent in living organisms, and materialists in terms of the properties of molecules, and above all in terms of the chemical genes. In the next chapter, we turn to a discussion of the ways in which living forms actually do come into being, and examine how these various theories fit the available facts. The coming into being of form is called morphogenesis, from the Greek words *morphē*, form, and *genesis*, coming into being.

Practically everyone agrees that an understanding of morphogenesis is essential for a deeper comprehension of the nature of life; and practically everyone also agrees that very little is currently understood about it. But one thing is clear: any satisfactory theory of morphogenesis has to take into account the fact that all biological forms have evolved. Morphogenesis is rooted in ancestral history. This is evident not only from the study of evolutionary history, but even in the very processes of embryonic development. For example, all of us, when we were embryos, passed through a stage in which we had gill slits (Fig. 1.1), which seems to recall in some way the embryonic development of the ancestral fishes from which we are descended—or ascended.

The conventional explanation of the evolutionary basis of morphogenesis is, of course, in terms of the inheritance of chemical genes. The hypothesis of formative causation takes a broader view of heredity, and sees the inheritance of organic form—including the forms of molecules themselves—in terms of the inheritance of organizing fields that contain a kind of inbuilt memory. According to this point of view, living organisms, such as badgers, willow trees, or earthworms, inherit not only genes but also habits of development and behaviour from past members of their own species and also from the long series of ancestral species from which their species has arisen.

CHAPTER 5

The Mystery of Morphogenesis

The Unsolved Problem of Morphogenesis

As plants and animals develop from fertilized eggs, their form and organization become increasingly complex. How this happens is still a mystery.

From a materialistic point of view, the source of an organism's form must already be present within the fertilized egg in some material manner. This theory was first elaborated in the seventeenth century in the doctrine of preformation. Preformationists maintained that the egg contained a tiny version of the adult organism, which then just grew and unfolded. This theory of preformation turned out to be wrong. Preformationism was then resurrected in a subtle form in the late nineteenth-century doctrine of the germ-plasm (pp. 78–81). In its modern form, the germ-plasm is identified with the chemical genes. The debate now centres on the question of whether or not the genes do in fact control and direct the processes of morphogenesis. Are genes alone sufficient? Or does development depend on non-material organizing principles as well? And if so, what are they, and how do they work?

In this chapter, we trace the history of this long debate and examine the forms in which it is still going on. It has been characterized again and again by mechanists' denial of the existence of the purposive organizing principles of vitalism; and then by the reinvention of these principles in new guises such as the germ-plasm, selfish genes, genetic programs, patterns of information, internal representations, and so on.

The discussion in this chapter enables the idea of morphogenetic fields to be seen in its biological context; then in the following chapter we explore the nature of these fields.

Organisms Are Not Preformed

According to the preformationists, fertilized eggs contained miniature organisms. Development was nothing but a growth and unfolding of these pre-existing material structures. This hypothetical process was called "evolution."

Preformationists in the seventeenth and eighteenth centuries differed among themselves as to whether these tiny organisms were contributed by the egg or the sperm; most favoured the sperm. Some thought they had proved it; they saw just what they were looking for. For example, one microscopist observed miniature horses in horse sperm, and similar tiny animals with big ears in the sperm of donkeys.[1] Likewise, with the eye of faith, human sperm could be seen to contain tiny homunculi (Fig. 5.1).

Although this theory gave a satisfyingly simple explanation for the development of individual organisms, it ran into appalling theoretical difficulties when the succession of generations was considered. For if a rabbit, for example, grows from a miniature rabbit in a fertilized egg, then it itself must contain miniature rabbits in its germ cells, and they must in turn contain nested within them an endless series of future generations. An early-eighteenth-century opponent of preformationism calculated that at least $10^{100,000}$ rabbits must have existed in the first rabbit, assuming that creation took place 6,000 years ago and that rabbits begin to breed at six months of age.[2]

In due course, the theory of preformation was completely refuted by empirical facts. When developing embryos were closely observed, new structures could be seen to appear that were not there before. For example, in 1768 C. P. Wolff showed that in chick embryos "the intestine is formed by the folding back of a sheet of tissue which is detached from the ventral surface of the embryo, and that the folds produce a gutter which in the course of time transforms itself into a closed tube."[3] By the mid-nineteenth century, as the further study of embryology provided overwhelming evidence for it, development was generally agreed to be *epigenetic*, in other words to involve the appearance of material structures that were not present before (Fig. 5.2).

Epigenesis is just what would be expected from the point of view of Platonic and Aristotelian theories, which have never supposed that all of an organism's form must be contained in the matter of the fertilized egg. By contrast, epigenesis is inevitably problematic from the mechanistic point of view. Somehow, more material form has to arise from less. Embryos have to pull themselves up by their own material bootstraps, as it were.

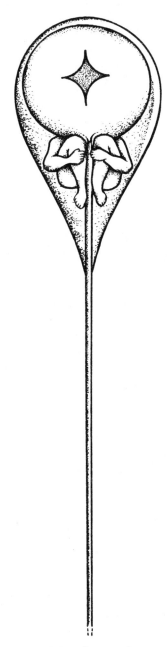

Figure 5.1 A human sperm containing a homunculus, as seen by an early eighteenth-century microscopist. (From Cole, 1930)

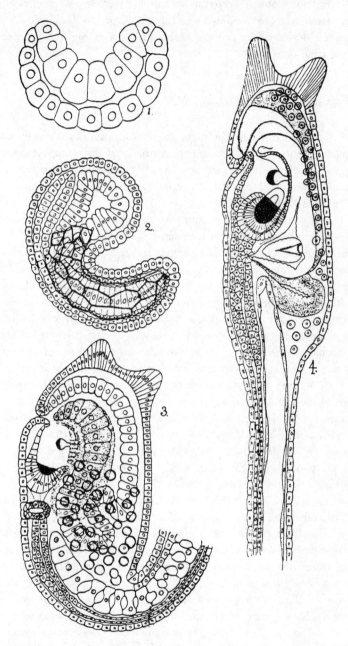

Figure 5.2 Stages in the embryonic development of a sea-squirt. (From Russell, 1916)

The same problem is presented by the phenomenon of regeneration. For in regeneration, more complex forms can arise from less complex ones, as in embryonic development: a whole willow tree, for example, can regenerate from a small cutting.

The Regeneration of Wholeness

According to the preformationists, the organism developed by a kind of inflation of its original miniaturized form. But if that were so, how could it regenerate parts of itself that were lost? For a crude analogy, think of an inflatable rubber doll: how could it possibly regenerate its arm or any other part of itself that was cut off?

If, on the other hand, various parts of organisms have the capacity to regenerate, then might not this same capacity account for their development in the first place? Hartsoeker put it tersely in 1722: "An intelligence that can reproduce the lost claw of a crayfish can produce the entire animal."[4]

The capacity to regenerate is in fact one of the most fundamental features of living organisms, and any theory of life has to try to explain it. All organisms have some regenerative powers, even if only when young or only in certain tissues. For example, we ourselves are continuously regenerating our blood, our intestinal lining, our skin; our wounds heal; broken bones knit themselves together; severed nerves grow out again; and if part of the liver is lost, new liver tissue develops to replace it.[5] Many of the lower animals have such a strong regenerative ability that they can duplicate whole creatures from isolated parts. A flatworm, for example, can be cut into pieces, and each piece—a head, a tail, a side, or a mere slice—can grow into a complete worm (Fig. 5.3). And many plants can form new plants from separated parts: thousands of cuttings can be taken from a willow tree, each of which can grow into a new willow.

Processes of regeneration reveal that in some sense organisms have a wholeness that is more than the sum of their parts; parts can be removed, and yet wholeness can be restored. A part of a flatworm is more than just a part of a material whole; it has a kind of implicit wholeness that goes beyond its actual material structure: if it is isolated from the rest of the worm, it can turn into a whole worm. One of the most striking ways in which living organisms differ from machines is their capacity to regenerate. No man-made device has this capacity. If a computer, for example, is cut into pieces, these cannot, of course, form whole new computers. They remain pieces of a broken computer. The same goes for cars, telephones, and any other kind of machinery.

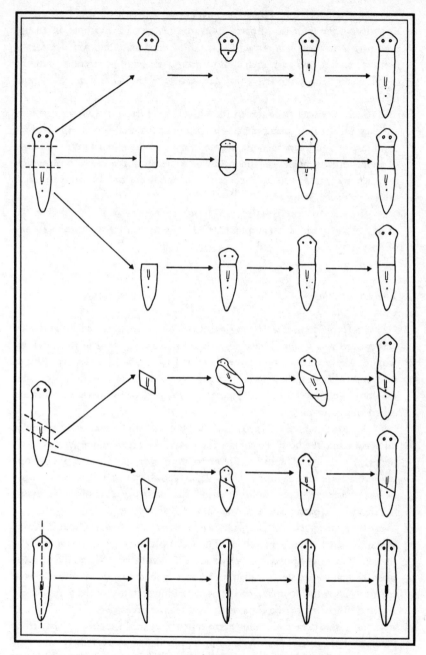

Figure 5.3 The regeneration of complete flatworms from pieces of worm (of the genus *Planaria*), cut as indicated on the left. (After Morgan, 1901)

Nevertheless, there are some physical systems with holistic properties that survive the removal of parts. If an iron magnet, for example, is cut up, each part is a whole magnet with a complete magnetic field. Or if a part is removed from a hologram, which is a physical record of interference patterns in the electro-magnetic field, this part can give rise to the entire original image.

Such physical analogies to the holistic properties of living organisms are examples of field phenomena. Fields are not material objects, but regions of influence. Do the epigenetic development and the regenerative capacities of living organisms depend on fields, or something like fields, with which they are associated? Or are they due to material objects that are present in the egg to start with?

We now follow this debate through another turn of the spiral in the theory of the germ-plasm and the vitalist reply to it in Hans Driesch's theory of entelechy.

The Germ-Plasm

Preformationism in its original form had to be abandoned; the theory was proved to be false. It was resurrected in a subtle form in the 1880s by August Weismann, who proposed that fertilized eggs contain material structures that, although they do not have the actual form of the adult organism, somehow give rise to it. These material structures were in what Weismann called the germ-plasm.

Weismann made a theoretical division of organisms into two quite different parts, the body, or somatoplasm, and the germ-plasm. He described the germ-plasm as a "highly complex structure" with the "power of developing into a complex organism."[6] This is the repository of all the specific causes of form observed in the adult organism: each particular part of the organism is caused by a particulate material unit, called a determinant.

By contrast, the somatoplasm is that part of the organism that is shaped and moulded by the germ-plasm. The germ-plasm is the active agent, and the somatoplasm responds passively to it. The germ-plasm affects the somatoplasm, but not vice versa. This process is represented in the familiar scheme in figure 5.4, which emphasizes the potential immortality of the germ-plasm, and the mortality of the organisms to which it gives rise.

In animals, the embryonic germ cells are separated relatively early from the rest of the body, and Weismann supposed that there is no transfer of information from the body to the germ cells; he thought there could be no modification of the germ-plasm as a result of what happens in the body.

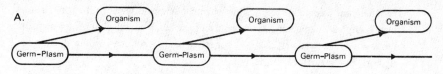

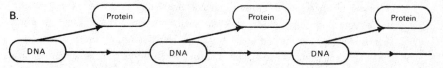

Figure 5.4 A: Weismann's scheme, indicating the continuity of the germ-plasm from generation to generation, and the transient nature of the adult organisms. B: The "central dogma" of molecular biology, in which Weismann's scheme is interpreted in terms of DNA and proteins.

Organisms inherit only the inherited characteristics of their ancestors; they cannot inherit any characteristics that their ancestors acquired for themselves through their adaptation to the environment or through the development of new habits. We return to the question of the "Lamarckian" inheritance of acquired characteristics in chapter 8 and consider some of the evidence in its favour. But according to Weismann, such inheritance was simply impossible.

In plants, the germ cells are not set aside from the rest of the body at an early stage of development, as they are in most animals. Pollen and egg cells are formed in flower buds, which arise on the growing shoots. Nevertheless, Weismann's theoretical principles were (and still are) assumed to apply to plants as well, in spite of their fundamental difference from animals.

Weismann believed that the determinants for each part of the body were parcelled out to various embryonic tissues in the course of development. Each received not a miniature version of the adult structure, as in old-style preformationism, but rather the material structure of the determinant somehow "directed" the formation of the adult structure.

This theory was taken up by Wilhelm Roux, the founder of the school of "developmental mechanics." Roux postulated that the early embryo is like a mosaic, containing parts that develop independently, although in harmony with each other.

This theory soon turned out to be inadequate. In the 1890s, the embryologist Hans Driesch discovered that when half a young sea-urchin embryo was destroyed, the remaining half did not give rise to half a sea-urchin, as it would have done if the embryo were like a mosaic of more or less independent parts; rather, it adjusted to the loss and went on to form a small but complete organism. Conversely, Driesch showed that if two young embryos were artificially fused together, they produced not a double sea-urchin, but a normal single one.[7]

This ability of embryos to adjust to damage is called regulation. It is closely related to regeneration. Many other examples of embryonic regulation have been discovered since the time of Driesch, not only in very young embryos, but in the developing organs of older embryos as well. For example, the developing wing buds of chick embryos have the ability to adjust themselves and produce normal wings in spite of quite severe damage.

Driesch pointed out that this ability of embryos to regulate shows that their different parts do not develop independently in a rigidly predetermined way. Rather, they respond and adjust to each other. They can change the way they develop if other parts are damaged; cells that would have given rise to one particular structure in a normal embryo can give rise to another one instead if need be. Thus Driesch refuted Weismann's original theory of determinants that were parcelled out progressively into the developing embryonic tissues.

However, towards the end of the nineteenth century, as a result of studies of chromosomes (threadlike structures within the cell nuclei), the cell nucleus was generally recognized to contain the material substance of heredity. Weismann consequently identified the germ-plasm with the chromosomes.[8] With the rediscovery in 1900 of the findings of Gregor Mendel and the subsequent development of genetics, Weismann's determinants came to be identified with the genes. And with the discovery of the structure of the genetic material, DNA, and the way it "codes" for the sequence of amino acids in proteins, the germ-plasm theory seemed to be fully confirmed. Well-defined biochemical substances could be written into Weismann's scheme for the germ-plasm and somatoplasm: DNA and protein (Fig. 5.4). This scheme represents what is known as the central dogma of molecular biology: the genetic material acts as a template for the synthesis of proteins, but never the reverse. Just as in Weismann's original scheme, this rules out on theoretical grounds the possibility of the inheritance of acquired characteristics. Neither the form, function, nor behaviour of the body can exert any specific influence on the genetic constitution, or genotype.

Darwin's theory of evolution accepted the idea of the inheritance of acquired characteristics, and Darwin himself in his theory of panspermia proposed that the germ cells were modified through the incorporation of "gemmules" from different parts of the body.[9] The neo-Darwinian theory differs from Darwin's in that it is firmly based on Weismannian principles. These are incorporated into the distinction between the genotype, or genetic constitution, on the one hand and the phenotype, or the organism as it actually appears, on the other. It is the genotype that evolves, and it is the genotype that determines the phenotype. Consequently, in principle, "a theory of development would effectively enable one to compute the adult organism from genetic information in the egg."[10]

Entelechy

In accounts written by contemporary biologists, vitalism is usually treated as if it were a kind of superstition which has been swept away by the advance of rational understanding.[11] The discrediting of vitalism is usually said to have begun with the first artificial synthesis of an organic chemical, urea, by Friedrich Wöhler in 1828, and to have proceeded faster and faster ever since. Jacques Monod, for example, has expressed this conventional view as follow:

> Developments in molecular biology over the past two decades have singularly narrowed the domain of the mysterious, leaving little open to vitalist speculation but the field of subjectivity: that of consciousness itself. There is no great risk in predicting that also in this area, for the time being still "reserved," such speculation will prove as sterile as in all the others where it has been practised up to now.[12]

However, most nineteenth-century vitalists did not in fact deny that living organisms contain organic chemicals that could be both analysed and artificially synthesized; even the great chemist Justus von Liebig argued that in spite of the fact that many organic substances could be synthesized in the laboratory, and that many more would be made in the future, chemistry would never be in a position to create an eye or a leaf. These, he believed, were due to a kind of cause which organized the chemicals into "new forms so that they gain new qualities, forms and qualities which do not appear except in the organism."[13]

Rather vague ideas such as these were common throughout the nineteenth century; it was only in the 1900s that Driesch worked out a vitalist theory in some detail. He started his career as a mechanist, in the school of developmental mechanics, but he came to the conclusion that the facts of embryonic regulation, regeneration, and reproduction show that something with an inherent wholeness acts *on* the living system, but is not a material *part* of it. Following Aristotle, he called this non-material causal factor entelechy. He regarded entelechy as purposive or teleological, directing physical processes under its influence towards ends or goals contained within itself.[14]

According to Driesch, entelechy guides the morphogenesis of the developing organism towards the characteristic form of its species. The genes are responsible for providing the material *means* for morphogenesis, the chemical substances to be ordered, but the ordering itself is due to entelechy. Similarly, in Driesch's view, the nervous system provides the means for the behaviour of an animal, but entelechy organizes its activity, using it as an instrument, as a pianist plays on a piano. Behaviour can be affected by damage to the brain, just as the music played by the pianist is affected by damage to the piano; but this only proves that the brain is a necessary means for producing behaviour, as the piano is a necessary means for the pianist.

Because entelechy "contains" the end towards which a process under its control is directed, if a normal pattern of development is disturbed, the organism can reach the same goal in a different way: it can regulate or regenerate.

Driesch proposed that development and behaviour are under the control of a hierarchy of entelechies, all derived from and ultimately subordinate to the overall entelechy of the organism. He regarded these entelechies as *natural* causal factors, not "metaphysical" or "mystical" entities, that act on physical and chemical processes, imposing order and pattern on changes that would otherwise be indeterminate. But he developed this theory when science was still under the sway of classical physics; it was generally believed that all physical processes are fully deterministic and in principle completely predictable. If this were the case, then there would be no scope for the action of entelechy, because the physical and chemical processes within organisms would already be fully determined by the laws of physics.

Thus, Driesch believed, for entelechy to be able to impose order on processes within living organisms, these processes must be physically indeterminate, at least on a microscopic scale. Since such indeterminism was not admitted by the physics of his day, Driesch was driven to suppose that entelechy *itself* introduced this indeterminism into organisms. He suggested that it did so by affecting the timing of physico-chemical processes. It suspended them, and then released them again as required for its purposes.[15]

This proposal seemed to be a fatal weakness in Driesch's system. In the context of the reigning orthodoxy, any interference with physical determinism was unthinkable, and Driesch's proposal seemed impossible in principle.

It is surely ironic that by the end of the 1920s, when vitalism seemed to most biologists to have been discredited, undreamt-of changes were occurring in physics. Heisenberg proposed the uncertainty principle in 1927, and as quantum theory developed it became clear that on the microscopic scale physical events are not fully determinate, but predictable only statistically in terms of probabilities. It was no longer necessary for entelechy to introduce indeterminism into living organisms for its ordering effects to be possible: indeterminism is inherent in their physical nature anyway.

Driesch readily agreed that some aspects of living organisms could be explained mechanistically; he was aware of the importance of enzymes and other proteins, and thought that genes would ultimately be understood in chemical terms. These opinions have been confirmed by subsequent discoveries. But he argued that development and behaviour would never be fully understood mechanistically, but would only be comprehensible in terms of purposive organizing principles. At least so far, this prediction seems valid. Very little is actually understood about morphogenesis in physical and chemical terms; and the organizing principles of vitalism, which were denied by the mechanistic theory, have returned in such guises as selfish genes and genetic programs. The central paradigm of modern biology has in effect become a kind of genetic vitalism.

Selfish Genes

Weismann's germ-plasm was supposed to have a more or less immutable structure, which determined the form of the organism. His germ-plasm-somatoplasm duality, like its descendent, the genotype-phenotype dichotomy, recalls the Platonic distinction between the changeless Form or Idea and the phenomenon in which it is reflected. Just as the phenomenon has no effect on the Idea, so the phenotype has no effect on the genotype. It is as if Weismann incarnated the Idea of the organism into the germ-plasm, which also possessed the controlling and organizing properties of the psyche or entelechy; Weismann thought of it as a "central directing agency."[16]

His notion that each "determinant" in the germ-plasm is responsible for a particular physical characteristic was at the same time atomistic in spirit. It is paralleled by the more recent notion that particular characteristics are determined by particular genes or sets of genes. In other words, inherited characteristics—for example the shape of a pigeon's foot, or its inborn homing ability—are genetically *determined:* there are foot-shape genes and homing genes. Or again in other words, there are genes "for" foot shape and genes "for" homing. This idea plays a central role in neo-Darwinian evolutionary theory, in which genes "for" particular characteristics are subject to selection pressure; genes are in competition, and some genes are more successful than others in terms of the number of copies that are propagated. Genes "for" characteristics that favour more copies of these genes surviving and reproducing will increase in frequency within the interbreeding population through natural selection. Genes "for" unfavourable characteristics will be selected against; they will decrease in frequency. The rates of change of these frequencies have been calculated mathematically by theoretical population geneticists. But in order to formulate the appropriate equations, it is necessary to make some simplifying assumptions. One of these is the Weismannian view of genes as independent determinants that can be selected more or less independently of each other.

This assumption underlies neo-Darwinian thinking and is carried to its furthest extreme in the school of sociobiology, which attempts to account for practically all aspects of animal behaviour and social life in terms of genetic determinants whose frequencies depend on the pressures of natural selection. The leading proponent of sociobiology, E. O. Wilson, has extended his analysis to human society, on the assumption that there are genes, subject to natural selection, "for" traits such as homosexuality, xenophobia, and altruism.

Finally, the genes have come to life. They are intelligent, but they are also selfish, ruthless, and competitive, like "successful Chicago gangsters." This is the theory of selfish genes, propounded by Richard Dawkins. He traces their descent from primitive "replicator" molecules in the primeval soup:

> The replicators which survived were the ones that built *survival machines* for themselves to live in. . . . Now they swarm in huge colonies, safe inside gigantic lumbering robots, sealed off from the outside world, communicating with it by tortuous indirect routes, manipulating it by remote control. They are in you and in me; they created us, body and mind; and their preservation is the ultimate rationale for our existence.[17]

Although organisms are regarded as "throwaway survival machines," there is nothing mechanistic about the selfish genes. They have powers to "create form," to "mould matter," to "choose," to engage in "evolutionary arms races," and even to "aspire to immortality." As Dawkins remarks, "DNA works in mysterious ways."[18]

The trouble with the selfish-gene theory is that it cannot possibly be true, as Dawkins himself acknowledges. DNA molecules cannot really be selfish or intelligent or mould matter or think. But he promotes the idea of selfish genes as a "thought experiment" and as a "powerful and illuminating" metaphor. Indeed, he shows clearly that it is more interesting to imagine that organisms are controlled by small living things inside them than that they are blind, unconscious mechanisms. Moreover, he demonstrates that this way of looking at organisms is implicit within neo-Darwinism; he describes it as "a neo-Weismannist view of life."[19]

Selfish genes bear little resemblance to the chemical molecules of DNA.

They have been endowed with the properties of life and mind, and have become more like miniaturized entelechies. DNA molecules have been given comparable organizing and controlling powers by means of another powerful metaphor: the genetic program.[20]

Genetic Programs

Whereas selfish genes are individualistic, recalling Weismann's atomistic determinants, genetic programs are more holistic, recalling his idea of the germ-plasm as a central directing agency. They play much the same role as Driesch's entelechies.

The notion of the genetic program has several major attractions. First, it appears to account for the fact that most hereditary characteristics—for example the form of a cauliflower—have no obvious connection with DNA or protein molecules. If the genes somehow *program* the development of the cauliflower, then the vast gulf between this complex living structure and the DNA molecules seems less disturbing, even though nothing is actually known about the nature of the cauliflower program. Secondly, the program is a more subtle concept than the idea of genes "for" particular characteristics. Genes are not atomistic determinants of separate features of the organism, but somehow many different genes co-operatively play a part. If they are like elements of a program, then their harmonious and co-operative activity is easier to understand. Thirdly, this notion suggests that development is purposive. Programs contain within themselves information about the end to which they are leading. Thus organisms can develop purposively towards these ends that are contained in their programs; likewise embryos can regulate and organisms regenerate because of the purposive and holistic properties of these hereditary organizing principles. Finally, the idea of the genetic program seems to fit in well with the jargon of information theory and with the linguistic metaphors in general usage within modern biology. DNA "codes information," which can be "transcribed" into RNA molecules, and then "translated" into a sequence of amino acids as protein molecules are synthesized.

The metaphor of the genetic program can hardly fail to suggest that development is organized by a pre-existing purposive principle that is either mindlike itself or designed by a mind. Computer programs are intelligently designed by human minds for particular purposes, and act upon and through the electronic machinery of a computer. The computer is a machine, but the program is not.

Perhaps morphogenesis is indeed organized by such a purposive directing principle. But if so, "genetic program" is a misleading name for it: it is not genetic, in the sense that it is not in the genes, and morphogenesis is not in fact programmed in any meaningful sense of this word.

If the genetic program were carried in the genes, then all the cells of the body would be programmed identically, because in general they contain exactly the same genes. The cells of your arms and your legs, for example, are genetically identical. Moreover, these limbs contain exactly the same kinds of protein molecules, as well as chemically identical bone, cartilage, and so forth. Yet they have different shapes. Clearly, the genes alone cannot explain these differences. They must depend on something else: on formative influences which act differently in different organs and tissues as they develop. These influences cannot be inside the genes; they extend over entire tissues and organs. At this stage the concept of the genetic program fades out, and is replaced by vague statements about "complex spatio-temporal patterns of physico-chemical activity not yet fully understood," or "mechanisms as yet obscure."

The idea that development is programmed is misleading because for a phenomenon to be programmatic "it is a necessary condition that in addition to the phenomenon itself, there exists a second thing, the programme, whose structure is isomorphic with, i.e., can be brought into a one-to-one correspondence with, the phenomenon."[21] This is indeed the case in the clear causal chain leading from the sequence of chemical bases in DNA molecules to the sequence of amino acids in peptides. But here the programming ends. The folding up of the peptides into the characteristic three-dimensional structure of proteins is not programmatic, for it has no isomorphic correspondence in the DNA. And in relation to morphogenesis itself, it is most unlikely that the overall sequence of events is isomorphic with the genes. For example:

> Studies of the development of the nervous system have shown that the notion of genetic programming is not only defective at the conceptual level, but also represents a misinterpretation of the knowledge already available from developmental studies. . . . We already know enough about its mode of establishment to make it most unlikely that the nervous system is pre-specified; rather, all indications point to stochastic [i.e., probabilistic] processes as underlying the apparent regularity of neural development.[22]

Nevertheless, in spite of its inadequacy, and in spite of the fact that many biologists now recognize that it is misleading, the genetic program continues to play a large conceptual role in modern biology. There seems to be a need for such an idea. This is just what vitalists and organicists have been saying all along.

Modern biology grew up in opposition to vitalism, the doctrine that living organisms are organized by purposive, mindlike principles (Fig. 5.5). Mechanists denied this.[23] But modern biology now has purposive mindlike

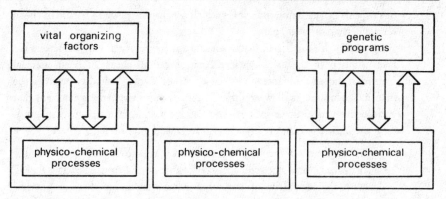

Figure 5.5 On the left, a diagrammatic representation of the vitalist theory: the physico-chemical processes within organisms interact with and are organized by vital organizing factors, such as entelechy. The mechanistic theory denies the existence of such vital factors, and asserts that life can be understood in terms of physico-chemical processes alone (centre). In the modern conception (right), these processes are organized by genetic programs or genetic information, which play much the same role as the organizing factors of vitalism.

organizing principles of its own: the genetic programs. Moreover, purpose is no longer denied but admitted. The old term *teleology,* with its Aristotelian associations, has been replaced by the new term *teleonomy,* the "science of adaptation." As Dawkins has pointed out, "in effect teleonomy is teleology made respectable by Darwin, but generations of biologists have been schooled to avoid 'teleology' as if it were an incorrect construction in Latin grammar, and many feel more comfortable with a euphemism."[24]

Thus the paradigm of modern biology, although nominally mechanistic, has in effect become remarkably similar to vitalism, with genetic "programs" or "information" or "instructions" or "messages" playing the role formerly attributed to vital factors such as entelechies.

Mechanists have always accused vitalists of trying to explain the mysteries of life in terms of empty words, such as entelechy, which "explain everything and therefore nothing." But the vital factors in their mechanistic guises have exactly this characteristic. How does a marigold grow from a seed? Because it is genetically programmed to do so. How does a spider instinctively spin its web? Because of the information coded in its genes. And so on.

The Duality of Matter and Information

All attempts to force the organizing principles of life into material objects such as genes have failed: they keep bursting out again. The concept of purposive organizing principles which are non-material in nature has been reinvented again and again.

In fact this duality of matter and non-material organizing principles has been implicit in the mechanistic theory of life all along. It is an essential feature of the machine metaphor. All machines involve a duality between the material components of which they are made and the purposive designs that were conceived in the minds of their designers and makers. As a contemporary theoretical biologist, Francisco Varela, has expressed it:

> What defines a machine organization is relations, and hence . . . the organization of a machine has no connection with materiality, that is, with the properties of the components that define them as physical entities. In the organization of a machine, materiality is implied but does not enter per se.[25]

This duality of form and matter is in fact inherent in all traditional philosophies of form, as we saw in the preceding chapter. In the modern context, it is usually conceived of in terms of the duality of matter and *information*. Information is what *informs;* it plays an *informative* role, as Norbert Weiner, the founder of cybernetics, emphasized in his concept of the primacy of information over matter and energy. He saw this distinction as essential for the doctrine of materialism: "No materialism which does not admit this can survive at the present day."[26] This may sound like a radical position, but in fact ever since the seventeenth century, the survival of materialism has depended on its combination with the Platonic notion of non-material organizing principles: the laws of nature (chapter 2).

But if biological "information" cannot be understood in terms of the material structures of the genes alone, then what is it? Is the information Platonic, somehow transcending time and space? Or is it immanent within organisms?

In the following chapter we consider the possibility that such information is immanent in morphogenetic fields, inherited in a non-material manner by organisms from their predecessors. But before doing so, we need to look in more detail at the reasons for thinking that it is not inherited materially in the genes and at the reasons why it cannot be fully explained in terms of the chemistry of developing embryos.

Why Genes Are Overrated

What genes are known to do is to code information for the sequence of chemical building blocks in RNA and protein molecules. Thus they help to provide a detailed understanding of the way in which organisms inherit their biochemical potentialities. What they are *not* known to do is to code for morphogenesis or for inherited patterns of behaviour. They are not "determinants" of the characteristics of the organism.

Genetics deals with hereditary *differences* between organisms. For instance, if a certain gene is present or absent, the structure of a fruit-fly may differ (Fig. 5.6). But the fact that the presence of mutant genes leads to differences in form does not prove that genes themselves determine the form.

To see the force of this point, consider the analogy of a radio set. A mutation in one of its transistors might cause the sounds that it is producing to become distorted; and a mutation in one of the components in its tuning

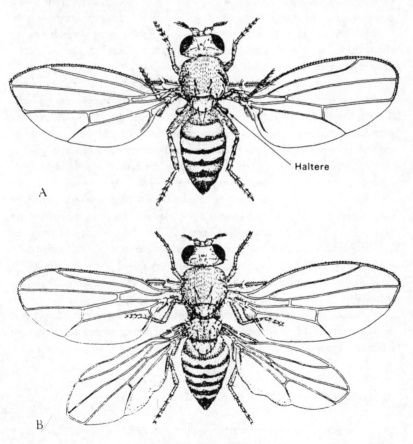

Figure 5.6 A: A normal specimen of the fruit-fly *Drosophila*. B: A mutant fly in which the third thoracic segment has been transformed in such a way that it duplicates the second thoracic segment. Such flies are known as bithorax mutants.

circuit might cause the set to pick up another radio station: an entirely different series of sounds would come out of the loudspeakers. But the fact that mutations in the set's components can cause differences in the sounds the set produces does not prove that these sounds are determined or programmed by the components of the set. These are necessary for the reception of the program, but the sounds are in fact coming from radio stations and are transmitted through the electro-magnetic field. The mutant component is not a component "for" a particular program or type of sound.

Many biologists are, of course, well aware that it is misleading to speak of genes "for" particular characteristics. Dawkins, for instance, has made it clear that if a geneticist speaks of genes "for" red eyes in the fruit-fly *Drosophila,* he implicitly means that "there is variation in eye colour in the population: other things being equal, a fly with this gene is more likely to have red eyes than a fly without the gene." However, he defends talking about genes "for" particular characteristics on the ground that it is "routine genetic practice."[27] And so it is.

The reason these habits of thought endure in spite of the fact that they are known to be misleading seems to be that they are almost unavoidable. They follow from the fundamental assumption made by Weismann, and by Mendelian geneticists and neo-Darwinian theoreticians, that heredity *must* be explicable in material terms. Hence all the hereditary information for the shape of a pigeon's foot or the web-spinning instincts of a spider *must* be in the genes: where else could it possibly be?

By contrast, the hypothesis of formative causation provides a different interpretation of the role of genes. It suggests that they do what they are known to do: namely to code information for the sequence of chemical building blocks in RNA and protein molecules. But it does not project onto the genes the ability to organize the whole organism. Rather, it looks for these hereditary organizing principles in fields that are inherited non-materially.

But is there any need for such a concept? Why cannot embryonic development be fully understood in terms of chemical patterns that both arise from and control the activity of the genes?

Chemical Theories of Pattern Formation

The genes an organism inherits underlie its capacity to make particular RNA and protein molecules. With the help of the ingenious techniques of genetic engineering, specific portions of DNA can now be transferred from one organism to another, enabling it to make proteins that it could not make before. For instance, the DNA that constitutes the human gene for insulin has been transferred to cells of the bacterium *Escherischia coli,* and this protein can now be made in commercial quantities by growing these modified bacteria and purifying the insulin they produce. Genes enable cells to make particular proteins.

In the course of morphogenesis, cells differentiate and different kinds of cells come to make different proteins. Although they all contain the same genes, different genes are *expressed.* For example, in the development of a chrysanthemum flower, at a particular stage, the enzymes responsible for the synthesis of the pigment molecules are made in the cells of the developing petals, and subsequently, through the activity of these enzymes, the coloured pigments appear in the petals. But to describe such chemical changes does not explain how they happen as they do or how morphogenesis is controlled.

Chemical changes *accompany* morphogenesis, and organisms could not develop without the production of appropriate molecules in appropriate quantities in appropriate cells at appropriate times. But how is the production of molecules related to morphogenesis? No one knows. Morphogenesis is generally supposed to happen automatically in a manner as yet obscure through the self-assembling properties of these material constituents. It is as if the delivery of the right building materials and machinery to plots of ground resulted in the spontaneous growth of houses of just the right form.

The main emphasis in research on morphogenesis over the last few decades has not in fact been on morphogenesis per se, but on the control of protein synthesis. How are the right proteins made in the right cells at the right times and in the right amounts? How is the expression of genes controlled as cells differentiate within the developing organism?

Clearly, patterning influences of some kind are at work within the developing tissues and organs. These are generally thought of as systems of "positional information," which "tell" cells where they are and thus enable them to respond appropriately by making the right proteins. What is the nature of this so–called positional information?

The most popular idea is that it is chemical, and depends on concentration gradients of specific chemical substances called morphogens. There has been very little success in actually detecting and identifying the putative morphogens;[28] the main progress has come from making mathematical models of the ways in which such chemical patterns could in theory be formed.

Many of these models depend on the principle that Ilya Prigogine has summed up in·the phrase "order through fluctuations."[29] In a system that is unstable, far from thermodynamic equilibrium, random fluctuations can be amplified through various kinds of positive feedback, and under certain conditions they spontaneously give rise to patterns. For example, in certain kinds of chemical reaction, where at least two substances are reacting catalytically with each other and diffusion is taking place, concentration patterns can appear as the chemicals react (Fig. 5.7). Prigogine has led the way in showing how such processes can be described mathematically in terms of non-equilibrium thermodynamics. He has pointed out that order can arise from "chaos" in comparable ways in many systems, ranging from patterns of convection in heated fluids to patterns of urban growth. For example, once towns and cities have started to develop in certain places, they tend to increase in size through the migration of population, leading in turn to an intensification of economic activity and more migration; but this growth is limited by a variety of other factors, including competition with other cities, especially those that are nearby.

Hans Meinhardt has summarized the principles involved in the making of such models of pattern formation in developing organisms as follows:

> Assuming that development is controlled by substances, any theory of development has to describe concentration changes of substances as a

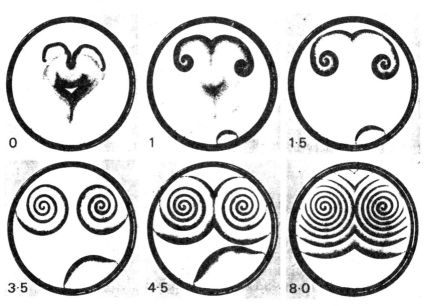

Figure 5.7 The formation of spiral chemical waves when the Belousov–Zhabotinskii reagent is allowed to stand in a shallow dish. The waves appear spontaneously as the chemical reaction proceeds, or they can be initiated by touching the surface with a hot filament, as in the pictures shown here. The numbers indicate how many seconds had elapsed after the initial photograph was taken. (Adapted with permission from *Being to Becoming* by Ilya Prigogine, copyright © 1980, W. H. Freeman and Co.)

function of other substances involved and as a function of spatial co-ordinates and time. Two conditions have to be fulfilled before a stable pattern can be generated. (1) A local deviation from an average concentration should increase further, otherwise no pattern would be formed, and (2) the increase should not go to infinity. Instead, the emerging pattern should reach a stable steady state.[30]

On the basis of such assumption, Meinhardt and his colleague Alfred Gierer and others have constructed a variety of mathematical models involving hypothetical "activator" and "inhibitor" substances. These have been used in computer simulations to show the kinds of patterns they can generate (Fig. 5.8). The interesting feature of some of these models is that they show self-regulatory properties such that the pattern can be restored after part of the model system is "removed." Meinhardt and Gierer have in fact suggested that they are models of morphogenetic fields. We will come back to them again in the next chapter in the context of other ideas about the nature of these fields.

If the hypothetical activators and inhibitors are actually identified in embryos and shown to play the kind of role such models suggest, then they will help to explain how the synthesis of different proteins in different cells is controlled. But they will still leave unexplained what the cells do with the proteins, how they take up their shapes, how they behave, how some of them move around within embryos, how tissues and organs take up their forms, and how organisms respond to their environment. How is the gulf between these hypothetical chemical gradients and the actual organism to be bridged? The answer suggested by the author of the concept of positional information,

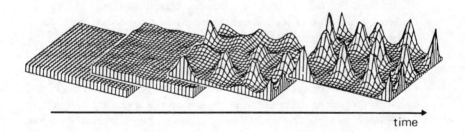

time

Figure 5.8 A computer model of pattern formation. A bristlelike pattern is formed in a non-growing field as a result of random fluctuations which lead to a local production of an "activator" which enhances further "activator" production, and at the same time to the production in these centres of an "inhibitor," which diffuses outwards and inhibits further centres from forming nearby. (From *Models of Biological Pattern Formation* by H. Meinhardt; Academic Press, 1982. Reproduced by permission.)

Lewis Wolpert, is that the cells "interpret that information according to their genetic program."[31]

As we have seen, the concept of the genetic program is misleading, if only because development is not in fact programmatic. A number of leading developmental biologists have in recent years suggested that this concept be abandoned.[32] Sydney Brenner, for example, has proposed that it be replaced by terms such as "internal representation" or "internal description."[33] He has summarized current thinking among developmental biologists as follows:

> At the beginning it was said that the answer to the understanding of development was going to come from a knowledge of the molecular mechanisms of gene control. I doubt whether anyone believes that any more. The molecular mechanisms look boringly simple, and they don't tell us what we want to know. We have to try to discover the principles of organization.[34]

What might such principles of organization be? This is just the question that organismic philosophers and biologists have been wrestling with for decades.

Organic Wholes

The organismic or holistic approach developed under the influence of Whitehead's philosophy of organism (pp. 34–35) and has been influential within biology since the 1930s. It has enabled holistic properties of organisms to be recognized without the need to adopt a vitalist position; and indeed it has offered an attractive way of "transcending" the vitalist-mechanist controversy.[35] Vitalists stressed the holistic, organic qualities of plants and animals, but they did not challenge the mechanistic orthodoxy of physics in relation to non-living things; they made a sharp distinction between the inanimate realm and the realm of life.

By contrast, mechanists assert that there is no difference in kind, only in degree, between the realm of biology and the realms of chemistry and physics. Organicists agree with the mechanists in this respect, and preserve their intuition of the fundamental unity of nature. But rather than regarding living organisms as inanimate machines, they regard physical and chemical systems, such as atoms, molecules, and crystals, as in some sense living; they are not mere inanimate material objects, but "structures of activity," or organisms.

The organismic approach is not reductionistic or atomistic in intent: it does not assume that atoms or subatomic particles have a privileged place in the scale of nature, and it does not attempt to explain all the properties of larger and more complex organisms in terms of the properties of their parts; at each hierarchical level of complexity, organisms behave as wholes, with an organic unity that is irreducible.

In general these hierarchies are "nested" in such a way that higher-level wholes are made up of parts which are themselves organisms at a lower level. Thus sugar crystals, for example, are organisms whose parts are sugar molecules, which are wholes made up of carbon, hydrogen, and oxygen atoms, which are wholes composed of electrons in orbitals around nuclei, and the nuclei are wholes made up of yet smaller organisms, the nuclear particles, which are themselves composed of entities such as quarks. Living organisms show a similar hierarchical arrangement, with organs, containing tissues, containing cells, containing organelles such as nuclei and mitochondria, containing complex molecules, and so on (Fig. 5.9).

Arthur Koestler has proposed the term *holon* for such organisms, which at the same time can be wholes made up of parts, and parts of higher-level wholes: "Every holon has a dual tendency to preserve and assert its individuality as a quasi-autonomous whole; and to function as an integrated part of an (existing or evolving) larger whole. This polarity between the Self-Assertive and Integrative tendencies is inherent in the concept of hierarchic order."[36] For such a nested hierarchy of holons, he proposed the term *holarchy*.

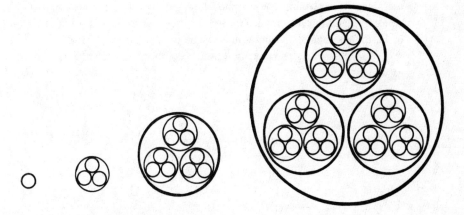

Figure 5.9 Successive levels in a nested hierarchy of morphic units, or holons. At each level, the holons are wholes containing parts, which are themselves wholes containing lower-level holons, and so on. This diagram could represent subatomic particles in atoms, in molecules, in crystals, for example; or cells in tissues, in organs, in organisms.

Another term with a meaning equivalent to holon is *morphic unit*.[37] The word *morphic* emphasizes the aspect of form, and the word *unit* emphasizes the unity or wholeness. Such units are formed by what L. L. Whyte has called "morphic processes" in which "well-formed terminal states can arise from less-formed initial ones."[38]

The organismic approach has encouraged a search for general principles that apply to organisms or "systems" at any level of complexity. This is the goal of general systems theory, which has been strongly influenced by cybernetics, the theory of communication and control, with its basic concepts of information-transfer and feedback. Many mathematical models have been constructed in the spirit of this systems approach, both within biology and in industry, commerce, and society in general.[39] Related "systems approaches" include game theory; and the metaphor of games, relying on the interplay of chance and rules, has been applied to biological evolution and to the development and behaviour of living organisms.[40]

In living systems theory, J. G. Miller has distinguished seven levels of living systems (cell, organ, organism, group, organization, society, and supranational system) and identified nineteen "critical subsystems" at each level: for example "reproducer," "boundary," and "ingestor." At the cellular level, for instance, these particular subsystems are represented by, respectively, chromosomes, cell membranes, and gaps in cell membranes.[41] Such a classification permits illuminating cross-level comparisons and insights.

However, the very generality of the systems approach has limited its usefulness in explaining the morphogenesis of any actual plants or animals. Among organismically minded biologists, the most fruitful idea has proved to be the concept of morphogenetic fields.

Morphogenesis remains mysterious. Can morphogenetic fields help us to understand it?

CHAPTER 6

Morphogenetic Fields

Fields of Different Kinds

Fields are non-material regions of influence. The earth's gravitational field, for example, is all around us. We cannot see it—it is not a material object; but it is nevertheless real. It gives things weight and makes things fall. It is holding us down to earth at this moment; without it we would be floating. The moon moves around the earth because of the curvature of the earth's gravitational field; the earth and all the other planets move around the sun because of the curvature of the sun's field. Indeed the gravitational field pervades the entire universe, curving around all matter within it. According to Einstein, it is not *in* space and time: it *is* space-time. Space-time is not a bland background abstraction; it has a structure, which actively shapes and includes everything that exists or happens within the physical universe.

There are also electro-magnetic fields, which are quite different in nature from gravitation. They have many aspects. They are integral to the organization of all material systems, from atoms to galaxies. They underlie the functioning of our brains and bodies. They are essential to the operation of all our electric machinery. We can see the things around us now, including this book, because we are connected to them through the electro-magnetic field, in which the vibratory energy of light is travelling. And all around us there are countless other vibratory patterns of activity within the field which we cannot detect with our senses; but we can tune in to some of them with the help of radio and TV receivers. Fields are the medium of "action at a distance," and through them objects can affect each other even though they are not in material contact.

We take all this for granted. All our lives go on within these fields all the time, whether or not we know how physicists model them mathematically. There is no doubt that these fields are physically real, however we model them or whatever we choose to call them. We know that these fields exist through their physical effects, even though we usually cannot detect them directly with our senses. For example, the spatial structure of the field around an iron magnet is invisible in itself, but its existence can be revealed by sprinkling iron filings around it (Fig. 6.1). This field, like other kinds of fields, has a continuous, holistic quality, and unlike a material object cannot be chopped up into separate parts. For example, if a magnet is cut in half, each half of it does not have half the original field, abruptly ending where the magnet was cut; it becomes a whole magnet, surrounded by a complete magnetic field.

In addition to these familiar kinds of fields, there are also, according to quantum field theory, various kinds of matter field—electron fields, neutron fields, and so on: microscopic fields within which all the particles of matter exist as quanta of vibratory energy.

None of these various kinds of fields is reducible to the others. It has been a longstanding hope of physicists to understand them all as aspects of a single unified field. In contemporary theoretical physics attempts are being

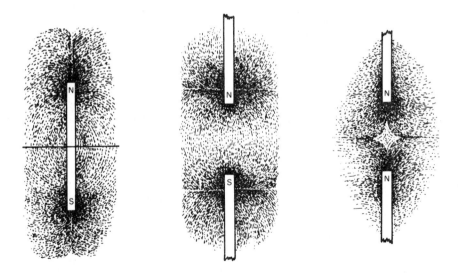

Figure 6.1 Left, the magnetic field around a bar magnet, revealed by sprinkling iron filings around it. Centre, the field between two magnets with north and south poles attracting each other; and (right) with their north poles in mutual repulsion.

made to derive them hypothetically from the original unified field of the cosmos, which then differentiated within itself into the known fields of physics by "curling up" in various ways as the universe grew and evolved. In the context of these new evolutionary field theories: "The world, it seems, can be built more or less out of structured nothingness."[1]

The nature of fields is inevitably mysterious. According to modern physics, these entities are more fundamental than matter. Fields cannot be explained in terms of matter; rather, matter is explained in terms of energy within fields. Physics cannot explain the nature of the different kinds of fields in terms of anything else physical, unless it be in terms of a more fundamental unified field, such as the original cosmic field. But then this too is inexplicable—unless we assume it was created by God. And then God is inexplicable.

We can, of course, assume that fields are as they are because they are determined by eternal mathematical laws, but then there is the same problem with these laws: how can we explain *them?*

We return to a discussion of the known fields of physics in chapter 7 and consider recent theories of the evolution of fields in chapter 17. First we explore the possibility that there are many more kinds of fields than those currently recognized by physics: the morphogenetic fields of all the various kinds of living cells, tissues, organs, and organisms.

Morphogenetic Fields

At the beginning of the 1920s, at least three biologists independently proposed that in living organisms morphogenesis is organized by fields: Hans Spemann, 1921; Alexander Gurwitsch, 1922; Paul Weiss, 1923. These fields were called developmental, embryonic, or morphogenetic. The idea was that they both organized normal development and guided the processes of regulation and regeneration after damage. Gurwitsch wrote of them as follows:

> The place of the embryonal formative process is a field (in the usage of the physicists) the boundaries of which, in general, do not coincide with those of the embryo but surpass them. Embryogenesis, in other words, comes to pass inside of the fields. . . . Thus what is given to us as a living system would consist of the visible embryo (or egg, respectively) and a field.[2]

Paul Weiss applied the field concept to the detailed investigation of embryonic development, and in his seminal book *Principles of Development* wrote of the fields as follows:

A *field* is the condition to which a living system owes its typical organization and its *specific* activities. These activities are specific in that they determine the *character* of the formations to which they give rise. . . . Inasmuch as the action of fields does produce spatial order, it becomes a postulate that the field factors themselves possess definite order. The three-dimensional heterogeneity of developing systems, that is, the fact that these systems have different properties in the three dimensions of space, must be referred to a three-dimensional organization and heteropolarity of the originating fields.[3]

The specific nature of the fields, according to Weiss, means that each species of organism has its own morphogenetic field, although fields of related species may be similar. Moreover, within the organism there are subsidiary fields within the overall field of the organism, in fact a nested hierarchy of fields within fields (cf. Fig. 5.9).

During the 1930s C. H. Waddington attempted to clarify the field concept with the idea of "individuation fields" associated with the formation of definite organs with characteristic individual shapes. In the 1950s he extended the field idea in his concept of the chreode,[4] or developmental pathway. He illustrated this by means of a simple three-dimensional analogy, the epigenetic landscape (Fig. 6.2). The development of a particular part of

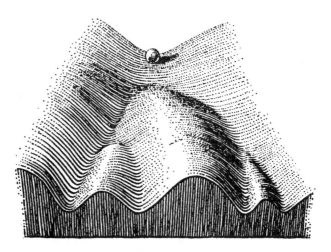

Figure 6.2 Part of an "epigenetic landscape," illustrating Waddington's concept of chreodes as canalized pathways of change. The chreodes correspond to the valleys, and lead to particular developmental end-points, which could, for example, be the sepals, petals, stamens, and pistils in a flower. (From *The Strategy of Genes* by C. H. Waddington; George Allen and Unwin, Ltd., 1957. Reproduced by permission.)

the egg is represented by the rolling of a ball downwards. It can follow a branching series of alternative paths, which correspond to the pathways of development of the different types of organs. In the organism these are quite distinct: for example, the heart and the liver have well-defined structures and do not grade into each other through a series of intermediate forms. Development is "canalized" towards definite end-points. Disturbances of normal development may push the ball away from the valley bottom and up the neighbouring hillside, but unless it is pushed over the top into another valley it will find its way back to the valley bottom—not to the point from which it started, but to a later position on the canalized pathway of change. This represents embryonic regulation.

The concept of morphogenetic fields and of chreodes within them differs from Driesch's idea of entelechies in that the field concept implies the existence of profound analogies between the organizing principles of the biological realm and the known fields of physics. By contrast, Driesch, as a vitalist, stressed the radical difference between the realm of life and the realms of physics and chemistry. However, there is no doubt that many features of entelechies were carried over into the concept of morphogenetic fields. Like entelechy, these fields were endowed with properties of self-organization and goal-directedness; and like entelechy, they were assumed to play a causal role, guiding the systems under their influence towards characteristic patterns of organization. For example, Weiss thought of the fields as complexes of organizing factors which "cause the originally indefinite course of the individual parts of the germ to become definite and specific, and furthermore, cause this to occur in compliance with a typical pattern."[5] And Waddington's concept of chreodes canalizing development towards particular goals strongly resembles the pulling or attracting of pathways of development towards ends given by entelechy. The ends or goals of the chreodes from the point of view of a developing system lie in the future, and Waddington described them in the language of dynamics as "attractors."[6] Modern mathematical dynamics is teleological in that it involves the idea of "basins" within which are "attractors" representing the states towards which dynamical systems are drawn.[7]

René Thom has developed Waddington's ideas in mathematical models in which the structurally stable end points towards which systems develop are represented by attractors or basins of attraction within morphogenetic fields.

All creation or destruction of forms, or morphogenesis, can be described by the disappearance of the attractors representing the initial forms, and their replacement by capture by the attractors representing the final forms.[8]

Thom himself has compared this approach to Driesch's: "Our method of attributing a formal geometrical structure to a living being to explain its stability may be thought of as a kind of *geometrical vitalism;* it provides a global structure controlling the local details like Driesch's entelechy."[9]

The field approach contrasts with the scheme of Weismann and his followers in that the field, rather than the germ-plasm, is central. It is the field, not the germ-plasm, that shapes the organism. But development does not depend on fields alone, of course; it is also affected by genes and environmental influences. This is represented in the diagram in figure 6.3, devised by Brian Goodwin, which clearly brings out the difference between the Weismannian approach and the idea of morphogenetic fields.

The Nature of Morphogenetic Fields

What exactly are morphogenetic fields? How do they work? Despite the widespread use of this concept within biology, there are no clear answers to these questions. Indeed, the nature of these fields has remained as mysterious as morphogenesis itself.

As might be expected, the fields have been interpreted in radically different ways, which reflect the three major philosophies of form. From the Platonic point of view, they represent changeless Forms or Ideas, which may in turn be thought of in a Pythagorean spirit as essentially mathematical. In an Aristotelian manner, they inherit most of the features of entelechies, and

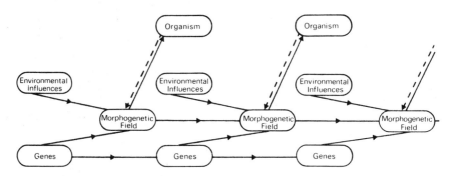

Figure 6.3 The influence of genes and the environment on morphogenetic fields (after Goodwin, 1984). The addition to Goodwin's diagram of the arrows (broken lines) indicating that the organisms affect the fields brings this scheme into close correspondence with the hypothesis of formative causation.

play a causal role in organizing the material systems under their influence. From a nominalist perspective, they merely provide convenient ways of describing the phenomena of morphogenesis, which are usually assumed to proceed entirely mechanistically. All these interpretations coexist within developmental biology, and sometimes the same author oscillates among them, even within a single paragraph.

The causal role of the fields, and the features they inherited from Driesch's entelechy, are usually left implicit. But interpretations in a Platonic or Pythagorean spirit have been made explicit.

Gurwitsch stressed the geometrical properties of the fields and treated them as ideal mathematical constructions. The source and extent of a field were not confined to the material of a developing organism, and its centre could well be a geometrical point outside the organism.[10]

Thom has tried to develop what could be described as a dynamic Platonism, in which not only forms themselves can be characterized mathematically, but also the ways in which forms change into each other. This is the basis of his catastrophe theory, in which the possible ways in which forms can change into each other are categorized in terms of a limited number of basic "catastrophes." His models of morphogenetic fields incorporate such catastrophes, and he thinks of these fields as mathematical objects that somehow determine biological forms. He compares them with the mathematical structures that, in physics, are taken to determine chemical forms:

> If sodium and potassium exist, this is because there is a corresponding mathematical structure guaranteeing the stability of atoms Na and K; such a structure can be specified, in quantum mechanics, for a simple object like the hydrogen molecule, and although the case of the Na or K atom is less well understood, there is no reason to doubt its existence. I think that likewise there are formal structures, in fact, geometric objects, in biology which prescribe the only possible forms capable of having a self-reproducing dynamic in a given environment.[11]

He argues that the reductionist attempt to "reconstruct a complex space out of simple elements" is quite incapable of giving any understanding of morphogenesis, and concludes that "the Platonic approach is in fact unavoidable."[12]

Brian Goodwin also emphasizes the mathematical nature of morphogenetic fields, which he sees in terms of "generative field equations." The development of organisms is not to be understood in terms of the germplasm, as Weismann supposed, or the DNA or the genetic program. "Rather, generation is to be understood as a process which arises from the field properties of the living state, with inherited particulars arising to stabilize particular solutions of the field equations so that specific morphologies are generated"[13] (Fig. 6.3). In other words, organisms take up the shapes that the stabilization of the field equations requires; and genes indirectly bring about their effects on form by stabilizing some solutions of the field equations rather than others. It is the hope of Goodwin and his colleague Webster that an understanding of these generative equations will enable a rational science of biological form to be constructed (pp. 67–69).

> What is required is the deduction of the correct relational order which generates the observed phenomena, and this order or organization is not directly observable, though it is real. This logical relational order is what defines the distinctive organizational properties of living organisms. . . . The appropriate mathematical description is provided by field equations. . . . An understanding of morphogenesis provides the basis for a rational taxonomy, one based on the logical properties of the generative process rather than a genealogical taxonomy based upon the accidents of history.[14]

But while from a Platonic or Pythagorean point of view the fields represent an objective mathematical reality, and are equally objective if conceived of in an Aristotelian spirit as immanent organizing principles, from a nominalist perspective they have no reality outside our minds. On occasion, some of the proponents of the field concept have indeed denied that they have any objective existence. Paul Weiss, for example, on the one hand thought of them as "physically real," but on the other hand thought that the field concept was nothing more than an abstraction produced by the mind. "Being an abstraction we cannot expect it to return more than what we have put into it. Its analytical and explanatory value, therefore, is nil."[15]

A similar ambiguity was shown by Waddington, who did so much to develop and promote the field concept within biology. On one occasion he wrote as follows:

> Any concept of a "field" is essentially a descriptive convenience, not a causal explanation. . . . The operative forces have to be experimentally identified separately in each instance. Only if the forces are always the same or of very few kinds, as they are in gravitational or electromagnetic fields, or if the maps are always the same, would the field concept be a unifying paradigm; and we know that none of these conditions is fulfilled.[16]

But if the fields have no causal role, and are merely a convenient way of talking about complex physical and chemical processes, then there seems to be little to distinguish this approach from a sophisticated version of the mechanistic theory. And indeed, among contemporary biologists morphogenetic fields are often conceived of in conventional physical or chemical terms. But when this approach is pursued far enough, sooner or later it leads away from explanations in purely material terms and back towards a mathematical or Platonic vision.

This is what has happened through the mathematical modelling of morphogenetic fields by Gierer, Meinhardt, and others (pp. 91–93). They start with a conventional mechanistic assumption:

> Since we do not yet know the biochemical or physical nature of the fields, we have to introduce an assumption as to the general class of physics the phenomenon belongs to. If we assumed that the basic phenomenon were magnetism, we would then try to understand it in terms of Maxwell's equations. The realistic assumption seems to be that morphogenetic fields have the same basis as other biological phenomena which have physical explanations so far: namely that they are primarily due to the interaction and movement of molecular compounds.[17]

Such processes can then be described by appropriate equations. However, as Gierer points out:

> Such equations are rather non-committing with respect to details of molecular mechanism. They represent an attempt to "demystify" morphogenetic fields, proposing that they are due to conventional molecular biology and nothing else; and yet, they impose stringent constraints on the construction of theories and models.[18]

These mathematical models are usually based on the assumption that there are self-activating chemical processes in certain regions, with inhibitory effects spreading over a wider area. The local activation is self-enhancing, so some slight initial advantage in a particular place can develop into a striking activation. But the production and spreading of inhibitory effects prevent an overall catalytic explosion, causing activation in one part of the area to proceed only at the expense of deactivation elsewhere, until a stable pattern is formed. Computer simulations based on such models show that they can generate a variety of simple patterns (Fig. 5.8), and that some of these are able to "regenerate" after damage.

But while such models may help to account for the *spacing* between different patterns of chemical activity in cells, particularly the production of different proteins, they explain neither the forms of the cells nor the structures they give rise to. For example, an understanding of the factors influencing the spacing of hairs on a leaf would not explain the shapes of the hairs. In a similar way, to return to one of Prigogine's examples (p. 92), a mathematical model of urbanization may shed light on the factors affecting the rate of urban growth, but it cannot account for the different architectural styles, cultures, and religions found in, say, Indian and Brazilian cities.

Diffusing chemicals are not the only possible physical factors in terms of which morphogenetic fields can be modelled: other candidates include chemical and electrical pulses,[19] electrical fields,[20] and the viscoelastic properties of gels.[21]

Although such models are based on assumptions about possible chemical or physical mechanisms, they are essentially mathematical in nature, and their explanatory value is inseparable from the mathematics. Just as classical physics brings together the Platonic and the materialist traditions, so these mathematical models attempt to provide a comparable synthesis, as Gierer has made explicit:

> Only a combination of knowledge of mathematics and matter is expected to lead to a satisfactory understanding of biological pattern formation. It is psychologically understandable that most biochemists and molecular biologists prefer the materialist and most mathematicians the formal aspect of the problem. Philosophically, it appears that the formal mathematical aspect is more fundamental than the structural one for understanding, but it is insufficient for experimental confirmation. However, it is worth noting that the relative explanatory value of mathematics versus matter is the subject of an age-old controversy traceable to Pythagoras and Plato in favour of mathematics, and Democritus (as well as Marx) in favour of the materialist notion, and perhaps not objectively resolvable.[22]

The Evolution of Morphogenetic Fields

The kinds of theories about morphogenetic fields that we have just considered have had a wide influence on contemporary research, and provide the most promising way of modelling the processes of morphogenesis. But for over sixty years, these fields have existed in a theoretical limbo. They seem to be new kinds of fields so far unknown to physics, but at the same time not to be new kinds of fields at all, or just words that refer to regularities which we can describe and model.

I believe that it is possible to go beyond these unsatisfactory ambiguities by taking into account what must be one of the most essential features of these fields: they have evolved. They have an inherently historical aspect. Organisms inherit them from their ancestors. How could these fields be inherited?

There seem to be only two kinds of answer possible. The first is in terms of a combination of genetics and Platonism, in the traditional spirit of the mechanistic theory. The second possibility is that *memory* is inherent in the fields.

The first of these approaches implies the existence of transcendent mathematical formulae for all possible living organisms. Richard Dawkins has produced a computer model of this Platonic realm, called Biomorph Land, in which the possible forms of organisms, called biomorphs, exist. Natural selection propels populations of organisms along trajectories of gradual genetic change into new biomorphs through an intermediate series of biomorphs. But all the possible biomorphs pre-exist quite independently of the actual course any particular evolutionary process might take; they are already mathematically specified in the computer program for Biomorph Land.[23]

Biological evolution, conceived of from the Platonic perspective, depends on the evolution of genetic systems which enable some of the possible forms of organisms to be reified in the physical world; but the formulae or biomorphs themselves do not evolve. They are like the eternal Forms of all possible species, and they exist in a transcendent realm that is quite independent of these organisms' actual existence. The morphogenetic field equations for *Tyrannosaurus rex,* for example, already existed before the earth came into being, or even before the cosmos was born. They were totally unaffected by the evolutionary appearance of this kind of dinosaur, and equally unaffected by its subsequent extinction.

On the other hand, if morphogenetic fields contain an inherent memory, their evolution can be conceived of in a radically different way. They are not transcendent Forms, but immanent in organisms. They evolve *within* the realm of nature, and they are influenced by what has happened before. Habits build up within them. From this point of view, mathematical models of these fields are only models; they do not represent transcendent mathematical realities that *determine* the fields.

The idea that morphogenetic fields contain an inherent memory is the starting point for the hypothesis of formative causation. The reason I am putting it forward is that I think it could lead towards a genuinely evolutionary understanding of organisms, including ourselves. I do not believe that the only available alternative, the traditional mechanistic combination of materialism and Platonism, can do so because it is rooted in a pre-evolutionary conception of the universe, a conception that physics itself is in the process of superseding.

The Hypothesis of Formative Causation

The hypothesis of formative causation, which the rest of this book explores, starts from the assumption that morphogenetic fields are physically real, in the sense that gravitational, electro-magnetic, and quantum matter fields are physically real. Each kind of cell, tissue, organ, and organism has its own kind of field. These fields shape and organize developing microorganisms, plants, and animals, and stabilize the forms of adult organisms. They do this on the basis of their own spatio-temporal organization.

The temporal aspect of morphogenetic fields is brought out most clearly in the concepts of chreodes and morphogenetic attractors. The morphogenetic fields relate developing organisms to future patterns of organization, towards which chreodes guide the developmental process.

So far, this proposal merely makes explicit what has been implicit in the concept of morphogenetic fields all along. What is new in the hypothesis of formative causation is the idea that the structure of these fields is not determined by either transcendent Ideas or timeless mathematical formulae, but rather results from the actual forms of previous similar organisms. In other words, the structure of the fields depends on what has happened before. Thus, for example, the morphogenetic fields of the foxglove species are shaped by influences from previously existing foxgloves. They represent a kind of pooled or collective memory of the species. Each member of the species is moulded by these species fields, and in turn contributes to them, influencing future members of the species.

How could such a memory possibly work? The hypothesis of formative causation postulates that it depends on a kind of resonance, called morphic resonance. Morphic resonance takes place on the basis of similarity. The more similar an organism is to previous organisms, the greater their influence on it by morphic resonance. And the more such organisms there have been, the more powerful their cumulative influence. Thus a developing foxglove seedling, for example, is subject to morphic resonance from countless foxgloves that came before, and this resonance shapes and stabilizes its morphogenetic fields.

Morphic resonance differs from the kinds of resonance already known to science, such as acoustic resonance (as in the sympathetic vibration of stretched strings), electro-magnetic resonance (as in the tuning of a radio set to a transmission at a particular frequency), electron-spin resonance, and nuclear-magnetic resonance. Unlike these kinds of resonance, morphic resonance does not involve a transfer of energy from one system to another, but rather a non-energetic transfer of information. However, morphic resonance does resemble the known kinds of resonance in that it takes place on the basis of rhythmic patterns of activity.

All organisms are structures of activity, and at every level of organization they undergo rhythmic oscillations, vibrations, periodic movements, or cycles.[24] In atoms and molecules, the electrons are in ceaseless vibratory movement within their orbitals; large molecules such as proteins wobble and undulate with characteristic frequencies.[25] Cells contain innumerable vibrating molecular structures, their biochemical and physiological activities exhibit patterns of oscillation,[26] and the cells themselves go through cycles of division. Plants show daily and seasonal cycles of activity; animals wake and sleep, and within them hearts beat, lungs breathe, and intestines contract in rhythmic waves.[27] The nervous system is rhythmic in its functioning, and the brain is swept by recurrent waves of electrical activity.[28] When animals move, they do so by means of repetitive cycles of activity, as in the wriggling of a worm, the walking of a centipede, the swimming of a shark, the flying of a pigeon, the galloping of a horse. We ourselves go through many such cycles of activity, for example in our chewing, walking, cycling, swimming, and copulating.

According to the hypothesis of formative causation, morphic resonance occurs between such rhythmic structures of activity on the basis of similarity, and through this resonance past patterns of activity influence the fields of subsequent similar systems. Morphic resonance involves a kind of action at a distance in both space and time. The hypothesis assumes that this influence does not decline with distance in space or in time.

The coming into being of form does not take place in a vacuum. All processes of development start from systems that are already specifically organized. For example, an embryo develops from a fertilized egg containing DNA, proteins, and other molecules that are organized in particular ways and are characteristic of the species. Such organized starting structures, or morphogenetic germs, enter into morphic resonance with previous members of the species. In other words, the developing embryo is "tuned in" to the fields of the species and thereby becomes surrounded by, or embedded within, the chreodes that shape its development, as the development of countless other embryos before it has been shaped.

Because all past members of a species influence these fields, their influence is cumulative: it increases as the total number of members of the species grows. Since these past organisms are similar to each other rather than identical, when a subsequent organism comes under their collective influence, its morphogenetic fields are not sharply defined, but consist of a composite of previous similar forms. This process is analogous to composite photography, in which "average" pictures are produced by superimposing a number of similar images (Fig. 6.4). Morphogenetic fields are "probability structures," in which the influence of the most common past types combines to increase the probability that such types will occur again.

Influence Through Space and Time

In Weismann's scheme, there is a one-way flow of influence from the germ-plasm to the somatoplasm (Fig. 5.3); in modern terms, there is a

Figure 6.4 Composite photographs of 30 female and 45 male members of the staff of the John Innes Institute, Norwich, England. (Reproduced by courtesy of the John Innes Institute.)

one-way flow from genotype to phenotype. A Platonic interpretation of fields as generative equations shares this idea of unidirectional influence (pp. 67–69): the fields, in combination with genetic and environmental factors, generate the adult organism. In no sense does the actual form of the organisms alter the field equations, which are assumed to transcend physical reality.

By contrast, the hypothesis of formative causation postulates a two-way flow of influence: from fields to organisms and from organisms to fields. This can be represented by including extra sets of arrows in Goodwin's diagram (Fig. 6.3).

The differences between the various possible theories of form are indicated diagrammatically in figure 6.5. A Platonic interpretation of the forms of organisms in terms of archetypal Ideas implies a one-way influence from the Idea to the organism, and the Idea itself undergoes no change (Fig. 6.5A). Indeed it cannot change, since it is transcendent, beyond both time and space. The Form is potentially present everywhere and always, and can be

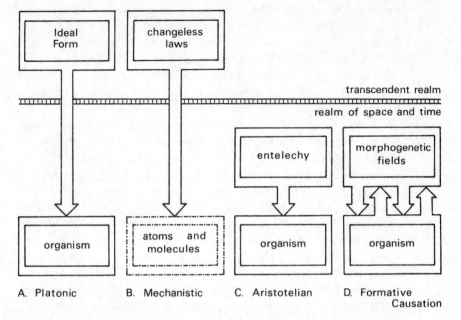

Figure 6.5 A diagrammatic comparison of different theories of form. In the Platonic theory, there is a one-way influence from the transcendent Form to the organism; in the mechanistic theory, there is a comparable one-way influence of the transcendent laws of nature on the atoms and molecules that constitute the organism. In the Aristotelian theory, the organizing entelechies are immanent within and around the organism, rather than transcendent, and in this respect resemble morphogenetic fields. However, entelechies, like Platonic Forms and transcendent laws, are fixed in nature and hence cannot evolve; they are not affected by what actually happens in successive generations of organisms. By contrast, according to the hypothesis of formative causation, morphogenetic fields are affected by what happens within organisms, and contain an inherent memory; they are evolutionary in nature.

reflected in the form of organisms anywhere and at any time in the universe, if the conditions are appropriate.

The mechanistic theory emphasizes the reality of atoms and molecules within organisms, but regards the way in which they interact as a consequence of universal laws (Fig. 6.5B). Like the Platonic Ideas, these laws are not material things which can be located in space and time; rather, they are potentially present and active throughout the universe: they always have been and always will be.

Aristotelian entelechies, by contrast, do not have a transcendent existence beyond space and time (Fig. 6.5C). They are associated with organisms and do not exist independently of them. But they remain the same: they do not evolve. Like Platonic Ideas or universal laws, they have a one-way influence *on* organisms; but their nature is not affected *by* the organisms.

Morphogenetic fields resemble entelechies in that they do not exist transcendently, in a free-floating manner, independently of actual organisms (Fig. 6.5D). But they are influenced by organisms and through morphic resonance are shaped by the fields of similar past organisms.

We are used to the idea of causal influences acting at a distance in space and time through fields: for example, when we look at distant stars we are subject to influences coming from far away and from thousands of years ago through the medium of the electro-magnetic field in which the light is travelling. But the idea of morphic resonance involves a different kind of action at a distance, which is harder to conceive of because it does not involve the movement of quanta of energy through any of the known fields of physics.

This raises the problem of the medium of transmission: how does morphic resonance take place through or across time and space? In answer to this question, we might imagine a "morphogenetic aether," or another "dimension," or influences passing "beyond" space-time and then re-entering. But a more satisfactory approach may be to think of the past as pressed up, as it were, against the present, and as potentially present everywhere. The morphic influences of past organisms may simply be *present* to subsequent similar organisms.

We are so used to the notion of immutable physical laws that we take them for granted; but if we pause to reflect on the nature of these laws, they are profoundly mysterious. They are not material things, nor are they energetic. They both transcend space and time and are, at least potentially, present in all places and at all times.

Although morphic resonance seems mysterious, the conventional theories seem no less so when we stand back and look at the remarkable assumptions they embody. The hypothesis of formative causation is not a bizarre metaphysical speculation that contrasts with a hard, empirical, down-to-earth theory of mechanism. The mechanistic theory depends on assumptions that are, if anything, *more* metaphysical than the idea of formative causation.

Morphic Fields

Morphogenetic fields, in the sense of the hypothesis of formative causation, will hereafter be referred to by the term *morphic fields*. This term is easier to pronounce; and it also serves to distinguish this new conception of morphogenetic fields from the conventional conceptions of them. The term is more general in its meaning than *morphogenetic field*, and includes other kinds of organizing fields in addition to those of morphogenesis; as we shall see later, the organizing fields of animal and human behaviour, of social and cultural systems, and of mental activity can all be regarded as morphic fields which contain an inherent memory.

Fields of Information

Information is a fashionable word. We live in an "information age" and our lives are surrounded by information technologies. Information plays a formative or in-formative role. But what is it? Both within and beyond the bounds of scientific discourse, the general usage of this word bears no close or well-defined relation to the technical conception of information in information theory. This mathematical procedure has a rather narrow range of applications, and has been of very limited value within biology.[29] When biologists speak of "genetic information," for example, usually they are using this word in a vague, non-technical sense, which is often interchangeable with an equally vague and non-technical sense of the word *program*.

> Information, the modern source of form, is seen to reside in molecules, cells, tissues, "the environment," often latent but causally potent, allowing those entities to recognize, select, and instruct each other, to construct each other and themselves, to regulate, control, induce, direct, and determine events of all kinds.[30]

The nature of this information remains obscure, and the use of alternative words such as *instructions* or *programs* does little to clarify it. Is it physical or is it mental? Is it essentially mathematical? Is it some kind of conceptual abstraction? If so, an abstraction from what?

In so far as information is used to explain the development and evolution of bodies, behaviour, minds, and cultures, it cannot be regarded as static, but must itself develop and evolve.

Morphic fields play a role comparable to information and programs in conventional biological thought, and they can indeed be regarded as *fields of information*. Thinking of information as contained in morphic fields helps to demystify this concept, which otherwise seems to be referring to something that is essentially abstract, mental, or mathematical, or at any rate non-physical in nature. It also draws attention to the evolutionary nature of biological information, for these fields contain a built-in memory sustained by morphic resonance.

The Appearance of New Fields

The morphic fields of any particular organism, say a sunflower plant, are shaped by influences from previous generations of sunflowers. But morphic resonance cannot explain how the *first* fields of this kind arose. In the context of biological evolution, the sunflower fields are clearly closely related to the fields of other related species, such as Jerusalem artichokes, and are no doubt descended from the fields of a long line of ancestral species. But how the fields of the sunflower genus, or of the family Compositae of which it is a member, or of the first flowering plants, or indeed the first living cells, came into being in the first place cannot be explained in terms of the hypothesis of formative causation. This is a question of origination, or creativity.

Fields of new kinds of organisms *must* somehow come into being for the first time. Where do they come from? Perhaps they do not come *from* anywhere, but somehow arise spontaneously. Perhaps they are organized by some "higher" kind of field. Or perhaps they represent a manifestation of pre-existing archetypes which until then were entirely transcendent. Perhaps, indeed, they arise from changeless Forms or mathematical entities, which through coming into being in the physical universe take on a life of their own. These possibilities are explored in more detail in chapter 18. But in considering the hypothesis of formative causation, it does not matter which, if any, of these answers is preferred. The hypothesis deals only with morphic fields that have already come into being.

Here again, we should bear in mind, the alternatives to the hypothesis of formative causation pose problems that are just as profound. If organisms are organized by changeless mathematical laws, by generative equations, or by whatever it is that mathematical models correspond to, we do not have to ask where these come from, because they are presumed to be eternal. But then we have the problem of changeless laws or equations which somehow already existed at the time the universe came into being. The generative equations for sunflowers, for example, would have had to exist in some sense before the first living cells arose on earth, and before the Big Bang itself.

Even if we recoil from such metaphysical speculations and adopt a thoroughly empirical approach, the fact remains that the hypothesis of formative causation makes some testable predictions that differ radically from those of conventional theories. And the reason why they differ is that the currently orthodox theories of science implicitly assume that the laws of nature are everywhere and always the same. Whether or not this assumption is recognized to be metaphysical, or even acknowledged at all, it is definitely present. It underlies the ideal of the repeatability of experiments, and is built into the foundations of the scientific method as we know it (chapter 2). The hypothesis of formative causation throws this assumption into question. It suggests that the invisible organizing principles of nature, rather than being eternally fixed, evolve along with the systems they organize.

CHAPTER 7

Fields, Matter, and Morphic Resonance

In this chapter, we first consider how physicists have conceived of the relationship of fields to matter. We then examine how the idea of morphic fields proposed by the hypothesis of formative causation relates to orthodox conceptions of fields, and go on to explore some of the consequences of this hypothesis at the levels of molecules and crystals, looking at some of the ways in which the theory could be tested experimentally. The chapter concludes with a discussion of the role of morphic resonance from a system's own past in the development and maintenance of its structure.

Aether, Fields, and Matter

Morphic fields, like the gravitational, electro-magnetic, and quantum matter fields known to physicists, are intimately related to matter. They interact with it and organize it. At first sight, this idea seems to imply a duality of fields and matter. But matter is no longer conceived of as a passive, inert substance; it is no longer made up of the hard billiard-ball atoms of nineteenth-century physics. It is now thought to consist of rhythmic processes of activity, of energy bound and patterned within fields.

In order to obtain a clearer conception of the relationship of morphic fields to their associated organisms, it is helpful to trace the development of the field concept within physics and its relationship to the concept of matter.

Modern field theories are rooted in the work of Michael Faraday, who through his investigation of magnetism came to the conclusion that "lines of force" extended around a magnet (Fig. 6.1). These were states of strain and were physically real.[1] But they were not made of ordinary matter. So what sort of reality had they? He was not sure, and suggested alternative interpretations. Either they have physical existence as states of a material medium "which we may call aether," or else they have physical existence as states of "mere space." He preferred the idea that the lines of force were modifications of space because it was linked to other speculations of his which treated material particles as point centres of converging lines of force, an interpretation that broke down the distinction between matter and force.[2] Indeed, he supposed that forces themselves were the sole physical substance, a substance filling all space, in which each point of the force field had a certain amount of force associated with it. Each point interacted with its neighbours, allowing for vibrations of force and for all kinds of *patterns* of force, including material bodies.[3]

However, these ideas of Faraday's were not adopted by his successors, and it was not until Einstein that his favoured conception of the field as a state of "mere space" was again taken seriously in physics. Maxwell adopted Faraday's less favoured view and regarded the field as a state of a material medium, the aether. The aether had something of the nature of a fluid, within which were rotating tubelike vortices. If there was a difference between the speed of rotation of neighbouring vortices, then forces were exerted and states of stress arose. But Maxwell was very guarded in his use of the fluid analogy:

> The substance here treated must not be assumed to possess any of the properties of ordinary fluids except those of freedom of motion and resistance to compression. It is not even a hypothetical fluid which is introduced to explain actual phenomena. It is merely a collection of imaginary properties which may be employed for establishing certain theorems in pure mathematics in a way more intelligible to many minds and more applicable to physical problems than that in which algebraic symbols are employed.[4]

But whatever its nature, the field was essential for Maxwell's description of electro-magnetic interactions at a distance because of the time delay in transmission. He took this delay to mean that physical processes must be taking place in the intervening space.[5]

In the late nineteenth century, Hendrik Lorentz abandoned the idea that the aether is a mechanical substance. He thought of it as immobile, and made a clear separation between aether and matter. He rejected contemporary attempts to conceive of the aether as a subtle form of matter, and thought of it as altogether different. Others took this process further; and by the end of the century, fields rather than matter were coming to be seen as primary. Instead of trying to explain fields in terms of matter, attempts were being made to explain matter in terms of fields. For example, Joseph Larmor wrote in 1900 that "matter may be and likely is a structure in the aether but certainly aether is not a structure made of matter."[6]

But then what was the aether? Lorentz continued to think of it as in some sense substantial. As late as 1916 he wrote: "I cannot but regard the aether, which can be the seat of an electro-magnetic field with its energy and its vibration, as endowed with a certain degree of substantiality, however different it may be from all ordinary matter."[7] For Lorentz, the aether served as a medium, albeit non-mechanical, and also as an absolute reference frame, with a role similar to that of Newton's absolute space.

But for Einstein the aether became "superfluous." In his special theory of relativity (1905), the electro-magnetic field permeates the vacuum of empty space, and this space is no longer absolute. The field has no mechanical basis whatsoever; nevertheless, it is the seat of complex processes and, like ponderable matter, has energy and momentum. It can interact with matter and in so doing exchange energy and momentum with it. But the field is independent of matter. It is not a state of matter. It is a state of space.[8]

In his general theory of relativity, Einstein extended the field concept to gravitational phenomena. The gravitational field, a space-time continuum curved in the vicinity of matter, replaced the Newtonian concept of a gravitational force acting at a distance. Gravitation is a consequence of the geometrical properties of space-time itself. But Einstein failed to formulate a unified field theory in which electro-magnetic effects could also result from these geometric properties. Many attempts have been made to define such a theory, and physicists are continuing these attempts (chapters 6 and 17).

The general theory of relativity deals with large-scale phenomena such as the movements of planets, and indeed embraces the very structure of the universe. The intimate interrelations of fields and matter in the realm of very small things such as atoms are the province of quantum theory.

Quantum theory started from the idea that atoms absorb and emit light in quanta, or units, of energy. Light waves have to be thought of in terms of "packets," and these quanta give light a particulate aspect. The "particles" of light are called photons.

A "quantum jump" took place in quantum theory in 1924 when de Broglie suggested that just as light waves have particulate properties, so should particles of matter display wavelike properties.[9] This led to an entirely new conception of electrons and other subatomic particles, which had previously been pictured as tiny billiard balls. Experiments soon showed that electrons did indeed behave like waves. This is now a matter of practical engineering. For example, the electron microscope operates like an ordinary microscope, but uses electrons instead of light waves. But de Broglie's theory is not confined to subatomic particles: *all* matter has a wavelike aspect, even whole atoms and molecules.

This theory provides the basis for the idea of quantum matter fields. These fields are different in kind from electro-magnetic fields, but are as real. The matter waves do not merely describe the behaviour of single particles such as electrons; they are taken to be aspects of a matter field in which the particles are quanta of excitation. Thus an electron is a particle in a matter field, just as a photon is a particle in the electro-magnetic field.

There are many kinds of matter fields, one for each type of particle: an electron is a quantum of the electron-positron field; a proton is a quantum of the proton-antiproton field; and so on. Different kinds of matter fields can interact with each other, and they can also interact with electro-magnetic fields. All these interactions are mediated by quanta.[10]

In these quantum matter fields, there is no duality of field and particle in the sense that the field is somehow external to the particle. Indeed, the essential physical reality has become a set of fields, and the fields specify the probabilities of finding quanta at particular points in space. The particles are manifestations of the underlying reality of the fields.

These fields are states of space, or of the vacuum. But the vacuum is not empty; rather, it is full of energy, and itself undergoes quantum fluctuations that create new quanta "from nothing," which are then annihilated again. A particle and its antiparticle can spring into "virtual existence" at a point in space and then immediately annihilate each other.[11]

Atoms and Organisms: Fields Within Fields

The result of all this is that particles of matter are quanta of energy in fields, which are states of space, or the vacuum. This is the modern foundation for the understanding of material reality. Yet this extraordinary theoretical vision has so far had very little effect on our understanding of living organisms. Biologists need know next to nothing about quantum matter fields, and molecular biologists deal with molecules that can for most purposes still be thought of as composed of ball-like atoms. Quantum physics has dissolved atoms into a complex system of quantized fields, but much of the old atomistic way of thinking has persisted in other areas of science; atoms still seem to provide a reassuringly firm foundation for biology, and even for much of chemistry.

While physicists have felt free to introduce scores of different kinds of matter fields in the context of subatomic particles, chemists have not introduced any comparable new fields at the molecular level. Certainly some of the properties of molecules have been interpreted in terms of the principles of quantum physics: chemical bonds, for example, can be understood in terms of shared electronic orbitals which embrace the bonded atoms.[12] But these are still aspects of the electron–positron field: they are not a qualitatively new kind of field. And in mechanistic biology there is no consideration of the possibility of new kinds of fields which are unknown to quantum physicists: the known fields of physics are assumed to provide an adequate foundation for all the phenomena of life. This assumption arose within the mechanistic world view of classical physics, and it has simply persisted in spite of all the changes ushered in by the quantum theory.

But if we adopt an organismic rather than an atomistic perspective, there seems to be no good reason why organisms at all levels of complexity should not have characteristic fields. Indeed, de Broglie's original idea of matter waves implied such a view: entire atoms and molecules were wavelike quanta, as indeed were all forms of matter.

It might not be absurd to think of an insulin molecule, say, as a quantum or unit in an insulin field; or even of a swan as a quantum or unit in a swan field. But this may be just another way of thinking about morphic fields: any particular insulin molecule is a manifestation of the insulin morphic field; any particular swan is a manifestation of the swan morphic field.

Morphic fields may indeed be comparable in status to quantum matter fields. If atoms can be said to have morphic fields, then these may well be what are already described within quantum field theory. The morphic fields of molecules may already be partially described by quantum chemistry. But the morphic fields of cells, tissues, organs, and living organisms have so far been described only in vague and general terms. Something is known of their properties from the study of developing plants and animals (chapter 6), but the ways in which these fields actually organize the processes of morphogenesis remain obscure.

Morphic Fields as Probability Structures

An essential feature of morphic fields is that they are intrinsically probabilistic; in other words they are not sharply defined but are "structures of probability." There are at least three reasons for thinking that this is so.

First, individual organisms, or systems, or holons (pp. 94–96) at every level of complexity show indeterminate or probabilistic features. In the biological realm, individual cells, tissues, organs, and organisms, even if they are genetically identical and develop under practically identical conditions, are never exactly the same. Their variability itself suggests that probabilistic processes play an important part in their development. Moreover, detailed studies of their functioning have provided many examples of an intrinsic indeterminism or probabilism: for example, random fluctuations in the electrical potential across the membranes of nerve cells affect their tendency to "fire," and this has important consequences in the functioning of the nervous system.[13] For these reasons alone, it seems natural to assume that morphic fields are probabilistic in nature;[14] in so far as they interact with, or rather underlie, the material structures of organisms, the probabilistic behaviour of these structures seems likely to reflect an inherent probabilism of the associated fields.

The hypothesis of formative causation provides a second reason for thinking that these fields are probability structures. Morphic fields are built up and sustained by morphic resonance from innumerable previous similar organisms (pp. 108–9). These organisms, although similar, are inherently variable. No two clover plants, for example, are exactly the same; nor indeed are any two leaves on the same clover plant. Morphic resonance from many past organisms gives rise to a morphic field which is a composite or average of the previous forms: it cannot therefore be sharply defined, but is a probability structure (cf. Fig. 6.4).

Thirdly, if morphic fields are regarded as in some sense akin to quantum matter fields, then this would also suggest that they are probabilistic in nature. The relationship of morphic fields to quantum matter fields is, of course, still obscure; perhaps they are a different kind of field altogether and have no relationship to the fields of quantum theory. But if so, it would be difficult to conceive of the ways in which these fields interact, as presumably they must. On the other hand, if they are similar in kind to quantum matter fields, not only would their interactions be easier to conceive, but it would be possible to look forward to the development of a unified theory that would embrace both.

The morphic field of an organism organizes the parts, or holons, within it; and the fields of these holons in turn organize the lower-level holons within them. For example, an organ field organizes tissues, and a tissue field organizes cells, and a cell field organizes subcellular holons such as the nucleus and the cell membranes. Both the holons and their associated fields are arranged in a nested hierarchy (Fig. 5.9).

At every level, the fields of the holons are probabilistic, and the material processes within the holon are somewhat random or indeterminate. Higher-level fields may act upon the fields of lower-level holons in such a way that their probability structures are modified. This can be thought of in terms of a restriction of their indeterminism: out of the many possible patterns of events that could have happened, some now become much more likely to happen as a result of the order imposed by the higher-level field. This field organizes and patterns the indeterminism that would be shown by the lower-level holons in isolation.

A Provisional Hypothesis

Obviously, the hypothesis of formative causation is still vague, and the nature of morphic fields remains ill-defined. If the fields were considered to be changeless, then the hypothesis would probably be untestable: it would not be experimentally distinguishable from a Platonic conception of these fields, or from theories that assume that there are no such fields, only complex patterns of physical interaction which are in principle (although not in practice) explicable in terms of the known fields of physics.

But according to the hypothesis of formative causation, morphic fields are *not* changeless. They are influenced by what has happened before. Their probability structures change, and such changes should be detectable experimentally. Thus it is possible to test this hypothesis even though much about the nature of the fields and the processes of morphic resonance remain unknown. Several possible experimental tests of the hypothesis will be discussed in this and the following chapters.

Before Faraday's investigations, magnetic effects were usually conceived of in terms of "effluvia" or "subtle fluids." Though Faraday's lines of force and Maxwell's aetherial vortices were better defined, their nature remained obscure. The modern theory of the electro-magnetic field represents a major advance in detailed understanding; but even so, the nature of the field and its quantized excitations are barely imaginable in terms of our direct experience.

We are still at an early stage in the understanding of morphogenesis, and morphic fields are almost as poorly defined as magnetic effluvia were before Faraday. Nevertheless such vague concepts can serve as a bridge to more definite ones. Theories of magnetism did not jump straight from the effluvia stage to the quantized electro-magnetic field; the modern understanding took over a century to develop.

The concept of morphogenetic fields has already been developing for over sixty years, and if the present interpretation in terms of morphic resonance is supported by experimental evidence, it will enable researchers to understand these fields more deeply. Even so, a more detailed and adequate theory of these fields, including their relationship to the known fields of physics, could take years or even decades to evolve.

The hypothesis of formative causation is inevitably preliminary and provisional. But even in its present form, the concept of morphic resonance enables a large range of phenomena in chemistry, biology, and psychology to be seen in a new light, and it gives rise to many predictions. The remainder of this book explores these consequences and implications.

Molecular Morphic Resonance

If morphic fields are associated with holons at all levels of complexity, then we should expect each kind of chemical molecule to be associated with a characteristic morphic field.

At first sight this notion may seem unnecessary. It is usually assumed that molecular structures are in principle completely comprehensible in terms of quantum theory and electro-magnetic fields. This means, in fact, that they are already conceived of in terms of fields. However, this assumption that the known kinds of fields are adequate to account for all chemical phenomena is supported by very little evidence.

Quantum mechanics is able to give a detailed account of the simplest of all chemical systems, the hydrogen atom. But with more complicated atoms and with simple molecules its methods are no longer so precise. The complexity of the calculations becomes formidable, and only approximate methods are feasible. Even the simplest of all molecular systems, the hydrogen molecule-ion, containing two protons and one electron, presents insuperable problems. Its properties can be calculated only by making a series of simplifying assumptions:

> One distinctive feature of this simple system is that it is a three-body problem, which can no more be exactly solved in quantum than in classical mechanics. . . . What the elementary textbooks on quantum chemistry fail to explain is that empirical spectroscopic evidence that might test these calculations is thin. Moreover, such tests as there may be tend to disappoint the calculators. But nobody need feel ashamed, for the complete calculation of the hydrogen molecule-ion, rotations as well as vibrations, depends on a succession of manifestly false assumptions.[15]

For complex molecules and crystals, even more drastic approximations and simplifying assumptions have to be made in order to apply a mathematical analysis. These calculations have given an increased understanding of some of the properties of molecules and crystals; but this is a very different matter from predicting their forms and properties from first principles. Only if this were possible could we conclude that these known principles are capable of explaining the facts of chemistry. But this has never been demonstrated. The question is in fact quite open, even in relation to relatively simple molecules and crystals.[16]

The Structure and Morphogenesis of Proteins

The conventional assumption that the known principles of physics are adequate to account for the structures and properties of molecules reaches enormous proportions in relation to complex molecules such as proteins. These are made up of chains of amino acids, called polypeptide chains, which spontaneously fold up into a characteristic three-dimensional conformation (Fig. 7.1). A protein can be *denatured*, in other words made to unfold into a flexible polypeptide chain that has lost its original conformation. If the denaturing treatment is gentle enough, it can usually be reversed. Unfolded polypeptide chains spontaneously refold into their original conformations: their normal shape is regenerated. A current text-book comments on the phenomenon as follows:

> This behaviour confirms that all the information determining the conformation must be contained in the amino acid sequence itself. . . . Although all the information required for the folding of a protein chain is contained in its amino acid sequence, we have not yet learned to "read" this information so as to predict the detailed three-dimensional structure of a protein whose sequence is known.[17]

The problem in understanding protein folding arises from the astronomical number of possible ways a polypeptide chain could fold up. Out of all the possible conformations, the protein adopts only one. Moreover, as the protein folds up, it cannot "explore" at random these possible conformations until it "finds" the one that is most stable energetically. Consider, for example, a chain of 100 amino acids such as might be found in a typical protein. Such a chain could have up to 10^{100} possible conformations if each amino acid could adopt, on average, ten conformations. "If all internal bond rotations interconverting these conformations occurred independently at the maximum rate of 10^{13} sec $^{-1}$, the average time to sample all conformations would be 10^{85} seconds, or 10^{77} years. Another estimate is 10^{50} years. Since proteins are often observed to refold within seconds, the conclusion that the folding process is not random is perhaps inevitable."[18]

Studies of protein structure have shown that the polypeptide chains (which are called the primary structure) arrange parts of themselves into

Insulin   Ribonuclease A

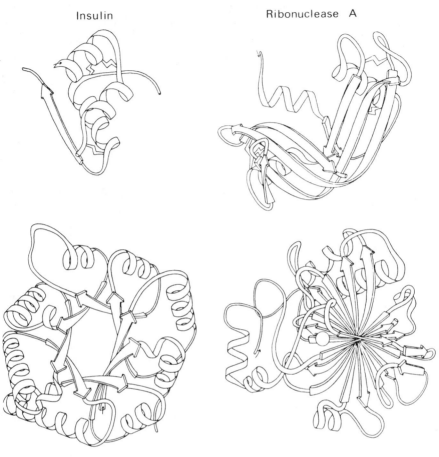

Triose phosphate isomerase   Carbonic anhydrase

Figure 7.1 Diagrammatic representation of the three-dimensional structure of four kinds of protein molecules. (From *Advances in Protein Chemistry 34* by J. S. Richardson, 1981. Reproduced by permission.)

helices or sheets (called α-helices and β-sheets). They are the proteins' secondary structure. These "structural clichés" in turn are often arranged in particular patterns, called domains, which are similar in many different proteins.[19] A protein may have several different domains, which can be thought of as modular or structural units from which the protein as a whole is constructed. The conformation of the protein as a whole is called the tertiary structure. Finally, individual protein molecules often assemble with others to produce aggregates of characteristic form. This is the quaternary structure (Fig. 7.2).

Thus protein structure appears to consist of a hierarchy of levels. One current theory proposes that protein folding occurs stepwise through these hierarchical levels. Another theory proposes that folding starts at "preferred points" of the polypeptide chain and then propagates from these "nucleation centres."[20]

Many attempts have been made to calculate the structure of proteins on the basis of the sequence of amino acids in their polypeptide chains. These models take into account known facts about protein structure and about the properties of different amino acids. On the basis of a variety of simplifying assumptions, these are then used to compute the conformations of the protein most likely to be stable from a thermodynamic point of view. But even so, they generate dozens or even hundreds of structures, all of which are equally stable, or in other words are "minimum-energy" structures. In the literature on protein folding, this is known as the "multiple-minimum problem."[21] In successful calculations, one of the predicted conformations does in fact turn out to correspond to the known structure of the protein. But why should this conformation be adopted rather than the others? Even if more detailed calculations were able to show that this conformation was slightly more stable thermodynamically than the others, such a slight difference might not be able to stabilize the protein in this form, since the molecule is subject to relatively large energy fluctuations as a result of thermal vibration.

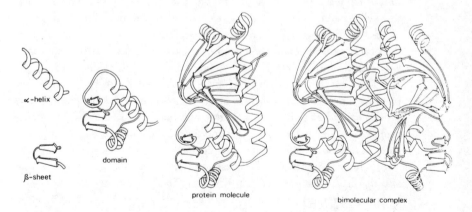

Figure 7.2 Hierarchical levels of protein structure. The bimolecular complex shown here is of the catabolite activator protein, which plays a role in the control of protein synthesis in the bacterium *Escherischia coli* by binding to DNA. (From *Advances in Protein Chemistry 34* by J. S. Richardson, 1981. Reproduced by permission.)

In any case, the assumption that the actual conformation of the protein is determined simply according to thermodynamic principles is not empirically testable. The protein just will not fold into the other theoretically possible structures, and therefore their energetic stability cannot be compared experimentally.[22]

One suggested explanation for the unique conformations of protein molecules is in terms of evolutionary speculation. Present-day proteins "are the rare survivors of a very long evolutionary process in which the vast majority of proteins had more random conformations, were less useful, and were therefore discarded through natural selection."[23] But natural selection cannot explain the rapidity of the folding process. In order to account for this, it has been suggested that the process of folding takes place in such a way that it leads only to one of the possible stable structures. Perhaps "the observed folded state is not the most stable thermodynamically of all those possible, but merely the most stable state of those which are kinetically accessible."[24]

The hypothesis of formative causation provides an interpretation of protein folding that does not contradict, but rather complements, these interpretations. There are morphic fields for structural clichés such as α-helices; these are organized by higher-level fields into domains, and the overall field of the molecule organizes the domains into the characteristic structure of the protein. Higher-level fields organize proteins into aggregates.

Fields at each of these levels canalize the folding process towards a characteristic end-point; the processes of folding follow chreodes (Fig. 6.2). Out of the many possible ways of folding and the many possible final forms, the fields stabilize particular folding pathways and final forms. In other words, the fields greatly increase the probability of these structures, rather than other possible ones, coming into being. Or, looking at this the other way round, they greatly reduce the randomness of the folding process.

The morphic fields are themselves stabilized by morphic resonance from innumerable past structures of the same kinds. The long evolutionary process has indeed stabilized those structures that have been useful and therefore favoured by natural selection; and the vast numbers of these past molecules have a powerful stabilizing effect on the fields by morphic resonance.

This interpretation means that not only the amino acid sequence but also the fields determine protein structure. This in turn means that "all the information required for the folding of a protein chain" is *not* contained in its amino acid sequence (p. 123). Think again of the building analogy. The information for the structure of a house is not all contained in the building materials, even if they are supplied in modular units. The same materials can be used to build houses of different form; and, conversely, houses of identical design can be built from different materials: stone instead of brick, for example.

If protein structure is organized by fields, we could expect that these fields might give rise to similar structures even if the amino acid sequences were different. It is, in fact, known that domains of very similar structure occur in quite different proteins, and that these domains may contain different sequences of amino acids. Even entire proteins may have a structure very similar to that of other proteins in spite of great differences in their amino acid composition. One example is provided by a family of protein-cleaving enzymes called serine proteases, which include digestive enzymes such as trypsin and enzymes involved in blood clotting such as thrombin. If any two of the enzymes in this family are compared, only about 40% of the positions in their amino acid sequences are occupied by the same amino acid. Yet the similarity of their three-dimensional conformations, as determined by x-ray crystallography, is very striking. Most of the detailed twists and turns in these polypeptide chains, which are several hundred amino acids long, are identical.[25]

The haemoglobins provide an even more extreme example. These red proteins are responsible for the colour of blood, and are found in a very wide range of animals, both vertebrate and invertebrate. Even peas and beans produce haemoglobin: it is present in their root nodules, which is why the nodules are pink inside. The three-dimensional structures of these various kinds of haemoglobin are extremely similar. However, their amino acid sequences are quite different. In all known haemoglobin sequences, only 3 out of a total of 140 to 150 amino acids are the same in the same positions.[26]

Such an extraordinary stability of structure in spite of differences in amino acid sequence is astonishing if we assume that all the information required for the folding of the protein chain is contained in the amino acid sequence. But it is much easier to understand on the basis of the field hypothesis.

Experiments on Protein Folding

As we have just seen, the problem of protein structure has not proved to be fully explicable in terms of the sequence of amino acids in the protein chain and the known laws of physics and chemistry. From the orthodox point of view, this is just because the calculations are so complex that they cannot yet be carried out satisfactorily. From the point of view of the hypothesis of formative causation, the structure of proteins is organized by morphic fields maintained by morphic resonance from protein molecules of the same kind in the past.

If this hypothesis could be applied to protein structure only speculatively, it would be of little value in molecular biology. But in fact it should be testable by experiments on protein folding.[27]

When a protein is artificially denatured, for example by being placed in a strong solution of urea, its molecules unfold into long, flexible chains. When the denaturing agent such as urea is removed, the molecules usually refold again, and their normal three-dimensional structure is regenerated (p. 123).

Very little is known about the way in which proteins fold up inside living cells, and it is not possible to say how closely the refolding of a protein under experimental conditions resembles the way in which it folds up under natural conditions. If the refolding processes follow exactly the same pathways in test tubes as they do in cells, the folding chreodes will be strongly stabilized by morphic resonance from all the many times the protein has folded up before. But if it is possible to make protein unfold into a denatured state which differs from any state in which the protein normally exists inside living organisms, then the refolding of the chain may follow a somewhat different pathway from usual. In other words, the refolding may follow different chreodes than in the normal folding process. In this case, the more often it is refolded in the laboratory, the more the previous refoldings will stabilize this chreode by morphic resonance.

This chreode will become more probable, and consequently the refolding process may take place more quickly. This increased rate of folding may be measurable experimentally.

Consider the following experimental design. Several kinds of enzyme molecule which have not already been used in folding experiments are selected.[28] They are made to unfold, and then allowed to refold under standard conditions. The rate of refolding is measured for each of the enzymes in a laboratory in one place, say London.

Then, in another laboratory, say in Berkeley, *one* of the enzymes is selected at random, and large quantities of it are made to unfold and refold under the same conditions that are being employed in London. The experimenters in London are not told which of the enzymes has been selected for this treatment in Berkeley.

Subsequently, the rate of refolding of all the enzymes is measured again in London under the same standard conditions that were used before. If it is found that this enzyme now refolds significantly faster than it did on the previous occasion, and if there is no comparable increase in the rate of refolding of the other enzymes, which serve as controls, then this result would be in accordance with the idea of morphic resonance. Of course, such a result, indicating an action at a distance between the refolding protein molecules in Berkeley and in London, would seem utterly bizarre from the point of view of all conventional theories, but would provide striking support for the hypothesis of formative causation.[29]

Morphic Resonance in Crystallization

Although the structures of many kinds of crystals have been described in detail, the ways in which these crystals take up their structures as they crystallize is very obscure. In the first place, just as in the case of protein structures, it is not possible to predict from first principles the way in which the molecules will pack themselves together in the crystal lattice. Even with quite simple molecules there are many possible lattice conformations that are equally stable thermodynamically, and there is no clear reason why one rather than any of the others is actually taken up during crystallization.[30] Again, there is no way of empirically testing the assumption that the actual lattice structure is uniquely stable from an energetic point of view. The molecules simply will not crystallize into the other theoretically possible lattice structures, and therefore their energies cannot be measured and compared.

The second difficulty arises in trying to understand the way in which the crystal grows as a whole. Somehow, as molecules in solution come close to the growing surface of the crystal, they "snap" into place in the growing aggregate. But the way in which they do this cannot be directly observed, and attempts to model the process mathematically are still very crude and have not been very successful so far.[31] Such models take into account only local effects on the molecules joining the growing crystal. But crystals as a whole show patterns of symmetry which cannot possibly arise from a sum of local effects. Consider snowflakes. These crystals generally have a sixfold symmetry, but each is unique (Fig. 7.3). Within a snowflake, the intricate structure of the six arms is very similar, and these arms are themselves symmetrical. Although the differences *among* snowflakes can be explained in terms of random variations, the symmetrical development *within* each snowflake cannot be explained in this way.[32]

> [This] must be the consequence of some co-operative phenomenon involving the growing crystal as a whole. What can that be? What can tell one growing face of a crystal . . . what the shape of the opposite face is like? Only the lattice vibrations which are exquisitely sensitive to the shape of the structure in which they occur (but which are almost incalculable if the shapes are not simply regular).[33]

Figure 7.3 Snowflakes. (From *Snow Crystals* by W. A. Bentley and W. J. Humphreys; Dover Publications, Inc., 1962. Reproduced by permission.)

From the point of view of the hypothesis of formative causation, the lattice structure is organized by a lattice morphic field, and a higher-level field organizes the structure of the crystal as a whole. The same lattice structure, for example that of water, can be organized into different types of crystal, as in sheets of ice, in snowflakes, and in various kinds of frost. The morphic field of the crystal as a whole is associated with the "lattice vibrations which are exquisitely sensitive to the structure in which they occur" and organizes the pattern in which the crystal grows.

Crystallization Experiments

The fields of crystals that have already occurred many times in the past are highly stabilized by morphic resonance, and changes in these fields will not be experimentally detectable. But this is not the case with newly synthesized chemicals that have never existed before. Thousands of new kinds of molecules are made every year by synthetic chemists in universities and in industrial laboratories. Before such a substance crystallizes for the first time, there will not be morphic fields either for its lattice structure, or for the form of the crystal as a whole. There can be no morphic resonance from previous crystals of this type if none have existed. But when it crystallizes for the first time, the lattice and the crystal fields come into being. The second time, the fields will be influenced by morphic resonance from the first crystals; the third time, from the first and second crystals; and so on. There will be a cumulative build-up of morphic resonance stabilizing the fields of subsequent crystals, which will tend to render further crystallization of this type more probable. Consequently, the compound should tend to crystallize more and more readily as more of the crystals are made.

It is in fact well known to chemists that newly synthesized compounds are usually difficult to crystallize: weeks or even months may elapse before crystals appear in a supersaturated solution. Moreover, generally speaking, compounds become easier to crystallize all over the world the more often they are made. This happens in part because chemists tell each other of the appropriate techniques. But the most common conventional explanation for this phenomenon is that fragments of previous crystals are carried around the world from laboratory to laboratory, where they serve as "seeds" for subsequent crystallizations. The folklore of chemistry has a rich store of anecdotes on this subject. The carriers of the seeds are often said to be migrant scientists, especially chemists with beards, which can "harbour nuclei for almost any crystallization process."[34] Or else seeds are thought to move around the world as microscopic dust particles in the atmosphere.

If morphic resonance plays a part in this phenomenon, the more often the new compounds are crystallized, the more readily they should tend to crystallize all over the world, even when migrant chemists are rigorously excluded and when dust particles are filtered out of the atmosphere. Experiments can easily be designed to test this prediction.[35]

Symmetry and Internal Resonance

According to the hypothesis of formative causation, crystal structures are stabilized by morphic resonance from other crystals of the same kind that existed in the past. But, in addition, the symmetry of crystals such as snowflakes seems explicable only in terms of some kind of resonance *within* the growing crystal: such an explanation seems necessary whether or not we take morphic fields into account (pp. 129–31). This raises a very general point about the morphogenesis of symmetrical structures: their symmetry seems to require some kind of resonant communication between the symmetrical parts. Consider, for example, your right and left hands. They are different from everyone else's, in both the pattern of lines on the palms and the pattern of ridges on the finger-tips. Yet they are very similar to each other,[36] just as the arms of an individual snowflake are similar to each other. This suggests that within the developing organism morphic resonance takes place between similar structures, in this case between the fields of the embryonic hands. The same applies to other symmetrical structures such as the right and left sides of the face: again, although these are not exactly the same, they are very similar, and their development must have been correlated by some kind of resonant phenomenon.

We may conclude that in general within developing organisms there is an internal resonance between the fields of symmetrical structures, and this self-resonance is essential to their symmetry. Since symmetry is such an important feature of natural forms at every level of complexity, an internal resonance between symmetrical structures within the same organism is likely to be an important general feature of formative causation through morphic fields.

Such morphic resonance between spatially symmetrical structures that are developing at the same time within the same organism is, however, only one kind of self-resonance. Another aspect of self-resonance which is just as fundamental is the morphic resonance from an organism's own past.

Self-Resonance

The specificity of morphic resonance depends on the similarity of the patterns of activity that are resonating. The more similar the patterns of activity, the more specific and effective will the resonance be. In general, the most specific morphic resonance acting on a given organism will be that from its *own* past states, because it is more similar to itself in the past, especially in the immediate past, than to any other organism. This self-resonance will therefore tend to stabilize and maintain organisms in their own characteristic form, as well as harmonizing the development of symmetrical structures within the same organism. In living organisms, this self-stabilization of morphic fields may go a long way towards explaining how they are able to maintain their characteristic forms in spite of a continuous turnover of their chemical constituents.

If resonance from a holon's own past states is of such importance, then how far in the past does a pattern of activity have to be to exert an influence by morphic resonance? The very notion of resonance implies a relationship between vibratory structures of activity, and the identity of such a structure cannot be defined instantaneously. Its "present" must involve duration, since vibrations take time; and the frequency of vibration cannot be characterized until several similar vibrations have taken place. The "present" must therefore consist of several cycles of vibration; hence the duration of the present depends on the characteristic vibratory frequencies of the organism. The slower these are, the longer the "present" will be.

This general principle is, of course, apparent in quanta of radiation and of matter, which because of their wavelike nature cannot be considered to be sharply located: they are more like a "smear" of probability. There is an inherent uncertainty in locating them at a particular point and assigning them a particular momentum.

In quantum matter fields, the vibration of the field itself underlies the quanta, or particles. The field, as the ground of the vibration, must endure or persist in time; indeed persistence in time, which implies a linkage of the present with the past, is inherent in the nature of the field. This linkage cannot take place through any kind of independently persisting material structure, since particles of matter are themselves manifestations of the field. So if a vibrating field is connected to its own past, which it must be if it is to persist, the linkage must be intrinsically temporal in nature. It must in fact depend on some kind of self-resonance.

Just as the position and momentum of a particle cannot be defined with certainty, neither can the exact duration of its present: it shades off into the past. These past patterns of activity into which it shades off become present again by morphic resonance, and by doing so maintain and stabilize the field as it persists in time.

If this interpretation is valid, then the persistence of matter itself, and indeed of radiation, depends on a continuous process of resonance of the fields with their own past states. The continuity of any self-organizing pattern of activity at any level of complexity—from an electron to an elephant—results from this self-resonance with its own past patterns of activity. All organisms are dynamic structures that are continuously recreating themselves under the influence of their own past states.

These causal influences from an organism's own past states must be capable of passing through or across not only time but space, or rather space-time. This requirement becomes obvious when we consider a moving organism, for example a galloping horse: its past patterns of activity with which it is in morphic resonance occurred in different places from the ones it now occupies. If it is in morphic resonance with its own past states, including those of only a few seconds ago, this causal influence must traverse the intervening space-time. Or, to look at it another way, its past patterns of activity, wherever and whenever they were, can become present by morphic resonance.

Thus morphic resonance from the patterns of activity of similar past organisms, and self-resonance from an organism's own past, can be seen as different aspects of the same process. Both involve formative causal connections across both space and time. Self-resonance, through its high specificity, stabilizes an organism's own characteristic pattern of activity, and resonance with similar past organisms stabilizes the general probability structure of the field. This is what enables an organism to come into being and gives it its potentialities. As it actualizes itself, its own particular structure will tend to be maintained by self-resonance within the overall probability structure of the field.

This interpretation has much in common with Whitehead's idea that there is a "prehension" from the "actual occasions" of organisms to their immediate or more remote predecessors. The more often a pattern of activity has been repeated, the stronger its influence will be. In Whitehead's words, "any likeness between the successive occasions of a historical route procures a corresponding identity between their contributions to the datum of any subsequent actual entity; and it therefore secures a corresponding intensification in the imposition of conformity."[37] However, Whitehead's philosophy is rather obscure in this respect; and although he definitely envisaged a process similar to what is here called self-resonance, it is not clear to what extent he thought of a comparable influence from *different* organisms in the past.[38]

In chapter 6 we considered the role of morphic fields in biological morphogenesis, and in this chapter their role in the morphogenesis of molecules and crystals. We have also examined some of the general features of the hypothesis of formative causation: the idea of morphic fields as probability structures, and the importance of self-resonance in the development and maintenance of the form of individual organisms. We now turn to a discussion of the possible role of morphic resonance in biological heredity, and then in chapter 9 consider the nature of animal memory and the light shed upon it by the idea of self-resonance from an animal's own patterns of activity in the past.

CHAPTER 8

Biological Inheritance

Genes and Fields

Living organisms inherit genes from their ancestors. According to the hypothesis of formative causation, they also inherit morphic fields. Heredity depends on both genes and morphic resonance.

The conventional theory attempts to squeeze all the hereditary characteristics of organisms into their genes. Development is then understood as the *expression* of these genes through the synthesis of proteins and other molecules. The words *hereditary* and *genetic* are treated as synonyms. Thus inherited characteristics, such as the ability of an acorn to grow into an oak tree or of a wren to build a nest, are usually referred to as genetic, or as genetically programmed.

What is in fact known to be inherited genetically is DNA. Some of the DNA codes for the sequence of amino acids in proteins; some codes for RNA such as that found in ribosomes; and some is involved in the control of gene expression. However, in higher organisms only a small percentage of the DNA (in humans, about 1%) seems to be involved in such coding and genetic control. The function, if any, of the vast majority is unknown, although some probably plays an important structural role in the chromosomes. Furthermore, the total amount of DNA that is inherited seems to bear very little relationship to the complexity of the organism. Among the amphibians, for instance, some species have one hundred times more DNA than others; and the cells of lily plants contain about thirty times more DNA than human cells.[1]

There is also a poor correlation between the genetic differences between species and the form and behaviour of these species. Thus, for example, human beings and chimpanzees have genes that code for almost identical proteins: "the average human polypeptide is more than 99% identical to its chimpanzee counterpart."[2] And direct comparisons of the DNA sequences believed to be of genetic significance show that the overall difference between the two species is only 1.1%. By contrast, comparisons of species that are very similar to each other, such as different kinds of fruit-flies in the genus *Drosophila,* often reveal considerably greater genetic differences than between humans and chimpanzees.[3]

From the point of view of the hypothesis of formative causation, DNA, or rather a small part of it, is responsible for coding for RNA and the sequences of amino acids in proteins, and these have an essential role in the functioning and development of the organism. But the forms of the cells, tissues, organs, and the organisms as a whole are shaped not by DNA but by morphic fields. The inherited behaviour of animals is likewise organized by morphic fields. Genetic changes can *affect* both form and behaviour, but these patterns of activity are inherited by morphic resonance.

Consider the analogy of a television set, tuned to a particular channel. The pictures on the screen arise in the TV studio and are transmitted through the electro-magnetic field as vibrations of a particular frequency. To produce the pictures on the screen, the set must contain the right components wired up in the right way, and also requires a supply of electrical energy. Changes in the components, such as a fault in a transistor, can alter or even abolish the pictures on the screen. But this does not, of course, prove that the pictures arise from the components or the interactions between them, nor that they are programmed within the set. Likewise, the fact that genetic mutations can affect the form and behaviour of organisms does not prove that form and behaviour are coded in genes or programmed genetically. The form and behaviour of organisms do not arise simply from mechanistic interactions within the organism, or even between the organism and its immediate environment; they also depend on the fields to which the organism is tuned.

To pursue this analogy, developing organisms are tuned to similar past organisms, which act as morphic "transmitters." Their tuning depends on the presence of appropriate genes and proteins, and genetic inheritance helps to explain why they are tuned in to morphic fields of their own species: a frog's egg tunes in to frog rather than newt or goldfish or chicken fields because it is already a frog cell containing frog genes and proteins.

Genetic mutations can affect morphogenesis in two main ways. First, they can lead to distortions or alterations in a normal morphogenetic process, just as "mutant" components in a TV set can lead to distortions or alterations in the form or colour of the pictures. Second, they can result in the suppression of entire morphogenetic processes or in their replacement by different ones. These are analogous to "mutations" in the tuning circuit of the television: the original transmission is no longer picked up; either the screen goes blank, or the set picks up a different channel.

Mutations

We now consider how the inheritance of morphic fields by morphic resonance is related to the known facts about genetic mutations.

A great many mutations affect normal processes of development, often in quite small ways, and normal pathways of morphogenesis are convention-

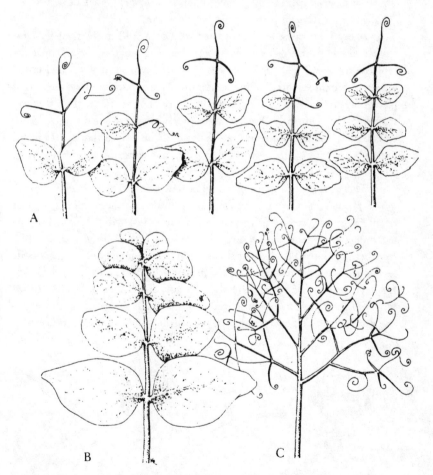

Figure 8.1 A: Normal pea leaves, bearing both leaflets and tendrils. B: Leaf of a mutant pea in which only leaflets are formed. C: Leaf of a mutant pea in which only tendrils are formed.

ally considered to be under the control of large numbers of "minor genes" and "modifying genes." But certain spectacular mutations are known in which entire structures are lost or are replaced by other structures. These are called homoeotic mutants. In pea plants, for instance, the leaves normally have leaflets near the base and tendrils at the tips (Fig. 8.1). One mutation, in a single gene, results in the replacement of all the leaflets by tendrils; another, in a different gene, has the opposite effect: all the tendrils are replaced by leaflets. Somehow these genetic mutations affect the tuning of the primordia in the embryonic leaf so that they all develop under the influence of either leaflet fields or tendril fields.[4] A similar metaphor is already implicit in the conventional interpretation of such genes as "switching on" or "switching off" entire pathways of development.

Many homoeotic mutations have been identified in the fruit-fly *Drosophila melanogaster*. For example, in antennapedia mutants, the antennae are replaced by legs (Fig. 8.2). These legs are of the type normally found in the second of the three pairs. Another mutation has the opposite effect: the second pair of legs is replaced by antennae.[5] In bithorax mutants, the third thoracic segment, which normally bears halteres (small balancing organs), is partially or completely transformed into a duplicate of the second thoracic segment, which bears wings. The resulting flies have four wings instead of two (Fig. 5.6).

There are several kinds of bithorax mutations.[6] These occur in nearby genes on the same chromosome, and their effects have been studied in great detail.[7] Recently some of these genes have been isolated and cloned by the techniques of genetic engineering, and the sequences of bases in their DNA have been analysed.[8] By means of sophisticated techniques involving antibodies stained with fluorescent dyes, the proteins that some of these genes code for have been localized in the early embryos; and it is now possible to see which of these proteins are produced in which segments, enabling the differing distribution of these proteins in normal flies and homoeotic mutants to

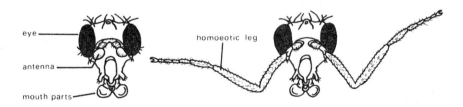

Figure 8.2 On the left, the head of a normal fruit-fly; on the right, the head of a mutant fly in which the antennae are transformed into legs as a result of a mutation within the antennapedia gene complex. (After Alberts, et al., 1983)

be compared.[9] The products of homoeotic genes clearly play an important role in determining which pathways of development the primordia in the embryonic segments will follow.

From the point of view of formative causation, the proteins coded for by these genes affect the tuning of the primordia, causing them to come under the influence of one field rather than another. Mutations in these genes alter the tuning of the primordia, just as "mutations" of components in the tuning circuit of a television can result in changes of channel. Sometimes the alteration is not complete, and a germ structure may be finely balanced between entering into morphic resonance with alternative fields. Homoeotic mutations often show variable penetrance, which is to say that not all the flies with the mutant genes have a mutant form; and they also show variable expressivity, which means that the mutant form may be expressed only partially: for example, antennapedia mutants sometimes have a normal antenna on one side of their head and a leg in place of an antenna on the other.

If homoeotic mutations bring about their effects by altering the normal patterns of vibratory activity in morphogenetic germs, then other factors that alter these patterns could have similar effects. Again, think of the television analogy. A change in tuning from one channel to another can occur because of a mutation in the tuning circuit; but it can also happen in response to a stimulus from the environment, such as a mechanical jolt to the tuning knob. The causes of the change in tuning are different, but their effect is the same.

It has been known for many years that the normal development of organisms can be disturbed by exposing embryos to toxic chemicals, X rays, heat, and various non-specific stimuli. The interesting thing is that many of the abnormalities that result from such treatments fall into definite categories, which are the same as abnormalities due to genetic mutations. The resulting organisms are called phenocopies. In the case of *Drosophila*, practically all of the homoeotic mutant forms can also appear in genetically normal flies as phenocopies.[10] Thus the mutant structures arise as a result of disturbances to the normal course of development, and these disturbances may be due *either* to genes *or* to influences from the environment; according to the hypothesis of formative causation they are a consequence of changes in the tuning of the germ structures within the embryo such that they become associated with different morphic fields—a leg field instead of an antenna field, for instance.

The difference between this hypothesis and the conventional genetic interpretation becomes clearer when we consider the inheritance of acquired characteristics. The possibility of this kind of inheritance is vigorously denied by orthodox geneticists, but according to the hypothesis of formative causation is not only possible but probable. We shall now see why this is so, and consider one of the ways in which the validity of the rival approaches can be compared by experiment.

The "Lamarckian" Inheritance of Acquired Characteristics

If plants of a particular species are grown under unusual conditions, for example at a high altitude, they generally develop somewhat unusually. The modified form they take up is an "acquired characteristic" which has come about in response to the environment. Likewise, if rats learn a new trick, this is an acquired behavioural characteristic, as opposed to an inherited instinct.

Until the late nineteenth century, it was almost universally believed that acquired characteristics could be inherited.[11] The early nineteenth-century zoologist Lamarck took this for granted, as did Charles Darwin.[12] This kind of inheritance is now commonly referred to as Lamarckian inheritance, but it could with equal justification be called Darwinian inheritance.

The idea of Lamarckian inheritance has the great advantage of making sense of many of the evolutionary adaptations of organisms. For example, camels, like many other animals, develop thick calluses on their skin as a result of abrasion. They possess such calluses on their knees just where the skin is subject to abrasion as they kneel down. This could well be an acquired characteristic; but, as a matter of fact, baby camels are *born* with thick pads at exactly the right places on their knees.

From a Lamarckian point of view, such calluses were acquired by ancestral camels as a result of their habit of kneeling, and then over many generations this acquired characteristic became increasingly hereditary, developing even in embryos before they have ever had a chance to kneel.

This idea is straightforward enough, and has a strong "common sense" appeal. However, it is denied dogmatically by neo-Darwinians, who, unlike Darwin, reject the possibility of such inheritance. From a neo-Darwinian point of view, camels are born with pads on their knees not because their ancestors acquired this characteristic as a result of their habits, but because it arose as a result of chance genetic mutations which just happened to produce pads in the right places. The mutant genes "for" knee pads were favoured by natural selection because it was of advantage to camels to be born with pads just where they and their ancestors would have acquired them anyway.

We return to a discussion of the evolutionary significance of the inheritance of acquired characteristics in chapter 16; we now consider the orthodox denial of this possibility and the light that is shed upon it by the hypothesis of formative causation.

The genetic theory of inheritance is rooted in the Weismannian assumption that the germ-plasm (genotype) determines the somatoplasm (phenotype), but not vice versa (pp. 78–81). It therefore rules out from first principles the possibility that there can be a Lamarckian inheritance of acquired characteristics.

Precisely because such a fundamental theoretical principle is at stake, this subject has been the focus of some of the bitterest controversies in the history of biology. In the West, Lamarckian inheritance has been treated as a heresy since the 1920s; but in the Soviet Union, from the 1930s to the 1960s the situation was the other way round. Under the leadership of T. D. Lysenko the inheritance of acquired characteristics became the orthodox doctrine; Mendelian geneticists were often persecuted, and sometimes liquidated.[13] These hostilities have by no means favoured an objective examination of the evidence.

From a Weismannian perspective, the theoretical reasons for rejecting the possibility of this type of inheritance have been strengthened by the discoveries of molecular biology. It is practically impossible to conceive of a mechanism whereby a learned pattern of behaviour in a rat, say, could cause specific modifications to the genes within the germ cells, such that the rat's progeny would be "programmed" to learn the same behaviour more easily.

Nevertheless, in spite of Weismannian theories, there is a large body of evidence that indicates that acquired characteristics *can* be inherited. Some of these experimental results are considered fraudulent, and this may well be the case with some of Lysenko's data. There was also evidence of fraud in a famous Lamarckian experiment carried out by Paul Kammerer, a story well told by Arthur Koestler in *The Case of the Midwife Toad* (1971). However, numerous other experiments carried out by dozens of biologists in the West before the 1930s[14] and by many Soviet biologists in the Lysenko period[15] provided evidence for an inheritance of acquired characters. These experimental findings are usually dismissed or simply ignored by geneticists and by neo-Darwinians. However, there is good evidence from more recent experiments, to which we now turn, that an inheritance of acquired characteristics does in fact take place.

The Inheritance of Acquired Characteristics in Fruit-Flies

In the 1950s a fascinating series of experiments was carried out with fruit-flies in Waddington's laboratory. The developing flies were subjected to abnormal stimuli, and as a consequence some developed in characteristically abnormal ways: they were phenocopies. In one experiment, young pupae, in which the larvae were metamorphosing into flies, were heated to 40°C for four hours. Some of the emerging flies had abnormal wings lacking cross-veins. In another experiment, eggs were exposed to fumes of ether for 25 minutes about three hours after they were laid. Some of the resulting flies were phenocopies of the bithorax type (Fig. 5.6). The abnormal flies were selected as parents of the next generation, which was again subjected to the abnormal stimulus, and so on. A higher and higher proportion of flies was abnormal in successive generations. After a number of generations, in one case as few as eight, these flies gave rise to progeny that showed the abnormal character even in the absence of the abnormal stimulus.[16] Matings between cross-veinless flies obtained in this way gave rise to strains that regularly produced cross-veinless flies when cultured at normal temperatures.[17] Similarly, bithorax-type flies appeared generation after generation without the ether treatment.

Waddington called this phenomenon genetic assimilation, which he defined as a "process by which characters which were originally 'acquired characters,' in the conventional sense, may become converted, by a process of selection acting for several or many generations on the population concerned, into 'inherited characters.' "[18] He explained it in terms of the selection of genes that gave the flies a capacity to respond to the environmental stress, which then went on producing the same abnormal pattern of development even in the absence of the stress. This seems at first glance to provide a neo-Darwinian interpretation of the inheritance of acquired characters, and the concept of genetic assimilation is now used in conventional evolutionary theory to account for otherwise puzzling examples of apparent Lamarckian inheritance, such as the pads on camels' knees.

But genetic assimilation does not depend on genes alone. Waddington's explanation of the continued appearance of abnormal flies in the "assimilated" strains in the absence of the environmental stimulus involved his idea of canalized pathways of development, or chreodes (Fig. 8.3). He attributed to these an autonomy that remained unexplained. "Developmental processes have some structural stability, so that once you have got a developmental process going in a certain direction it tends to go on there independently of changes in the environment."[19]

This is just the kind of effect that would be expected on the basis of morphic resonance. The larger the number of abnormal flies that appeared in the population, the more would the abnormal chreodes be stabilized by morphic resonance and the greater would be the probability of the abnormal development. The interpretation does not deny the role of genetic selection in Waddington's experiments, but it suggests that an increasing proportion of flies would show the abnormal character in successive generations even in the *absence* of selection by the experimenter of abnormal flies as the parents of the next generation.

And this is what happens. Waddington's experiments did not include a control line in which *unselected* flies were allowed to mate at random in

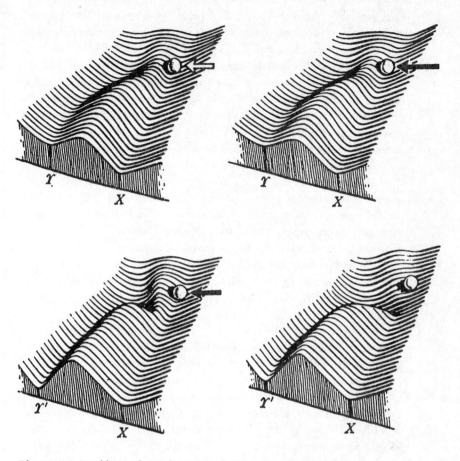

Figure 8.3 Waddington's explanation of "genetic assimilation" in terms of chreodes. The upper left drawing represents the original stock of fruit-flies; normal development follows the chreode leading to the normal adult form, X. A developmental modification, Y, for example the formation of four-winged flies, involves following a different chreode. The developing system can be forced to cross a threshold or col into the Y chreode by an environmental stress, represented by the white arrow (upper left). A genetic mutation can have a similar effect, represented by the black arrow (upper right). The two lower diagrams represent alternative models of genetic assimilation. In the one on the left, in Waddington's words, "the threshold protecting the wild type is lowered to some extent, but there is an identifiable major gene which helps push the developing tissues into the Y path. On the right, the genotype as a whole causes the threshold to disappear and there is no identifiable 'switch gene.' Note that in both the genetic assimilation diagrams there has been a 'tuning' of the acquired character, i.e., the Y valley is deepened and its end-point shifted from Y to Y'." (From *The Strategy of Genes* by C. H. Waddington; George Allen and Unwin, Ltd., 1957. Reproduced by permission.)

each generation. Mae-wan Ho et al. have recently repeated Waddington's experiment, treating the eggs of successive generations with ether. But, unlike Waddington, they allowed all the flies to mate at random; they did not pick out the abnormal flies in each generation to serve as the parents of the next. In fact, in their experiments the abnormal flies were at a disadvantage in mating, and natural selection worked against them. They found that the proportion of bithorax-type flies progressively increased from 2% in the first generation to over 30% in the tenth (Fig. 8.4).[20] In other words, in each generation most of the parent flies were normal in appearance, and yet more and more abnormal flies were produced in successive generations.[21]

Anticipating the objection that there must have been some subtle genetic selection going on in favour of abnormal development, they did a parallel experiment with an inbred strain of flies. There was very little genetic variability in this strain, hence little scope for selection. But here too the proportion of bithorax-type flies increased progressively.

When flies from these populations were returned to normal conditions, without ether treatment, they continued to give rise to a considerable proportion of abnormal progeny. This proportion gradually declined over several generations (Fig. 8.4).

From a conventional point of view, ether would not be expected to have any specific effects on the genes, and certainly not to cause specific mutations leading to the appearance of four-winged flies. Nor do the results obtained by Ho and her colleagues suggest that such genetic changes had occurred. By crossing control flies with flies from ether-treated populations, they found that the marked tendency of ether-treated flies to give rise to abnormal progeny was inherited through the mothers, not the fathers. (Waddington himself found a similar maternal effect in one of his experiments.[22]) They interpret this to mean that the ether treatment somehow changed the cytoplasm (the organized cell structures outside the nucleus), rather than the genes. Cytoplasm is inherited only from mothers, whereas genes are inherited from both parents. Somehow the ether-induced modifications of the cytoplasm persisted for several generations after ether treatment had ceased. Nothing of this kind would be expected on the basis of conventional genetic theory.

If the ether treatment in fact modified the cytoplasm in some way, then the developing flies would be specifically tuned to past flies with similarly modified cytoplasm, and this specificity would enhance morphic resonance from abnormal predecessors. As the experiment proceeded, there would be a cumulative influence from the increasing numbers of abnormal flies, increasing the probability of the pathway of development following the abnormal chreode (Fig. 8.3).

If the chreode leading to the bithorax form became more probable as the experiment proceeded, then we might expect that even normal flies of

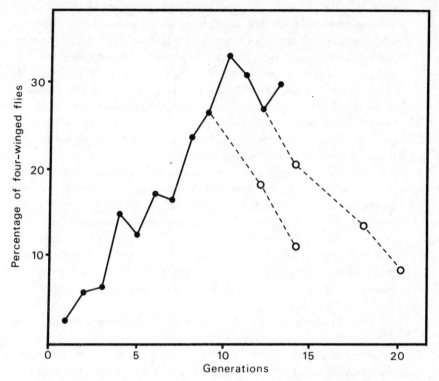

Figure 8.4 The effects of exposing successive generations of fruit-fly eggs to ether on the percentage of bithorax-type flies in the population. The dotted lines show what happened when the ether treatment was discontinued in a subpopulation of the flies; the percentage of abnormal flies declined in successive generations. (After Ho, et al., 1983)

the same strain, whose parents had not been exposed to ether, would show an increasing tendency to produce abnormal progeny in response to ether treatment. The data of Ho's group suggested that this was indeed the case. After the experimental flies had been treated with ether for six generations, the effect of the same treatment on control flies was examined. In the first generation 10% of the progeny were abnormal, and in the second generation 20%.[23] This compares with 2% and 5% in the first and second generations of the experimental line (Fig. 8.4).

Thus, after many of the flies had already responded to ether by developing abnormally, there was a much greater tendency for new batches of flies to do so. This is just what would be expected from the point of view of the hypothesis of formative causation.

In further experiments of this type, it should be possible to confirm whether or not characteristics acquired in response to a stimulus such as ether have an increased probability of appearing in genetically similar organisms whose parents have not been exposed to the abnormal stimulus. These abnormal organisms should show an increased probability of appearing not only in the same laboratory, but in other laboratories hundreds of miles away. This would provide a good way of testing the hypothesis of formative causation. If the development of abnormal organisms in one place caused a higher proportion of organisms to develop the same abnormality elsewhere in response to the same stimulus, this result would of course be inexplicable from the point of view of all orthodox theories.

For decades, the debate about Lamarckian inheritance has centred not so much on empirical evidence for or against such inheritance as on the question of whether or not such inheritance is theoretically possible. According to the genetic theory of inheritance, it is impossible because the genes cannot be specifically modified as a result of characteristics that organisms acquire in response to their environment or through the development of new habits of behaviour. Lamarckians have generally assumed that some such genetic modification must take place, but have been unable to suggest how this could happen.

The hypothesis of formative causation provides a new approach, which does not fit into either of these standard positions. Acquired characteristics can be inherited not because the genes are modified, but because this inheritance depends on morphic resonance. This means that the inheritance can take place without any transfer of genes at all. For example, as we have just seen, fruit-flies in one place may inherit a tendency to develop abnormally, in response to ether, from fruit-flies of the same strain hundreds of miles away without inheriting any modified genes from them, and indeed in the absence of any known means of communication.

Dominant and Recessive Morphic Fields

We now consider the implications of the hypothesis of formative causation for the understanding of the phenomenon of genetic dominance.

The great majority of mutations are *recessive*. That is to say that if a mutant organism is crossed with a normal one, often called the wild type, the progeny are normal. The normal type is *dominant*. Some of the second generation produced by crossing the hybrids with each other show the mutant character, but again the majority are normal.

The study of this kind of phenomenon by Mendel laid the foundation for the science of genetics. In one of his classic experiments, he crossed normal pea plants with a variety that produced wrinkled seeds. The first generation had normal seeds. In the second generation about three-quarters of the progeny had normal seeds and about a quarter wrinkled seeds. This is called Mendelian segregation and is explained in terms of Mendelian determinants, or genes (Fig. 8.5). The normal plants have two copies of the gene "for" normal seeds. The abnormal plants have two copies of a mutant form of the gene, resulting in wrinkled seeds. Such alternative forms of the same gene are known as alleles. Each parent contributes one copy of each of its genes to the progeny. Consequently the hybrid peas have one allele "for" normal seeds and one "for" wrinkled seeds. Normal seeds are produced because the normal-seed gene is dominant and the wrinkled-seed gene is recessive. In the second generation, through the random combination of genes from the egg and pollen cells, on average one plant will have two round pea alleles, and one will have two wrinkled pea alleles, for every two plants that have one copy of each allele. The latter will have round seeds, as of course will the plants with round pea alleles; thus there will be an approximate ratio of three normal-seeded plants to one wrinkled-seeded plant.

This is elementary genetics, and is well known to every student of the subject. But the very familiarity of these concepts conceals a deep problem. *Why* are normal wild-type genes almost always dominant? The problem becomes apparent when we consider it in an evolutionary context. New features of organisms arise by mutation. But the great majority of mutations are recessive. If these mutants are favoured by natural selection, the mutant type becomes more common, and eventually becomes predominant; what were originally mutants become the normal or wild type. As this happens, the genes that were originally recessive become dominant. So dominance cannot be an intrinsic property of genes, because dominance itself evolves.

The evolution of dominance is usually explained in terms of the natural selection of more dominant versions of the mutant genes, and also in terms of the selection of large numbers of minor genes which provide a "genetic background" that favours the dominance of the favourable mutant. Here is a typical text-book account:

> If certain phenotypic properties are favoured, then, clearly, the determinant or determinants conferring them will also be favoured. What is more, it will then be a further advantage if the elements in question find expression in all the individuals carrying them. In other words, dominance will be favoured from the point of view of adaptation. This means that certain alleles will be preferred to their less dominant isoalleles, and, other things being equal, it also means that background genotypes facilitating expression will be favoured over those which do not.[24]

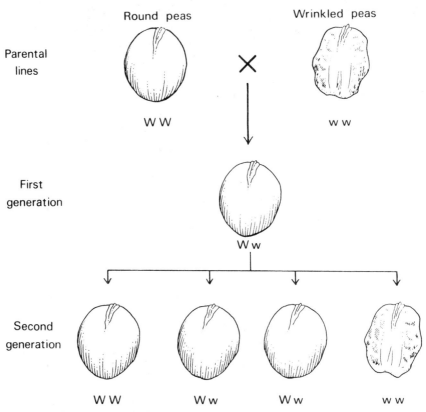

Figure 8.5 A standard example of Mendelian inheritance in peas. The gene W is dominant and leads to the development of round seeds. An alternative form of this gene, w, is recessive and leads to the production of wrinkled seeds when the W gene is absent. Only one copy of the gene is present in the egg and pollen cells; hence half the egg and pollen cells of the first-generation hybrids carry the W gene and half the w gene, resulting in an approximate ratio of one WW plant to two Ww plants to one ww plant in the second generation. Since W is dominant, this means there are approximately three times as many plants with round seeds as with wrinkled seeds.

This theory is inevitably speculative, and is also practically untestable, because the "background genotype" is too complex to be analysed genetically.

The hypothesis of formative causation provides an alternative explanation of dominance. The types most common in the past—the normal, wild types—stabilize the wild-type fields by morphic resonance. Mutant organisms, which are much rarer, are stabilized by much weaker fields simply because there have been so few of them. In hybrids, genes and proteins from both parental types are present, and the hybrids enter into morphic resonance with both the normal and the mutant types. The normal fields are stronger because of the far larger number of past organisms contributing to them, and they consequently "swamp" the mutant fields. A normal pattern of development will be far more probable: in other words it will be dominant. This is a dominance of fields, not a dominance of genes.

If a mutant type is favoured by natural selection, it becomes more and more common. Consequently, increasing numbers of organisms contribute by morphic resonance to the stabilization of these fields, and the mutant pattern of development becomes increasingly probable. From a genetic point of view, this increasing dominance of the mutant morphic fields would be interpreted as an increasing dominance of the mutant genes. Such changes in dominance as a result of morphic resonance from increasing numbers of mutant organisms could be investigated experimentally; some possible experimental designs are outlined in *A New Science of Life*.[25]

If two *species* are crossed, the hybrids will enter into morphic resonance with the fields of both. If both species are stabilized by morphic resonance from comparable numbers of past individuals, the fields will be of comparable strength; neither will be dominant, and the developing hybrid will be influenced to a similar extent by both. The hybrids can therefore be expected to exhibit features of both the parental species, and to be intermediate between them. This is in fact generally the case: think, for example, of mules, which are hybrids between horses and donkeys. The same is true in plants.

The Morphic Fields of Instinctive Behaviour

According to the hypothesis of formative causation, not only is the form of organisms shaped by fields, but so is their behaviour. Behavioural fields, like morphogenetic fields, are organized in nested hierarchies. They co-ordinate the movements of animals primarily through imposing rhythmic patterns of order on the probabilistic activities of the nervous system.[26] Behavioural fields are of the same general nature as morphogenetic fields: they are morphic fields and are stabilized by morphic resonance.[27]

In all animals certain patterns of motor activity are innate: for example the way in which mammals and birds scratch themselves (Fig. 8.6). And the instincts of animals are, of course, inherited from their forebears. For instance, young spiders hatch from their eggs with an inborn capacity to spin webs characteristic of their species; they set to work and spin them even if they are raised in complete isolation and have never seen another spider or a web. But even when animals *learn* new patterns of activity, they do so within an inherited framework of potentialities, and there is no sharp dividing line between instincts and learned behaviour, which depends on inherited capacities. For example, human babies do not have an innate ability to speak particular languages; they have to learn. But the *capacity* to learn human language is inherited, and this capacity is absent from other species.

The study of instinctive behaviour by ethologists has led to three major conclusions, often referred to as the classic concepts of ethology. First, instincts are organized in a hierarchy of systems, which are superimposed upon one another. Each level is activated primarily by a system at the level above it. Second, the behaviour that occurs under the influence of the major instincts often consists of chains of more or less stereotyped patterns of behaviour called fixed action patterns. Third, each pattern of behaviour requires a specific stimulus in order to be activated. The stimulus, or releaser, may come from within the body, or from the environment, in which case it is called a sign stimulus. One example is provided by European robins. During the breeding season, territory-holding males threaten other males that come too close. The fixed action pattern of aggressive behaviour is released mainly by the sign stimulus of the red breast, as shown by some simple experiments. Males attack crude models of robins with red breasts, or even a mere bundle of red feathers, but respond much less to accurate models without red breasts[28] (Fig. 8.7).

These features of inherited behaviour fit very well with an interpretation in terms of hierarchically organized morphic fields. The fixed action patterns can be thought of as chreodes; sign stimuli, such as the red feathers

Figure 8.6 The scratching behaviour of a dog and a European bullfinch. The inborn habit of scratching with a hind limb crossed over a forelimb is common to most reptiles, birds, and mammals. (After Lorenz, "The Evolution of Behavior," *Scientific American*, Dec. 1958)

Figure 8.7 Two models exposed to European robins during the breeding season. They attacked the tuft of red feathers on the right far more than a stuffed robin with a dull brown breast (left), showing that attacking behaviour is released mainly by the sign stimulus of the red breast. (After N. Tinbergen, *The Study of Instinct*, Oxford University Press, 1951)

attacked by robins, play the role of morphogenetic germs. They do so by setting up, through the senses, characteristic rhythmic patterns of activity within the nervous system, which enter into morphic resonance with particular behavioural fields—in the case of male robins responding to the sign stimulus of red feathers, the fields of attacking behaviour.

Behavioural chreodes canalize behaviour towards particular end-points (often called consummatory acts), and like morphogenetic chreodes have an inherent capacity to adjust or regulate the process so that the end-point is achieved in spite of fluctuations and disturbances. Ethologists have in fact observed that many fixed action patterns show a "fixed" component and an "orienting" component that is relatively flexible. For example, a greylag goose will retrieve an egg that has rolled out of the nest by putting its beak in front of the egg and rolling it back towards the nest (Fig. 8.8). As the egg is being rolled, its wobbling movements are compensated by appropriate side-to-side movements of the bill.[29] These compensatory movements occur in a flexible way, in response to the movements of the egg, within the framework of the fixed pattern of rolling; if the egg is suddenly removed they cease, but the movement of the bill towards the chest, once initiated, proceeds to completion.

Figure 8.8 A classic example of a fixed action pattern: a greylag goose rolling an egg back into its nest. The goose invariably attempts to roll the egg using its bill in this manner, rather than by using its foot or wing, or by using its bill in some other way. (After N. Tinbergen, *The Study of Instinct*, Oxford University Press, 1951)

The similarities between behavioural and morphogenetic chreodes, with their inherent regulatory capacities, are shown most clearly in patterns of behaviour that involve the building of structures, such as nests. For example, female mud wasps of a *Paralastor* species in Australia build and provision underground nests in an extraordinarily elaborate way. First they excavate a narrow hole about 3 inches long and ¼ inch wide in a bank of hard, sandy soil. Then they line this with mud. The mud is made by the wasp from soil near the nest; she releases water from her crop onto the soil, which she then rolls into a ball with her mandibles, carries into the hole, and uses to line the walls. When the hole has been fully lined, the wasp begins to construct a large and elaborate funnel over the entrance, building it up from a series of mud pellets (Fig. 8.9A). The function of this funnel appears to be the exclusion of parasitic wasps, which cannot get a grip on the smooth inside of the funnel; they simply fall out when they try to enter.

After the funnel is completed, the wasp lays an egg at the end of the nest hole, and begins provisioning the nest with caterpillars, which are sealed into cells, each about ¾ inch long. The last cell, nearest the entrance, is often sealed off empty, possibly as a protection against parasites. The nest hole is then sealed with a plug of mud, and the wasp destroys the carefully constructed funnel, leaving nothing but a few scattered fragments lying on the ground.

This is a sequence of fixed action patterns, governed by behavioural chreodes. The end-point of each of these chreodes serves as the sign stimulus or germ structure for the next. As in morphogenesis, the same end-points can be reached by a different route if the normal pathway of activity is disturbed: the behavioural equivalents of regulation and regeneration take place under the influence of the behavioural fields.

The ways the wasps react to damage of the funnel while it is under construction illustrate these general principles. First, in experiments carried

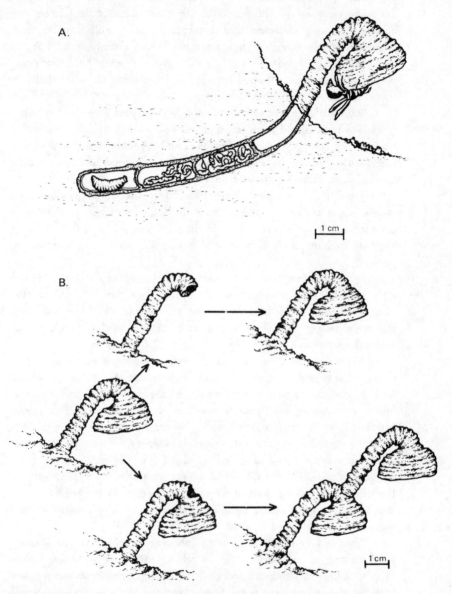

Figure 8.9 A: The nest, stocked with food, of the wasp *Paralastor*. B: The repair of funnels by *Paralastor* wasps. Above, the construction of a new funnel after the old one had been removed by the experimenter. Below, the extra stem and funnel made by the wasp in response to a hole above the normal funnel. (After S. A. Barnett, *Modern Ethology*, Oxford University Press, 1981)

out in the wild, funnels that were almost complete were broken off while the wasps were away collecting mud. However much of the funnels was missing, the wasps recommenced construction and rebuilt them to their original form; the funnels were regenerated. If they were broken off again, they were again rebuilt. This process was repeated seven times with one particular wasp, which showed no signs of reduced vigour as it rebuilt its funnel for the seventh time.[30]

Second, the experimenter stole almost completed funnels from some wasps and transplanted them to other nest holes where funnel construction was just beginning. When these wasps came back with pellets of mud and found the instant funnels, they examined them briefly inside and out, and then finished constructing them as if they were their own.

Third, the experimenter heaped sand around funnel stems while they were being constructed. These are normally about an inch long. If a nearly completed one was buried until only about ⅛ inch was showing, the wasp went on building it up until it was once again about an inch above the ground.

Finally, various holes were made in the funnels at different stages of construction. If these were made at an early stage, or if they involved removal of material from the bells of the funnels, the damage was detected at once, and the damaged area was repaired with strips of mud until the funnel assumed its previous form.

The most interesting behaviour occurred in response to a type of damage that would probably never happen under natural conditions: a circular hole made in the neck of the funnel after the bell of the funnel had been built. The wasps on their return soon noticed these holes and examined them carefully from the inside and the outside, but they were unable to repair them from the inside because the surface was too slippery for them to get a grip. After some delay the wasps started adding mud to the outside of the hole. This is just the type of activity that occurs when they start constructing a funnel over the entrance hole of the nest. The holes in the neck of the funnel thus came to act as a sign stimulus for the entire process of funnel construction, and a complete new funnel was made (Fig. 8.9B).

Thus behavioural fields, like morphogenetic fields, have an inherent goal-directedness, and they enable animals to reach their behavioural goals in spite of unexpected disturbances, just as developing embryos can regulate after damage and produce normal organisms, and just as plants and animals can regenerate lost structures.

Let us now consider how behavioural fields are transmitted.

The Inheritance of Behavioural Fields

Hereditary behaviour, like hereditary form, is influenced by genes, but it is neither "genetic" nor "genetically programmed." Under the hypothesis of formative causation, its characteristic patterns are organized by morphic fields, which are inherited by morphic resonance from past members of the same species.

Each of these behavioural fields organizes a particular pattern of behaviour. The fixed action patterns described by ethologists, such as the attacking behaviour of robins and the funnel-building activities of *Paralastor* wasps, are organized by such morphic fields.

The expression of the behaviour organized by these fields can be influenced by mutations in many different genes, but the effects of the genes on the behaviour may be very indirect. Some result in abnormal sense organs, nervous systems, or musculatures, and these can, of course, affect the ways in which an animal behaves. Other mutations affect behaviour by general effects on the vigour of animals.[31] But such mutations do not in themselves determine the *patterns* of behaviour; they merely affect the ways in which these patterns can be expressed.

In fact, studies on the inheritance of fixed action patterns have shown that there are indeed numerous genetic mutations that affect the performance of these patterns in various minor ways, but any given behavioural pattern "still appears in a clearly recognizable form if it appears at all."[32] In terms of the TV analogy, these mutant organisms are like faulty TV sets containing "mutant" components. These could bring about all kinds of distortions in the sound or pictures, or alterations in the colours on the screen. Nevertheless, in spite of these disturbances, the program being received by the set remains clearly recognizable. The set remains tuned to the same channel.

In addition to these various mutations, which affect the expression of a given behavioural field, we might expect to find other kinds of mutations, analogous to homoeotic mutations of form (p. 138–39), in which entire fixed action patterns disappear or are replaced by others. In terms of the TV analogy, these mutations affect the *tuning* of the set in such a way that an entire channel is lost or another program is received instead. And such mutations do indeed exist. They affect the appearance or non-appearance of entire fixed action patterns, just as homoeotic mutations affect entire organic structures.

One of the few examples studied in any detail concerns the nest-cleaning behaviour of honeybees in America in response to the American foulbrood disease, which kills larvae within the honeycomb. In one strain, called Brown, the worker bees uncap the cells in which larvae have died and remove the corpses from the comb. In another strain, van Scoy, they do nothing about the dead larvae. This leads to further infection. Colonies of the Brown strain, as a consequence of their hygienic behaviour, are more resistant to foulbrood disease than van Scoy colonies.

Crosses between queens from the one strain and drones from the other gave rise to hybrid queens that built up hybrid colonies. These colonies turned out to be unhygienic, showing that the hygienic behaviour is recessive. Further genetic analysis revealed that two recessive genes were involved in hygienic behaviour: one "for" uncapping the cells, and one "for" removing the corpses from uncapped cells.[33] From the point of view of formative causation, these fixed action patterns are not coded in the genes, but rather the genes somehow affect the tuning of the bees' nervous systems such that morphic fields for these patterns of behaviour come into play or they do not. Rather than genes "for" these patterns of behaviour, there are morphic fields for such patterns.

When two species are crossed, the instinctive behaviour of the hybrids often exhibits elements of the instincts of both parental types. Sometimes the parental patterns of behaviour conflict, as for example in crosses between two species of lovebird. Birds of one species make their nests from strips of leaves, which they tear off and carry to the nest in their bills. Birds of the other species carry strips of leaves to the nest tucked in among their feathers. Hybrids behave in a most confused manner, attempting to tuck in strips of leaves among their feathers, but so ineffectively that the strips drop out. They learn eventually that the only way they can carry them to the nest successfully is in their bills, but even then they still make attempts to tuck them in.[34]

In many cases, the hybrid behavioural pattern is intermediate between the parental patterns. This is shown particularly clearly in calls and songs, which have the advantage that they can be recorded and represented quantitatively. For example, female gibbons make impressive sounds in the morning in the presence of their mates, known as great calls. Different species make different calls. In the jungles of central Thailand, two species live in the same area and occasionally interbreed. The females of both species make great calls of similar length (14 to 21 seconds) and within a similar range of pitch. But one species utters on average 8 notes per call, the other averages 73. Hybrids between the two species, both in the wild and in zoos, make calls of an intermediate type[35] (Fig. 8.10).

The conventional theory and the hypothesis of formative causation interpret the known facts about the inheritance of behaviour in quite different ways; but the facts themselves do not enable us to decide which is the better interpretation. However, the two hypotheses lead to different predictions in

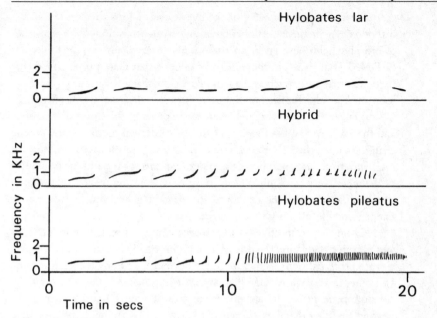

Figure 8.10 Sound spectrograms of the great calls of gibbons in the wild in Thailand. The calls of the hybrids are intermediate between those of the parent species. (After Brockelman and Schilling, 1984, copyright © Macmillan Magazines Ltd.)

cases in which animals acquire new patterns of behaviour. According to the conventional theory, such acquired abilities should have no hereditary effect on the offspring. By contrast, morphic resonance should produce a tendency for the new pattern of behaviour to be learned more readily by other members of the breed, even in distant parts of the world. These predictions are distinguishable empirically, and some possible experiments are discussed towards the end of the following chapter.

Morphic Resonance and Heredity

In this chapter, we have seen how the idea of morphic resonance sheds new light on the phenomena of heredity, which can be seen to depend both on genes and on morphic fields inherited by morphic resonance. The form and behaviour of organisms are not coded or programmed in the genes any more than the TV programs picked up by a TV set are coded or programmed in its transistors, or in any of its other material components.

From the point of view of the hypothesis of formative causation, the orthodox genetic theory of inheritance involves a projection of the properties of morphic fields onto genes, an attempt to squeeze them into the molecules of DNA. Thus there are thought to be genes rather than morphic fields for particular structures, such as the legs of fruit-flies, and for patterns of behaviour, such as the funnel-building activities of *Paralastor* wasps (Fig. 8.9). Genes, rather than morphic fields, are assumed to be dominant or recessive; and the evolution of dominance is believed to depend on ill-defined genetic changes, rather than on the cumulative building up of habits by morphic resonance from large numbers of similar organisms in the past. The possibility of the inheritance of acquired characteristics is denied on theoretical grounds, just because it cannot be explained in terms of genes. But in fact it may happen because of morphic resonance.

From this perspective, in the absence of the concept of morphic fields and morphic resonance, the role of genes is inevitably overrated, and properties are projected onto them that go far beyond their known chemical roles. Likewise, in relation to the phenomena of learning and memory, to which we now turn, properties are projected onto nervous systems which go far beyond anything that they are actually known to do. Brains, like genes, have been systematically overrated.

CHAPTER 9

Animal Memory

Morphic Resonance and Memory

The hypothesis of formative causation provides a radical reinterpretation of the nature of memory. It proposes that memory is inherent in all organisms in two related ways. First, all organisms inherit a collective memory of their species by morphic resonance from previous organisms of the same kind. Second, individual organisms are subject to morphic resonance from themselves in the past (pp. 132–34), and this self-resonance provides the basis for their own individual memories and habits.

As we have just seen in chapter 8, according to this hypothesis patterns of behaviour are organized by nested hierarchies of behavioural fields, just as patterns of morphogenesis are organized by nested hierarchies of morphogenetic fields. These behavioural fields organize the activities of the nervous system by imposing spatio-temporal patterns on its inherently indeterminate or probabilistic functioning (pp. 149–51). Behaviour is not determined only by the "wiring" of the nervous system and by physico-chemical processes going on within it, but depends on the organizing activity of these fields.

Just as appropriate genes are necessary for normal morphogenesis, so an appropriate nervous system is necessary for normal behaviour. Chemical or physical disturbances of the nerves can affect behaviour, just as disturbances of genes and proteins can affect morphogenesis. But behaviour is no more programmed in the nervous system than morphogenesis is programmed in the genes.

According to the hypothesis of formative causation there is only a difference of degree, not of kind, between inherited and learned behaviour. Both depend on morphic fields stabilized by morphic resonance. In instinctive behaviour, such as the building of nests by *Paralastor* wasps (Fig. 8.9), the influence of many other similar insects predominates; whereas in learned behaviour, such as the learning of the way out of a maze by a rat, resonance from an animal's own past is more important. Usually both play a part: instinctive behaviour involves an element of adaptation to the animal's particular circumstances, and learned behaviour takes place within the framework of potentialities provided by the species's morphic fields.

Learning inevitably involves memory, for the influence of past experience on present behaviour would not be possible if the experience were not in some way retained. There is of course no need for memory to involve consciousness: we ourselves are influenced by many unconscious memories which are expressed in our habits. We remember how to swim, write, or ride bicycles, but these habit memories are not conscious. There is no reason to assume that the habit memories we see at work in animals are any more conscious than our own.

Memory is conventionally believed to be explicable in terms of physico-chemical modifications of the nervous system, the "traces" of past experience. Attempts to locate such traces within the brain and to analyse them have so far been unsuccessful; but from the point of view of the mechanistic theory memory *must* depend on material traces of some kind. This is an a priori assumption:

> Memories are in some way "in" the mind, and therefore, for a biologist, also "in" the brain. But how? The term memory must include at least two separate processes. It must involve, on the one hand, that of *learning* something new about the world around us; and on the other, at some later date, *recalling,* or remembering that thing. We infer that what lies between the learning and the remembering must be some permanent record, a *memory trace,* within the brain.[1]

By contrast, through formative causation, memory depends on morphic resonance between patterns of activity within the nervous system now and similar patterns of activity in the past. It need not depend on physico-chemical modifications of the nerves. Memory need not be stored in material memory traces if it results from morphic resonance; the past can exert a direct influence on the present.

In this chapter, we first examine the evidence for memory storage within the brain; we then consider different types of learning, comparing the orthodox mechanistic interpretations with those of morphic resonance. Finally, we discuss what kinds of experiments could be done to find out which of these alternative approaches is in closer accordance with the way memory really works.

Are Memories Stored Inside the Brain?

The traditional idea of memory storage within the brain goes back to classical times. Stimuli falling on the sense organs produce disturbances in the brain, which cause the perception of the stimuli. The disturbances leave behind traces, minute changes in the structure of the brain. As a result of these changes, brain activity becomes more likely to follow the same paths again in response to stimuli that are similar or whose traces are intermingled or "associated" with those of the first stimulus.

In the seventeenth century, Descartes proposed a hydraulic version of this theory, based on the assumption that nerves are hollow and conduct a flow of "animal spirits": sensory nerves contain delicate threads attached to valves within the brain, the opening of which releases animal spirits, which pass through the nerves to appropriate muscles. Descartes in fact invented the concept of the reflex: animal spirits are "reflected" in the brain and pass back to the muscles[2] (Fig. 9.1). The memory traces "are nothing else than the circumstances that the pores of the brain through which the spirits have already taken their course on presentation of the object, have thereby acquired a greater facility than the rest to be opened again the same way by the spirits which come to them; so that these spirits coming upon the pores enter therein more readily than into the others."[3] This idea has the great attraction of simplicity, and is echoed in the modern theories of synaptic modification.

Pavlov's famous researches on conditioned reflexes greatly strengthened the traditional concept of memory traces. Pavlov himself was reluctant to claim that reflex arcs depended on specifically localized traces within the cerebral cortex because he found that the conditioning could survive considerable surgical damage to the brain.[4] But some of those who followed him were less cautious. In the first few decades of this century it was assumed by many biologists that *all* psychological activity, including the phenomena of the human mind, could ultimately be reduced to simple associations and chains of reflexes. The path of the reflex circuits was supposed to run from the sense organs to the sensory areas of the brain, thence through associative areas to the motor cortex, and finally to the motor cells, which conducted impulses to the muscles.[5] These definite pathways of connection were often conceived by analogy with a telephone system, with the nerve fibres as wires and the brain as the exchange where appropriate connections were made.

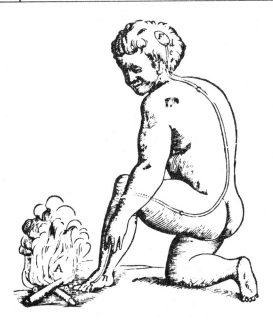

Figure 9.1 The diagram of a kneeling man by a fire used by Descartes to illustrate his idea of reflex action. (As reproduced by Boakes, 1984)

Modern theories usually rely on computer analogies. The central model is of coding, storage, and retrieval. Incoming nerve impulses from the sense organs are usually said to "encode" the external stimulus, and these then change the properties of other nerve cells within the brain so that these changes encode or "represent" the stimulus, but in some different way. These changes constitute the process of memory storage. Retrieval is the supposed process by which the stored pattern is revived as needed.

The complexity of computers means that this model can be developed in a more sophisticated way than the telephone model, but it still depends on definite memory traces, even if there are "back-up" storage systems. In computers the "traces" are carried either in the solid-state memory or on discs or tapes. If in a real computer the memory store is destroyed, then the memory is, of course, lost.

Much effort has been expended in an attempt to locate memory traces within the brain, and vast numbers of animals have been used up in the process. The classic investigations on the subject were made by Karl Lashley with rats, monkeys, and chimpanzees. For over thirty years he tried to trace conditioned reflex paths through the brain and to find the locus of specific memory traces, or "engrams." To do this, he trained the animals in a variety of tasks, ranging from simple conditioned reflexes to the solution of difficult problems. Either before or after the training, nerve tracts within the brain were surgically cut or portions of the brain were removed, and the effects on initial learning or post-operative retention were measured.

He first became sceptical of the supposed path of conditioned reflex arcs through the motor cortex when he found that rats trained to respond in particular ways to light showed no reduction in accuracy of performance when nearly the entire motor cortex was cut out. Likewise, most of the motor cortex of monkeys was removed after they had been trained to open various latch boxes. This operation resulted in a temporary paralysis, but after eight to twelve weeks they recovered sufficiently to be able to make the movements required to open the latches. They were then exposed to the puzzle boxes, and opened them promptly without random exploratory movements.

He then showed that learned habits were retained if the associative areas of the brain were destroyed. Habits also survived a series of deep incisions into the cerebral cortex which destroyed cross-connections within it. Moreover, if the cerebral cortex was intact, removal of subcortical structures such as the cerebellum did not destroy the memory either.

Lashley started as an enthusiastic supporter of the reflex theory of learning, but was forced by his results to abandon it:

> The original programme of research looked toward the tracing of conditioned-reflex arcs throughout the cortex, as the spinal paths of simple reflexes seemed to have been traced through the cord. The experimental findings have never fitted into such a scheme. Rather, they have emphasized the unitary character of every habit, the impossibility of stating any learning as concatenations of reflexes, and the participation of large masses of nervous tissue in the functions rather than the development of restricted conduction paths.[6]

In reviewing the types of human memory loss that follow brain damage, he came to a similar conclusion:

> I believe that the evidence strongly favours the view that amnesia from brain injury rarely, if ever, is due to the destruction of specific memory traces. Rather, the amnesias represent a lowered level of vigilance, a greater difficulty in activating the organized pattern of traces, or a disturbance of some broader system of organized functions.[7]

Lashley did not consider the possibility that memories might not be stored inside the brain at all. He suggested that rather than localized traces, there must be *multiple* memory traces throughout an entire functional area of the brain. He thought that this indicated that "the characteristics of the nervous network are such that when it is subject to any pattern of excitation, it may develop a pattern of activity, reduplicated through an entire functional area by spread of excitations, such as the surface of a liquid develops an interference pattern of spreading waves when it is disturbed at several points." He suggested that recall involved "some sort of resonance among a very large number of neurons."[8] These ideas have been carried further by his former student Karl Pribram in his proposal that memories are stored in a distributed manner analogous to the interference patterns in a hologram.[9]

Analogous experiments have shown that even in invertebrates such as the octopus, specific memory traces cannot be localized. Observations on the survival of learned habits after destruction of various parts of the brain have led to the seemingly paradoxical conclusion that "memory is both everywhere and nowhere in particular."[10]

The conventional response to such findings is that there must be multiple or redundant memory-storage systems distributed throughout various regions of the brain: if some are lost, back-up systems can take over. This hypothesis, invented to account for the failure of attempts to find localized memory traces, follows naturally from the assumption that memories *must* be stored somehow inside the brain; but in the continuing absence of any direct evidence, it remains more a matter of faith than of fact.

There is, however, good evidence that *changes* can occur in the brains of young animals as a result of the way they grow up. In one experiment, for example, young rats were raised either in solitary confinement in featureless cages, or in groups in larger cages furnished with a variety of playthings, which were regularly replaced with novel ones. After various periods of time, rats from both environments were killed and their brains examined. Those that had grown up in the enriched environment had bigger brains than those in solitary confinement, and the individual nerve cells and synapses were larger.[11] These results show that the development of the nervous system was influenced by its activity.

In somewhat more refined experiments with young monkeys, the effect of depriving them of the use of one eye (by stitching the eyelids together) has been studied in considerable detail. In normal adults, both the right and the left visual cortex of the brain receive nervous input from both eyes. Thus in the left visual cortex there are two orderly maps of the right half of the visual field, one received from the right eye and the other from the left; likewise in the right visual cortex there are two maps of the left half of the visual field. The input from the two eyes is segregated in a pattern of alternating cortical strips about 0.4 millimetres wide. But young monkeys that had one eye stitched up became blind in that eye after several weeks, and the strips connected with it became very narrow while those connected with the other eye expanded to take up almost all the space. Similar results have been obtained in experiments on kittens. The changes appear to be due to competition between the nerves connected to the two eyes: the inactive nerves linked to the closed eye made fewer connections with the cortical cells than the electrically active nerves from the other eye.[12] As in the case of the rats raised in an enriched environment, these results show that the way in which the nervous system developed depended on the activity of the nerves within it.

It does not seem surprising that changes in the functioning of the nervous system are associated with changes in the nerve cells themselves; we all know that changes in other tissues such as muscles occur as a result of use or disuse. Body-builders show us how far such changes can go. The fact that such changes occur in developing brains emphasizes once again that the nervous system is dynamic in its structure.

Perhaps the most careful and thorough attempts to demonstrate the occurrence of changes in the brain that might be connected with the formation of memory traces have been made with young chicks. A day after hatching, they underwent simple forms of training, the effects of which were studied by injecting radioactive substances. Greater amounts of these substances were incorporated into nerve cells in a particular region of the forebrain, especially in the left hemisphere, in chicks that learned to respond to the stimulus than in control chicks, which did not learn.[13] In other words, these experiments have shown that nerve cells in this region underwent more active growth and development when learning took place than when it did not.[14] As we have already seen, in the growing brains of young rats, kittens, and monkeys, nerve cells that were active developed more than those that were inactive. But this greater development of active cells does not prove that they contain specific memory traces. In fact, in experiments with chicks, when the region of the left forebrain associated with the learning process was removed a day after they were trained, the chicks could still remember what they had learned. Therefore these cells which were somehow involved in the learning process were *not* necessary for memory retention. Yet again, the hypothetical memory traces have proved to be elusive, and once more it has been necessary for those who searched for them so assiduously to postulate further unidentified "storage systems" somewhere else in the brain.[15]

Not only have the hypothetical memory traces proved to be spatially elusive, but their physical nature has also remained obscure. The idea of specific RNA "memory molecules" was fashionable in the 1960s but has now been more or less abandoned. The theory of reverberating circuits of electrical activity, giving a kind of echo, may help to account for short-term memory over periods of seconds or minutes, but cannot plausibly explain long-term memory. The most popular hypothesis remains the old favourite that memory depends on modifications of the synaptic connections between nerve cells in a manner still unknown.

If memories are somehow stored in synapses, then the synapses themselves must remain stable over long periods of time: indeed, the nervous system as a whole must be stable if it is to act as a memory store. Until recently this was generally assumed to be the case, even though it has long been known that there is a continuous process of cell death within the brain. But recent evidence suggests that the nervous systems of mature animals may be more dynamic than previously supposed.

Studies on canaries' brains, and in particular those parts involved in the learning of song, have shown that not only do many new connections between nerve cells continue to develop, but many new nerve cells appear. In males the number of neurons increases as the birds mature in the spring, but then decreases by about 40% by the autumn. As the new mating season approaches, the number of nerve cells increases again, and so on. Such changes have also been found in other parts of canaries' brains, and there is now evidence that in adults of other species too there is a turnover of neurons in the forebrain, the "seat" of complex behaviour and learning, with new cells being formed while others die.[16]

Brains also appear to be more functionally dynamic than was once thought. Recent studies on monkeys have shown that sensory areas of the brain that "map" different parts of the body are not "hard wired" or anatomically frozen, but are unexpectedly fluid. In one series of experiments the regions of the sensory cortex connected with touch sensations from the monkeys' hands were localized. The "map" in the brain was found to be subdivided into regions for each of the five fingers and for other surfaces of the hand. After one or more of the fingers was amputated, it was found that sensory input from the remaining adjacent fingers gradually shifted over a period of weeks into the missing finger's hitherto exclusive brain region (Fig. 9.2). The increased areas of brain connected to the adjacent fingers were associated with an increase in the acuity of sensation in these fingers.[17]

The dynamism of the nervous system is also shown when the brain is damaged. For example, if a portion of the sensory cortex is injured, the appropriate sensory "map" which used to be in the injured region can shift to the region surrounding it, albeit with some loss in acuity. This movement of the "map" probably does not depend on a growth or movement of nerve cells, but rather on a spatial shift of nerve-cell activity.[18]

This dynamism in the structure and functioning of the nervous system

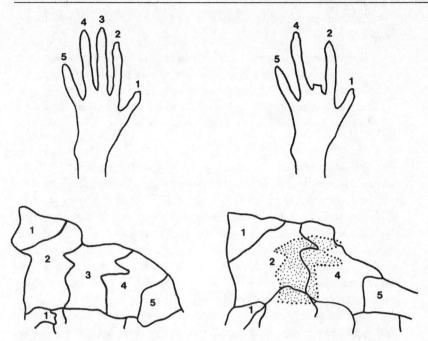

Figure 9.2 Brain maps of the area in the cortex of adult owl monkeys where the tactile inputs from the hand are received. Within several weeks of the amputation of the third finger, the area of the cortex in which it used to be represented is taken over by expanded areas representing the adjacent fingers. These brain maps were worked out by microelectrode analysis. (After Fox, "The Brain's Dynamic Way of Keeping in Touch," in *Science* 225:820–821, 24 August 1984; copyright 1984 by the AAAS)

presents great difficulties for the concept of memory traces. At the molecular level too, as Francis Crick has recently pointed out, there is a dynamism that makes the long-term storage of memory traces problematic. The time span of human memory is often years or tens of years. "Yet it is believed that almost all the molecules in our bodies, with the exception of DNA, turn over in a matter of days, weeks, or at the most a few months. How then is memory stored in the brain so that its trace is relatively immune to molecular turnover?" Crick has suggested a mechanism whereby "molecules in the synapse interact in such a way that they can be replaced by new material, one at a time, without altering the overall state of the structure." His ingenious hypothetical scheme involves protein molecules which he endows with a number of unusual properties. There is as yet no evidence that such molecules exist.[19]

In summary, in the words of a recent text-book, *Molecular Biology of the Cell:*

> Despite various scraps of physiological and biochemical evidence, a vast mass of psychological data, and a few general principles, we still understand almost nothing about the cellular basis of memory in vertebrates—neither the detailed anatomy of the neural circuits responsible nor the molecular biology of the changes that experience produces in them.

An interpretation of memory in terms of morphic resonance offers a new approach to these problems. If memories depend on morphic fields, then they need not be stored within the brain at all, but rather may be given by morphic resonance from the organism's own past. After damage to parts of the brain, these fields may be capable of organizing the nerve cells in other regions to carry out the same functions as before. The ability of learned habits to survive substantial brain damage may be due to the self-organizing properties of the fields—properties which are expressed in the realm of morphogenesis in regeneration and embryonic regulation.

This alternative to the conventional interpretation of memory can be distinguished from it by experiment. If the hypothesis of formative causation is correct, then it should be possible for the habit memories of one organism to influence another by morphic resonance, facilitating the acquisition of the same habits. Such an effect would not, of course, be expected on the basis of mechanistic theories of memory storage.

We now consider how the interpretation of memory in terms of morphic resonance from an animal's own past applies in the context of learning. We start with the simplest and most basic type of learning: habituation.

Habituation

If a stimulus is harmless and not followed by anything of interest, less and less response occurs as it is repeated. This is known as habituation. We ourselves become habituated in many different ways: we cease to notice the contact of our clothes with our skin; we usually become unaware of background noises, background smells, or background objects; we get used to new environments and settle down in new situations.

Animals too become accustomed to their environments. They generally react to the appearance of something new precisely because they are not used to it, often with alarm or avoidance. But if the new stimulus is harmless they soon cease to respond. Probably everyone has observed this kind of habituation taking place in pets, as well as in wild mammals and birds.

Habituation also occurs in lower animals such as snails, and even in single-celled organisms. *Stentor,* for example, an inhabitant of marshy pools, is a trumpet-shaped cell covered with rows of fine, beating hairs called cilia. Ciliary activity sets up currents around the cell, carrying suspended particles to the mouth, which is at the bottom of a tiny vortex (Fig. 9.3). The response of these creatures to various stimuli was studied in detail by H. S. Jennings over eighty years ago and is described in his classic work *The Behaviour of the Lower Organisms* (1906). When the object it is attached to is slightly jarred, "like a flash it contracts into its tube. In about half a minute it extends again, and the cilia resume their activity." If the same stimulus is repeated, it does not contract but continues its normal activities. This is not due to fatigue, since the animal still responds to a new stimulus, such as being touched. If this new stimulus is repeated, once again it does not react.

Habituation implies some sort of memory that enables harmless and irrelevant stimuli to be recognized when they recur. This may well depend on the resonance of the organism with its own past patterns of activity, especially those in the recent past. These past patterns would include its return to normal following the response to the harmless stimulus. Repeated irrelevant stimuli are assimilated into the organism's own background resonance; they become, as it were, part of itself. Conversely, any new kind of stimulus stands out precisely because it is new and unfamiliar.

In more complex organisms habituation involves the nervous system, and has been studied in great detail in the giant marine slug *Aplysia,* which grows to be a foot long. Normally the gill is extended, but it is withdrawn if the slug is touched (Fig. 9.4). This reflex soon ceases to take place if weak and harmless stimuli are repeated. (With stronger stimuli, the slug responds in an octopuslike way by releasing a brilliant purple ink which conceals it within an opaque cloud.)

The nervous system is very similar from slug to slug; identifiable cells occur in predictable places. The sensory and motor cells involved in the gill reflex have been located; only four motor cells have been found to be responsible for the withdrawal response.[20] (In higher organisms the "wiring diagrams" are much more complex than in slugs, and much more variable from individual to individual.) Electrical measurements from single nerve cells have shown that as habituation occurs, the sensory cells cease to excite the motor cells. This happens because they release fewer and fewer packets or "quanta" of chemical transmitter at the junctions or synapses with the motor cells.[21] This change in the functioning of the sensory cells persists for minutes or hours, depending on how often the stimulus is repeated. With

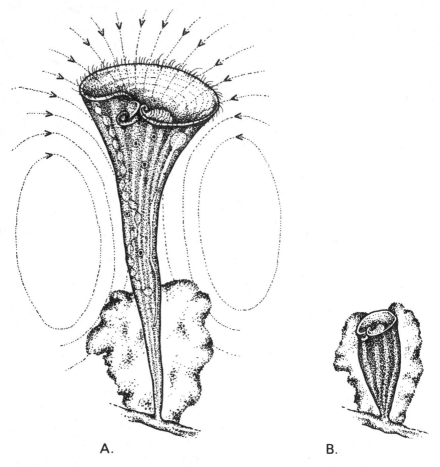

Figure 9.3 The single-celled organism *Stentor raesilii,* showing the currents in the water around it caused by the beating of its cilia. In response to an unfamiliar stimulus, it rapidly contracts into its tube (B). (After Jennings, 1906)

four training sessions of ten stimuli each, a profound habituation is produced which lasts for weeks. This means that some kind of memory of the stimulus can affect the sensory cells for long periods.

Since habituation can occur in a single cell such as *Stentor,* it is not surprising that cells within the nervous system of *Aplysia* can exhibit it as well. But there is no need to assume that it depends on physical or chemical memory traces within these cells. It may be due to morphic fields maintained by resonance with the organism's own past. These fields, modified by morphic resonance from the previous activity of the nervous system in

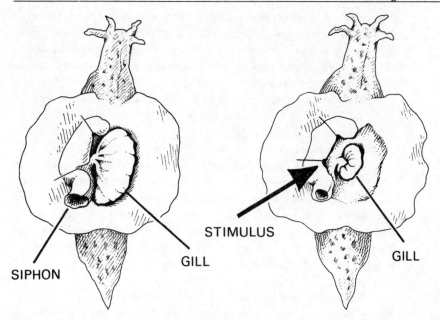

SIPHON

GILL

STIMULUS

GILL

Figure 9.4 *Aplysia,* a marine slug. On the left, its gill and siphon are extended. When the siphon is touched, both the siphon and the gill contract in a defensive reflex (right). (After Kandel, "Nerve Cells and Behavior," *Scientific American*, July 1970)

response to the harmless stimuli, organize the physical and chemical activities of the cells, including the release of chemical transmitters at the synapses. There may well be changes within the cells as a result of their activity, but this does not mean that the memory is stored within them as a material trace.

In higher animals, behavioural fields may embrace many millions of nerve cells. But here again, there is no need for habituation to involve memory traces; it may depend, as in *Stentor* and *Aplysia,* on morphic resonance with the organisms' own patterns of activity in the past.

Learning

From the point of view of the hypothesis of formative causation, inherited behavioural units or fixed action patterns are associated with particular morphic fields, such as the fields of attacking behaviour in robins or web-spinning behaviour in spiders. Morphic resonance from countless past members of the species gives these fields their probability structures, which organize the general expression of instinctive patterns of behaviour. However, actual experience within the framework of a behavioural field influences the way the actions are performed in similar circumstances on subsequent occasions, owing to self-resonance. Thus an animal acquires its own particular way of behaving instinctively.

There are many examples of such learning within the framework of instinct. Most young animals move rather clumsily at first, but become far better co-ordinated as time goes on. In part this improvement is due to maturation of the nervous system and the body in general; but it is due partly to practice as well.[22] An animal learns to carry out an inherited action pattern in a way that is appropriate to its particular body and environment.

Many bees and wasps instinctively go out on foraging or hunting expeditions, yet show a remarkable ability to memorize the terrain around their nests and are able to home by means of a variety of landmarks.[23] This kind of spatial learning is very widespread in the animal kingdom and makes possible a detailed adaptation of instinctive behaviour to the place where it is performed.

Perhaps the most dramatic type of instinctive learning is imprinting. Young birds such as chickens, goslings, and ducklings show an inherited behaviour pattern of following, and normally follow their mother. In his famous studies on geese, Konrad Lorenz got broods of newly hatched goslings to treat him as a mother figure and to follow him. Indeed, young goslings can imprint on almost anything that moves, including objects such as balloons.[24] After a period of imprinting of only fifteen to thirty minutes, the young birds recognize and approach the moving object when they are exposed to it again as long as seventy hours later. This ability to recognize the moving object is conventionally attributed to memory traces; but morphic resonance provides a direct connection. The object is recognized because through the senses it sets up specific patterns of activity within the nervous system, and these enter into morphic resonance with those previously set up by the same object.

The kind of learning on which experimental psychologists have concentrated is called associative learning. In Pavlovian conditioning, an automatic or unconditioned response such as the salivation of a hungry dog at the sight of meat can through repeated association with another stimulus, such as the ringing of a bell, become linked to it: a conditioned reflex is established, and the dog will salivate when the bell is rung even if no meat is provided.

The other main kind of associative learning depends on an animal's own activities. This was called operant conditioning by B. F. Skinner and the behaviourist school of psychology, and is also known as instrumental learning. For instance, if a cat by trial and error finds how to open a door and reaches a supply of food as a consequence, then it will sooner or later associate opening the door with receiving food; a conditioned response is established.

In conventional mechanistic terms, associative learning depends on the formation of new patterns of nervous connection within the brain. From the perspective of formative causation, by contrast, it results from the establishment of higher-level morphic fields that embrace previously separate patterns of activity within the nervous system. Such higher-level fields jump into being: they synthesize previously disparate parts and emerge as wholes. And indeed, associative learning often seems to involve definite discontinuities; it occurs in steps or stages. In trial-and-error learning, for instance, animals quite suddenly seem to grasp a connection, and we ourselves are familiar with jumps in learning: new patterns of connection suddenly "dawn on us" or "come in a flash." (The origin of new fields will be discussed in chapter 18.)

This may happen even without overt trial-and-error behaviour, by insight. Ethologists commonly use this word in connection with the behaviour of higher animals when they solve problems more rapidly than would be expected by trial and error. The classic example was provided by Wolfgang Köhler's studies of chimpanzees over sixty years ago. Presented with a banana too high to reach, they would, after some time, pile up boxes to make a stand for themselves so they could reach it, or would fit two sticks together in order to pull the banana down. Often they arrived at the solution quite suddenly, although they benefited from previous experience of playing with the boxes and sticks and showed considerable trial-and-error learning when actually building a stable pile of boxes.[25]

Such examples suggest that processes are going on that we can hardly avoid regarding as mental.[26] At the moment of insight a potential pattern of organized behaviour comes into being. This can be regarded as a new morphic field. If the behaviour pattern is repeated, the field will be increasingly stabilized by morphic resonance. This behaviour will become more probable, more habitual, and in our own experience increasingly unconscious.

The Transmission of Learning by Morphic Resonance

The nineteenth- and early-twentieth-century literature abounds in anecdotal accounts of the apparent hereditary transmission of acquired behaviour, especially in dogs. For example, a man who owned a young untrained pedigree Dobermann, in order to test its powers of perception, asked a friend to walk up to him in the street and pretend to attack him. When the friend lifted his hand to do so, the dog, barking viciously, jumped at the man. In itself this incident might show only that the dog had an instinctive tendency to spring to the defence of its master. But what happened seemed rather more remarkable:

> To me, who have trained dogs "on the man," the most interesting thing was the *way* in which the dog acted. It was exactly the same as is usually observed in highly trained police dogs when attacking criminals—a distinct behaviour which is well known to everyone who comes in contact with police dogs.[27]

Of course, such an observation could be dismissed on the ground that Dobermanns are used as police dogs because they have such an instinctive tendency anyway and police training only intensifies it. But such an argument is more difficult to apply to the response of naive gun dogs to the report of a gun; for dogs could not have had an instinctive response to the sound of gunfire before the invention of guns.

> The very careful, critical physiologist S. Exner relates how a young hunting dog, never before employed at a chase, as soon as he heard the first report of a gun started to search for a partridge which was not hit at all and which, for this reason, the dog could not very well have seen tumbling down to earth.[28]

Charles Darwin himself took a great interest in such stories, and published an account in *Nature* of a mastiff's violent antipathy to butchers and butchers' shops, presumed to be due to mistreatment at the hands of a butcher, which was apparently transmitted to at least two generations of its offspring.[29]

However, it was not until the 1920s that attempts were made to investigate experimentally the hereditary transmission of acquired habits. Some experiments provided evidence that such a transmission did in fact occur.[30] Pavlov, for instance, trained white mice to run to a feeding place when an electric bell was rung. The first generation required an average of 300 trials to learn, the second only 100, the third 30, and the fourth 10.[31] He later announced that attempts were being made to repeat these experiments but that they were "very complicated, uncertain, and moreover difficult to control."[32] No further results were published. (On the present hypothesis, the results would not be exactly repeatable because subsequent mice would be influenced by morphic resonance from those in the first experiment.) His last statement on the subject was that "the question of the hereditary transmission of conditioned reflexes and of the hereditary facilitation of their acquirement must be left completely open."[33]

The most thorough of all the investigations on the hereditary transmission of learning was begun at Harvard in 1920 by William McDougall. His own experiments, together with their sequels in Scotland and Australia, lasted for over thirty years and must be one of the longest series of experiments in the history of experimental psychology. McDougall used standard white laboratory rats and trained them in a water maze. They were put into a tank of water from which they could escape only by swimming to a gangway and climbing up it. There were two such exits, one on either side of the tank. One exit was illuminated, and if they chose this one they received an electric shock as they left the water. The other exit was quite safe. The next time they were put into the tank, the gangway that was previously illuminated was now in dim light, while the other exit was lit up and electric shocks were given there. The rats had to learn that it was painful to leave by the illuminated exit but safe to take the other one.

The first generation of rats made on average over 165 errors before learning to take the dim exit. Subsequent generations learned more and more quickly, until by the thirtieth generation the rats made an average of only 20 errors. McDougall showed that this striking improvement was not due to genetic selection for more intelligent rats, because even if he selected the most stupid rats in each generation as parents of the next, there was still a progressive increase in the rate of learning.[34] He interpreted these results in terms of Lamarckian inheritance, in other words in terms of the modification of the rats' genes.

This conclusion was unpalatable to many biologists. The only recourse was to repeat McDougall's experiments. When F.A.E. Crew did so in Edinburgh, the very first generation of his rats learned very quickly, with an average of only 25 errors, and some got it right the first time.[35] His rats seemed to be taking up where McDougall's left off. Neither he nor McDougall was able to account for this effect.

In Melbourne, W. E. Agar and his colleagues also found that the first generation they tested learned far quicker than McDougall's original rats. They continued to test fifty successive generations of rats over a period of twenty years, and like McDougall found a progressive increase in the rate of learning in subsequent generations. But, unlike McDougall, they also repeatedly tested control rats which were not descended from trained parents. These too showed a similar improvement.[36] The investigators very reasonably concluded that the progressive increase was not due to Lamarckian inheritance; if it had been, the effect should have appeared only in the progeny of the trained rats. But then why did the improvement occur? This effect has never been satisfactorily explained. But it is just what would be expected on the basis of morphic resonance.

Other experimental psychologists have, to their surprise, found very similar results. They were not, of course, looking for such improvements; they cropped up in the course of experiments carried out for a different purpose. For example, at the University of California R. C. Tryon bred rats with the intention of establishing "bright" and "dull" strains. He used a special kind of maze into which rats were released automatically, greatly reducing any influence of handling by the experimenter.[37] As expected, he found that the offspring of "bright" parents were more often "bright" than "dull," and the offspring of "dull" ones more "dull" than "bright." But he also found something he did not expect: *both* strains became progressively quicker at learning the maze.[38]

In general, according to the hypothesis of formative causation, an acceleration in learning should occur whenever animals are trained to do new tricks or adjust to new conditions: an increase in the average rate of learning or adjustment should occur as the training is repeated again and again, other things being equal. However, other things are rarely if ever equal, if only because trainers themselves tend to improve with experience. Nevertheless, there seems to be a wealth of anecdotal evidence that such changes do occur. Over the last few years, I have received fascinating accounts from dog owners, horse trainers, falconers, cattle ranchers, and dairy farmers about progressive improvements in the ease with which new generations of animals could be trained or became adapted to new methods. They all felt that only some of this improvement could be explained in terms of their own experience; some real change seemed to be happening in the animals as well. I do not know how general such observations are. It would be interesting to carry out a systematic survey among experienced trainers and farmers who have adopted new methods of animal husbandry to find out what changes of this type, if any, they thought had occurred.

Comparable changes may well be occurring all the time in psychology laboratories, but they are rarely, if ever, systematically documented. I have explored this possibility with two of the most ingenious experimental psychologists in Britain. When they devised new tricks for rats to perform, both found that in general the first rats tended to learn very slowly, but that as the series of experiments proceeded, new batches of rats usually picked up the tricks faster and faster. However, both believe that these improvements reflected improvements in their own performance as experimenters.[39] There is no doubt that experimenters can influence the performance of animals they are working with, and such "experimenter effects" have been well documented.[40]

Obviously, it is quite generally the case that animal trainers affect the animals they train, and also that trainers tend to improve with experience. But it may also be that the animals themselves improve as a result of morphic resonance from their predecessors. These influences are complementary rather than mutually exclusive.

Clearly these two kinds of influence need to be distinguished in experiments specifically designed to test the hypothesis of formative causation. One possible design is as follows. Several new tricks are devised for rats to perform, and appropriate pieces of apparatus are constructed in duplicate. The duplicate set is sent to a second laboratory, where experimenters are asked to test rats with each of the tasks and to record their rates of learning. They are requested to do the same thing again six months later with fresh batches of rats. Meanwhile in the first laboratory, *one* of these tasks is selected at random and thousands of rats are trained to perform it.

Experimenters at the second laboratory are not told which task has been selected. If they find a striking increase in the rate of learning in this task but not in the others, this could not in any obvious way be attributed to experimenter effects; rather the result would support the idea that morphic resonance from the trained rats in the other laboratory is effecting the accelerated rate of learning.

Such experiments would by their very nature not be exactly repeatable because of morphic resonance from previous experiments, but they could be replicated indefinitely with new species of experimental animals or with new sets of tricks.

The Case of the Blue Tits

The best-documented example of the spontaneous spread of a new habit concerns the opening of milk bottles in Britain by birds. They open the caps on bottles that are delivered to doorsteps early in the morning and drink as much as two inches of milk from the bottles (Fig. 9.5). Occasionally birds are found drowned head first inside the bottles. The bottles are usually attacked within a few minutes of delivery, and there are even reports of parties of tits following the milkman down the street and drinking from bottles on the cart while he is busy delivering milk.

The first record of this habit was from Southampton in 1921, and its spread was recorded at regular intervals from 1930 to 1947 (Fig. 9.6). It has been observed in eleven species, but most frequently in great tits, coal tits, and blue tits. Once discovered in any particular place, the habit spread locally, presumably by imitation.

Tits do not usually venture more than a few miles from their breeding place, and a movement of as much as fifteen miles is exceptional. Hence new appearances of the habit more than fifteen miles from where it had previously

Figure 9.5 A blue tit opening a milk bottle by tearing the foil cap. (After Hinde, 1982)

been recorded probably represented new discoveries by individual birds. A detailed analysis of the records showed that the spread of the habit accelerated as time went on, and that it was independently discovered by individual tits at least 89 times in the British Isles.[41]

The habit also appeared in Sweden, Denmark, and Holland. The Dutch records are particularly interesting. Milk bottles practically disappeared during the war, and became reasonably common again only in 1947 or 1948. Few if any tits that had learned the habit before the war could have survived to this date, but nevertheless attacks on bottles began again rapidly, and "it seems certain that the habit was started in many different places by many individuals."[42]

Hinde and Fisher have pointed out that bottle opening is related to the instictive behaviour of tits: "The initial discovery of the bottle as a source of food may be a logical consequence of the feeding habits of tits. They appear to have an inborn tendency to inspect a great variety of conspicuous

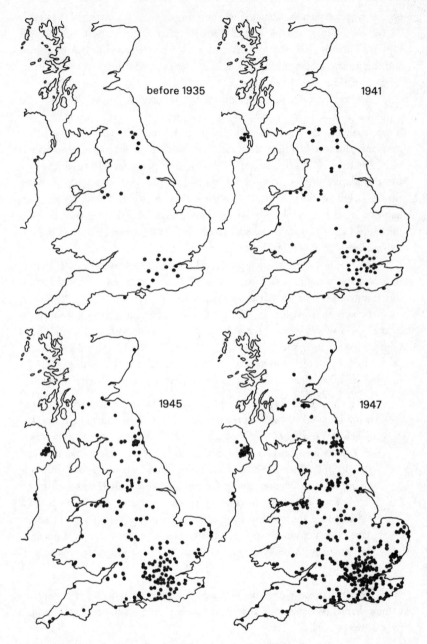

Figure 9.6 Distribution of recorded opening of milk bottles by tits up to and including the years indicated. (After Fisher and Hinde, 1949. Reproduced, by permission, from the monthly magazine *British Birds*.)

objects which contrast with their surroundings, and to test their palatability."
As for the opening of the bottles, "the hammering action with which foil
caps are punctured is very similar to a motor pattern used in opening nuts,
and the tearing action often used on cardboard tops is similar to a movement
used in tearing bark from a twig."[43]

An explanation in terms of formative causation complements this
suggestion. These instinctive motor patterns, themselves organized by
morphic fields, did not automatically give rise to the bottle-opening habit;
they only did so when they were embraced within a higher-level behavioural
field, the field of bottle opening. The field was progressively reinforced by
the cumulative effects of morphic resonance from previous milk-drinking
tits, and consequently enabled both the discovery and the passing on of the
habit by imitation to take place ever more readily. Morphic resonance would
therefore help to explain the spread of the bottle-opening habit as well as
its rapid reappearance in Holland after the war.

The case of the tits is just one example of a rapid evolutionary change
in behaviour in response to human activity. Many others have been observed,
but few have been systematically documented.

Behavioural evolution is currently occurring on a large scale in parts
of the world where towns have only recently been established. In Papua New
Guinea, for instance, the first town was founded in the 1870s, and even today
there are only sixteen towns with a human population exceeding 4,000. A
number of local species of birds have already adapted to living in them.

> Such shifts in behaviour do not occur instantly and newly acquired
> behaviours take time to spread. In several cases the timetable is known.
> It was not until 1971–74 that a flock of black kites settled in the town
> of Lae to feed on road kills, although the town was founded in the
> 1930s. Only since 1983 have brahminy kites learned to feed on road-
> killed toads in Port Moresby. Goldie's lorikeet, which was formerly a
> rare species occurring in primary forest, arrived in the 1970s to feed
> on casuarina seeds in highland towns, where it is now the most abun-
> dant bird. Lemon-bellied flycatchers expanded from the savanna into
> the town of Wau between 1976 and 1978, 50 years after Wau was
> founded.[44]

Such situations would provide a good opportunity to study the spread
of new habits of behaviour and to investigate the possible role of morphic
resonance.

As we have seen in this chapter, the hypothesis of formative causation
and the mechanistic theory provide radically different views of memory and

learning in animals. From a mechanistic point of view, memory depends on memory traces, as yet unidentified, which function in a manner that remains obscure. The inheritance of instinctive patterns of behaviour is different in kind from the ability of individual animals to acquire new habits of behaviour: inherited behaviour is programmed in the genes, and acquired patterns of behaviour cannot be inherited because there is no known way in which they could modify the genetic program.

By contrast, according to the hypothesis of formative causation, behaviour is organized by morphic fields associated with the activities of the nervous system. The inheritance of instincts and the building up of an animal's own habits both depend on morphic resonance, and there is no radical difference in kind between them. For this reason, habits acquired by some animals can facilitate the acquisition of the same habits by other, similar animals, even in the absence of any known means of connection or communication. There is already evidence that suggests that such effects actually occur; and the spread of the habit of opening milk bottles by tits highlights the possibility that these effects may be of considerable evolutionary significance.

We now consider the possible role of morphic resonance in human learning.

CHAPTER 10

Morphic Resonance in Human Learning

The Acquisition of Physical Skills

Generally speaking, we learn things from people who can already do them. In the case of physical skills, such as swimming or piano-playing, the skills are transmitted to us by people whom we imitate. In this transmission words play a subordinate role: it is notoriously difficult to learn such skills from books. By beginning to do these things we tune in to the morphic fields of the skills, and our learning is facilitated by morphic resonance not only with our teachers, but also with many other people unknown to us who have previously practised these skills.

In traditional societies, a great variety of crafts and skills have remained more or less the same for many generations—for example those involved in hunting, cooking, agriculture, weaving, and potting. The same is true of the traditional crafts such as those of blacksmiths, goldsmiths, and carpenters. Even in our own society, most trades are still learned through a system of apprenticeship. All such skills, according to the present hypothesis, involve nested hierarchies of morphic fields which are strongly stabilized by morphic resonance from countless people in the past.

In the case of long-established skills such as these it would be difficult to investigate empirically the role of morphic resonance in the facilitation of learning. Even in skills of relatively recent origin, such as cycle-riding, there are no objective data that would enable us to compare average rates of learning of children today with, say, those at the beginning of the century. According to the hypothesis of formative causation, other things being equal, the rate of learning should be faster now than it was then because many millions of people have learned to ride bicycles in the intervening period. And indeed anecdotal evidence suggests that children do tend to learn to cycle more readily than they used to. But of course other things have not remained equal: bicycle designs have altered, special children's bicycles are now common, motivation and opportunities have changed, and so on.

In a few instances, detailed data on the performance of physical skills at various times in the past are available, most notably in the form of records for various athletic events. These generally show an increasing level of performance. The best-known example is the running of the mile. Since Roger Bannister first broke through the four-minute "barrier" in 1954, many others have run the mile as fast and the top performances have continued to improve: for example, in 1967 the record was held by Jim Ryun at 3 minutes 51.1 seconds, and in 1985 by Steve Cram at 3 minutes 46.3 seconds. A similar pattern of improvement is shown in practically all other athletic events. Not only have records repeatedly been broken, but the *average* levels of performance in international competitions have also improved.

Here too many factors may have played a part: improved nutrition, better training methods, psychological factors, greater motivation, selection of athletes from a larger population of potential competitors, and so on. Any role that morphic resonance may have played cannot be teased apart from these other influences.

Experimental tests for the effects of morphic resonance need to be designed in such a way that other factors are kept as constant as possible. Several experimental designs and actual experimental results are considered towards the end of this chapter.

The Learning of Languages

Human babies have an inherited disposition to learn languages; the young of other mammalian species do not. In conventional terms this is thought of as a kind of programming in the DNA. From the present point of view it is due to morphic resonance from innumerable people in the past. This resonance underlies the general tendency to acquire language, and also facilitates the acquisition of particular languages, such as Swedish and Swahili, by resonance from previous speakers of these languages.

Languages have hierarchically ordered structures which, as René Thom has pointed out, can be thought of as a hierarchy of chreodes, or "canalized pathways of change" (Fig. 6.2). The sentence chreodes organize the phrases, the phrase chreodes the nouns, verbs, adverbs, conjunctions, and other parts of speech in the phrases, the word chreodes the syllables, and the syllable chreodes in turn organize the lowest-level chreodes, the phonemes.[1]

Such hierarchical patterns of organization are found in all languages; the grammatical ways in which the words are arranged and interrelated constitute the syntax of a language. However, the syntax alone does not confer meaning. It is perfectly possible to construct grammatically correct sentences that are meaningless; and of course meaningful sentences are not necessarily true. Beneath the grammatical structures of sentences, which Noam Chomsky has called the "surface structure," are further levels of organization that give a sentence its "deep structure"; and the deep structures of sentences are connected with further levels of organization and interrelationship which are the basis of meaning.

There are a variety of theories about the structure of language, and a consideration of the patterns of organization that give rise to meaning inevitably leads beyond the realm of linguistics into the territories of psychology and philosophy. These invisible and inaudible structures that lie below or beyond the surface structure of the language have been extremely difficult to characterize. After all, what kinds of things can these structures be? They are certainly patterns of organization, but what is their nature? From a mechanistic point of view they are assumed to be somehow connected with patterns of organization of nervous activity within the brain. The hypothesis of formative causation does not contradict this assumption, but rather complements it by enabling one to think of these structures in terms of nested hierarchies of morphic fields which act on and through the patterns of electrical activity in the nervous system.

An understanding of these structures is not merely a matter of academic or philosophical interest, but is of great practical importance, for example in attempts to program computers to translate from one language to another automatically and in the development of "artificial intelligence." These attempts have so far met with rather limited success. One reason for the slow progress in these fields, despite a vast investment of money and effort, could be that these attempts are based on a fundamental misconception. They take for granted the mechanistic theory of the organization of language and intelligence, which may well be wrong. The organization of language and learning may depend on morphic fields, which these computer models do not take into account.

Chomsky has argued that the rapid learning of language by children, including their grasping of grammatical rules, cannot possibly be explained in a behaviourist manner in terms of stimuli and conditioned responses. One of the most striking features of the use of language is its "creativity": by the age of five or six children are able to produce and understand an indefinitely large number of utterances that they have not previously encountered.[2] Chomsky thinks of this as an organic rather than mechanical process: "Language seems, to me, to grow in the mind, rather in the way that the physical systems of the body grow."[3]

In order to account for the remarkable ability of children to pick up languages, he has proposed that the basic organizing structures of languages are innate: children inherit them.[4] But since babies of any race seem to have the capacity to learn any human language, he has been driven to the conclusion that these inherited structures must be common to all languages: they represent what he calls a universal grammar, and he regards one of the tasks of linguistics as the determination of the universal and essential properties of all human languages. He regards these as "genetically programmed."

This notion of a universal grammar is the most controversial aspect of Chomsky's system. It is by no means clear that all languages share common generative organizing principles and deep structures of the kinds that Chomsky proposes in general terms on theoretical grounds—especially since many of these have never been specified.

A general morphic resonance from all of past humanity would indeed reinforce any organizing fields and chreodes that are in fact common to most if not all languages, and this would be in harmony with Chomsky's proposal. However, it is not necessary from this point of view to suppose that the grammatical structure of all languages depends on a single universal grammar. The general morphic resonance gives young children a general tendency to learn language, but as they begin to speak a particular language, such as Swedish, they enter into morphic resonance with the people they hear speaking it; their learning of its particular grammar and vocabulary is facilitated by this resonance. Speaking this language tunes them in, on the basis of similarity, to speakers of the same language, including many millions of speakers in the past.

Chomsky has pointed out that his theory makes a prediction that could be tested in principle, if not in practice. If an artificial language that violated the universal grammar were constructed, "it would not be learnable under normal conditions of access and exposure to data."[5] The same result would be expected on the basis of the hypothesis of formative causation (assuming for the purposes of argument that the universal grammar can indeed be specified). However, this hypothesis makes a prediction that differs from Chomsky's: an artificial language constructed in *accordance* with the universal grammar but differing in important respects from all natural languages, past and present, would be much more difficult to learn than any natural language. This would be because the chreodes are not stabilized by morphic resonance from past speakers of the language, for there would have been no such people.

The hypothesis of formative causation also predicts that languages spoken by very many people in the past should on average be easier to learn than those spoken by very few people, other things being equal. Why, then, cannot we all learn languages such as Mandarin and Spanish very easily, since so many millions have spoken them? In the case of adults, the situation is obviously complicated by the fact that the deep-seated habits associated with our own language strongly interfere with the acquisition of others. The more strongly these habits shape our speaking, listening, and understanding, the less easily will we be able to acquire the new patterns of the language we are learning. Perhaps this is why people with a "good ear" and an unusual ability to imitate the speaking of others are often particularly gifted in the learning of foreign languages. Through skilful imitation, they tune in more effectively to past speakers of the language than most of us can, and their further learning of it is facilitated to a greater extent by morphic resonance.

In the case of babies, where there are no established habits to interfere with the acquisition of language for the first time, there could indeed be differences in the ease with which they acquire common and rare languages: other things being equal, English, for example, might be easier to learn than a rare tribal language from the Amazon, just because so many more people have spoken it. In practice, of course, this would be very difficult to investigate because other things are unlikely to be equal; any effect of morphic resonance would be difficult to separate from differences in genetic constitution, cultural environment, methods of child-rearing, and so on. Nevertheless, it might not be impossible to test this prediction.

Education and IQ

On the present hypothesis, skills such as reading, writing, and arithmetical calculation depend on the ordering and patterning activity of morphic fields, just as physical skills and the speaking and understanding of languages do. The learning of reading, writing, and arithmetic should be facilitated by morphic resonance from those who have practised them before us. Moreover, it should become easier on average to learn such skills as more and more people acquire them.

The spread of modern education and the establishment of schools all over the world have meant that many hundreds of millions of people now know how to read and write, whereas only a century ago these abilities were the preserve of a small minority. Has the spread of these skills in itself facilitated their acquisition by successive generations of schoolchildren? Other things being equal, it should on average have become easier over the years for children to learn all the things that are taught in schools, including in recent years the ability to program computers. But, as usual, other things are not equal. A whole range of factors, psychological, social, economic, political, and technological, influence children's interests, motivations, and opportunities to learn. These factors operate in addition to any cumulative influence there might be from morphic resonance. In any case, very few quantitative data are available that would enable one to measure such changes.

The use of standardized intelligence tests is one of the very few methods of data collection that permit changes over time to be assessed. These tests have been conducted on a large scale over a period of decades. The hypothesis of formative causation predicts that, other things being equal, the average performance in these tests should have improved, both because the requisite mental skills have become easier to acquire owing to the cumulative effects of morphic resonance and because so many people have already taken these standardized tests: morphic resonance from past takers of these tests should in itself facilitate the taking of them. This would not necessarily mean that the average "intelligence" of the population is going up, but only that the ability to do standard intelligence tests is increasing. An improvement in the average performance in such tests has in fact occurred.

Within a given population and at a particular time, the average performance is by definition set at 100, and the intelligence quotient, or IQ, is calculated in relation to this mean. But if the actual scores of populations on standardized tests are compared over age groups or over years, changes in the average IQ over time can be calculated. In 1982, considerable interest and controversy were aroused by the claim that the average IQ of Japanese had been rising in the present century and was now about eleven points ahead of the average IQ in the United States.[6] But detailed studies of test scores in America soon revealed that these too had risen in recent decades at about the same rate as the Japanese scores, and a similar increase has now been found in at least twelve other countries.[7] Over the period from 1932 to 1978 the average IQ of Americans increased by 13.8 IQ points, or an average of 0.3 points per year.[8] These findings reawakened interest in earlier evidence, based on a comparison of U.S. Army mental tests carried out in the first and second world wars, which indicated that there had been large-scale gains in IQ between 1918 and 1943.[9]

There has been much discussion of the significance of these findings. In order to explain an average gain of 0.3 points per year in Japan since World War II, environmental changes of the most radical sort were invoked: massive urbanization, a cultural revolution from feudal to Western attitudes, the decline of inbreeding, and huge advances in nutrition, life expectancy, and education.[10] But these explanations suffered a setback when it was found that a similar increase had occurred in the United States over the same period, where there had not been such dramatic environmental changes as in Japan;

changes of a comparable magnitude occurred earlier in the United States, nearer the beginning of the century. A number of leading scholars in the field have recently proposed two possible explanations of these gains in IQ: increased sophistication in performance on standardized tests, and a "rising level of educational achievement."[11] Studies on the effects of repeated testing with parallel forms of IQ test have in fact shown that subjects can gain up to five or six IQ points through practice, but not much more than this.[12] This leaves enhanced "educational achievement" as a likely explanation.

However, the plausibility of both these explanations is thrown into question by the fact that in the period from 1963 to 1981 there has been a *decline* in the average performance of American high-school students on the standard Scholastic Aptitude Test (SAT), taken by over a million students a year. This consists of several subtests, and the greatest decline occurred in scores on the verbal SAT. An official advisory panel appointed to examine this decline found that about half of it could be explained in terms of the broadening of the sample of candidates; but the other half reflected a downward trend in test results of the general population and showed up in all socioeconomic groups of students. Various personal traits contribute to SAT scores: the advisory panel listed these as intellectual ability, study habits, motivation, self-discipline, and acquired verbal and writing skills. The panel suggested that these traits could have been influenced by causes such as less demanding school standards, student absenteeism rates of over 15%, the erosion of the nuclear family, and the influence of television.

But why have average IQ scores gone up while average SAT scores have gone down? One of the leading researchers in the field, J. R. Flynn, has argued that social factors, the influence of TV, and so forth, may have pulled down SAT performance, especially in the verbal test, while exerting much less influence on the skills involved in IQ tests. But this would not explain the *increase* in IQ scores: "The combination of IQ gains and the decline in Scholastic Aptitude Test scores seems almost inexplicable. . . . IQ gains of this magnitude pose a serious problem of causal explanation."[13]

The cumulative effects of morphic resonance might well help to explain an increase in IQ; but then they should also have given an increase in SAT scores, other things being equal. But as we have seen, other things were not equal; a variety of factors seems to have been working in the other direction, and may have offset any possible influence of morphic resonance.

As this discussion shows, it is practically impossible to reach any firm conclusions on the basis of this kind of evidence. In order to test the hypothesis it will be necessary to carry out specially designed experiments.

Some Experimental Tests

Two types of experimental test for the effects of morphic resonance in human learning are possible. First, tests that involve the acquisition of a new skill during the time span of the experiment itself. Such tests could involve the solving of novel puzzles, for example, or the playing of new kinds of video games. The average rate at which groups of inexperienced subjects can learn them is monitored at regular intervals in one country. Meanwhile, these puzzles are solved or the video games are played by thousands of people in another country. The average speed of learning for subjects exposed to them for the first time in the first country should increase as more and more people learn them elsewhere.

Experiments of the second type involve long-established skills. They attempt to detect the influence of morphic resonance from the many people in the past who have already learned these skills. Several experiments of this type have already been carried out.

Tests with Nursery Rhymes

In October 1982, coinciding with the announcement of a $10,000 prize to be awarded by the Tarrytown Group of New York for the best test of the hypothesis of formative causation, the British magazine *New Scientist* launched a competition for experimental designs that could be used in the testing of the hypothesis.[14] The results of this competition were announced in March 1983. The winner, Richard Gentle, proposed an ingenious test involving a Turkish nursery rhyme.[15] He suggested that English-speaking people could be asked to memorize two short rhymes in Turkish under standard conditions, one a traditional nursery rhyme, known to millions of Turks over the years, the other a new rhyme made by rearranging the words in the genuine nursery rhyme. The subjects would not be told which was which. After equal periods spent in memorizing each of the rhymes, they would be tested to find out which one they remembered better. If the learning of the genuine rhyme was facilitated by morphic resonance from millions of Turks, then it should be easier to memorize than the newly constructed rhyme.

I took up Gentle's suggestion, but used Japanese rather than Turkish rhymes. A leading Japanese poet, Shuntaro Tanikawa, kindly supplied me with three rhymes for this purpose: one is a genuine nursery rhyme known to generations of Japanese children, and the other two were specially composed to resemble it in structure. One of these is meaningful and the other meaningless in Japanese.

In a series of experiments with groups in Britain and America who learned the rhymes by chanting each of them a fixed number of times (without knowing which was which), 62% of those tested found the genuine rhyme easiest to recall half an hour later. This result was far above chance expectation: if the rhymes were of equal difficulty, by chance about 33% of those tested would have been expected to recall the genuine rhyme better than they recalled the new ones. In another experiment in which people were supplied with the rhymes in a written form 52% found the genuine rhyme easiest to learn, again a highly significant result. This method of learning was not as effective as chanting, which is of course much closer to the way that Japanese children learn the genuine rhyme. There were no consistent differences in the ease with which the two newly composed rhymes were learned.

These results, although encouraging, are open to the criticism that the new rhymes may be intrinsically harder to learn than the traditional nursery rhyme, in spite of the poet's efforts to make them of comparable difficulty. This argument gains additional force when we consider the history of nursery rhymes: presumably they are subjected to a process resembling natural selection, and maybe easy-to-remember rhymes are more likely to survive. If, on the other hand, one of the new rhymes had been memorized better than the nursery rhyme, it could have been argued that for some reason it was intrinsically easier. Consequently, this type of experiment cannot give conclusive results one way or the other.

The same difficulty of interpretation would arise in connection with other comparable experiments. One suggested test concerned the memorizing of passages of the Koran in Arabic. Many millions of Muslims have memorized such passages, and large numbers have learned the entire Koran by heart. In traditional Muslim schools in India, for example, it is not uncommon to find boys of twelve or even younger who have achieved this, even without any detailed understanding of the Arabic language. Morphic resonance may facilitate this process of learning, just because so many people have already learned to chant the Koran from memory. If so, it should be much easier to learn to chant passages of the Koran than comparable newly composed Arabic passages. But it would be impossible to establish that any newly composed passages were indeed of comparable difficulty. (And, of course, from an Islamic point of view, no newly composed passages could be compared with the divinely inspired originals.) The same difficulties would arise in connection with possible tests involving the memorization of Sanskrit mantras, recited by Brahmins over the centuries; or the Creed in Latin, learned by hundreds of millions of Roman Catholics; or even often-memorized passages by poets such as Shakespeare and Goethe. In no such case could it be established that any newly composed passages were truly comparable.

To overcome this problem, a different experimental design would be necessary. This could be carried out using several new rhymes, in Japanese for example, of similar metre and sound structure, and of similar difficulty. The rate at which they could be memorized would then be determined by testing people in, say, the United States. Then one of these rhymes, chosen at random, would be learned by many people in Japan. Subsequently, new sets of subjects in the United States would be asked to memorize the rhymes, and the rates at which they did so would again be measured. The one that had been learned by the Japanese should now be easier to memorize than before, but the other rhymes, which serve as controls, should not. Such an experiment might be practicable if one of the rhymes were used in a popular song in Japan, or even in an advertising jingle.

Tests with Hebrew and Persian Words

Three prizes were offered in the Tarrytown competition for the best test of the hypothesis of formative causation: a $10,000 first prize, a second prize of $5,000 provided by a Dutch foundation, and a third prize of $1,500 by Meyster Verlag, the publishers of the German edition of *A New Science of Life*.[16] The winners were selected by an international panel of judges (Professors David Bohm of London University, David Deamer of the University of California at Davis, Marco de Vries of the Erasmus University, Rotterdam, and Michael Ovenden of the University of British Columbia). The prizes were awarded in New York in June 1986. Two entrants tied for first place. Both of them had carried out similar experiments, quite independently, involving words written in foreign scripts, Hebrew in one case and Persian in the other.

These tests were based on the idea that words that have been read by millions of people may be associated with morphic fields that facilitate the perception of the patterns of the words. Hence people who are entirely unfamiliar with a foreign language and its script may find it easier to recognize or learn *real* words in this language than false words made up of letters in meaningless sequences. These non-words will not have been written or read by millions of people in the past, and hence the perception of their patterns will not be stabilized by morphic resonance. Note that such tests depend entirely on the visual patterns of the words: they do not involve hearing the words, nor do they involve any attempt to pronounce them; they are carried out in ignorance of the phonetic values of the letters.

Gary Schwartz, professor of psychology at Yale University, selected 48 three-letter words from the Hebrew Old Testament, 24 of them common and 24 rare. He then scrambled each word to produce a meaningless anagram containing the same three letters. This gave 96 words in all, half real and the other half false.

Over 90 students who were ignorant of Hebrew were shown these 96 words one by one, projected on a screen in a random order. They were asked to guess the *meaning* of each word and to write down the first English word that came to mind. They were then asked to estimate on a 0-to-4 scale the *confidence* they felt in their guess. The subjects were not told the purpose of the experiment, nor that some of the words were scrambled.

A few of the subjects did in fact correctly guess the meanings of some of the Hebrew words. Schwartz excluded these subjects from his analysis on the ground that they could possibly have had some knowledge of Hebrew in spite of the fact that they said they had none. He then examined the replies of the subjects who had always guessed the wrong meanings. Remarkably, on average they reported feeling more confident about their guesses when they were viewing the real words than the scrambled words, even though they did not know that some of the words were real and others were fake. This effect was roughly twice as strong with the common words as with the rare words. These results were highly significant statistically.

After Schwartz had tested his subjects in this way, he informed them that half the words were real and the other half scrambled. He then showed them all the words again, one by one, asking them to guess which were which. The results were no better than chance: the subjects were unable to do consciously what they had already done unconsciously. Schwartz interprets the greater confidence subjects felt about their wrong guesses of the meanings of the real words in terms of an "unconscious pattern recognition effect."

Alan Pickering, a psychologist at Hatfield Polytechnic in England, used two pairs of real and scrambled Persian words, written in Persian script (which resembles Arabic). He tested eighty students, showing each of them only one of the words. They were asked to look at the word for ten seconds and then to draw it after the viewing period was over. These reproductions of the actual and false words were subsequently compared by several independent judges in several different ways. The judges were not told the purpose of the experiment, nor did they (nor even Pickering himself) know which words were real and which scrambled.

The real words were reproduced more accurately than the false words. For example, in one method of judging, which involved comparing pairs of answers (paired at random) with the corresponding real and false words, on average in 75% of the pairs the real words were judged to be reproduced better than the false. This effect was highly significant statistically, with odds against chance of over 10,000 to 1. Pickering, like Schwartz, concluded that his results were in good agreement with the idea of morphic resonance.

An obvious alternative explanation is that real words tend to have certain aesthetic or other qualities that false words do not, for reasons that have nothing to do with morphic resonance. This argument is, however, very vague and ill-defined, which becomes apparent when we try to apply it in detail. Schwartz found a greater "unconscious pattern recognition effect" with common Hebrew words than rare ones. If this effect is due not to morphic resonance but to aesthetic or other properties of the real words, then why should *common* words be more aesthetic when written down than *rare* words? Do words become more frequently used within a language *because* their written forms are pleasing to the eye? Or do common written words become easier to recognize, owing to morphic resonance, *because* they are frequently used? Or are common words more recognizable *both* because of morphic resonance *and* for aesthetic reasons unconnected with morphic resonance?

These experiments clearly provide a promising starting point for further research using words in these and other languages.

Tests with Morse Code

Morse code was invented by Samuel Morse in the mid-nineteenth century for use in telegraphy. It has been learned and used by large numbers of people over the years, and is still in use today, for example by thousands of telegraph operators. Does morphic resonance from all these people make it easier to learn?

An experiment to test this possibility has been devised and carried out by Arden Mahlberg, an American psychologist who received the third prize in the Tarrytown competition. He constructed a new version of this code by reassigning the dots and dashes to different letters of the alphabet, and, using subjects who did not know Morse code, compared their ability to learn this new code with their learning of the genuine Morse code. The presentation of the material to be learned and the subsequent testing were carried out using letters with their associated dots and dashes in a written form. (The letters S and O were excluded because many people who do not know Morse code are nevertheless familiar with the code for S.O.S.) The subjects were exposed to the new code and the genuine Morse code one after the other, in random order, for equally brief periods.

In his first trials, Mahlberg found that, on average, subjects learned the real Morse code significantly more accurately than the new code.[17] In subsequent tests with new subjects, he found that the average accuracy of learning of the new code progressively increased until it was learned almost as well as the real Morse code. He suggests that the original difference could have been due to morphic resonance from past practitioners of Morse code, resulting in a significant facilitation of the learning of this code compared with the newly constructed code. But as the tests were repeated with fresh subjects, those in the subsequent tests were influenced by morphic resonance from those who were tested before them. This effect, owing to the high specificity of resonance with previous subjects tested in the same way under the same conditions, swamped the more subtle effects of morphic resonance from users of the real code, and led to a progressive equalization of scores of the two codes. He offers this as a tentative explanation of the results, and emphasizes the need to explore these possibilities in further experiments.

One way in which the experimental design could be improved would be to use *sounds* for the dots and dashes, rather than presenting them in written form. This method would correspond far more closely to the habitual experience of telegraph operators. For this purpose, a microcomputer could relatively easily be programmed to show a standard sequence of letters on the screen and at the same time beep the dots and dashes.

The first code invented by Morse, called early Morse code, was quite arbitrary in the way that dots and dashes were assigned to letters of the alphabet, and was not designed for ease of learning. Subsequent modifications were introduced in American Morse code and in International Morse code to reduce errors in transmission, mainly by assigning shorter signs to the most frequently used letters.[18] But these changes did not involve any systematic combining of letters with signs intended to make them easier to learn. Mahlberg took these factors into account in designing his new code, and seems justified in his assumption that the new code should not have been intrinsically harder to learn. This is clearly an essential feature of the experimental design; and in the construction of new codes for use in further experiments care will need to be taken to ensure that they are not intrinsically harder or easier than the real Morse code, at least as far as can be judged on the basis of modern cognitive psychology.

A Possible Test with Russian Typewriters

The first commercially successful typewriters were made by Remington in the 1870s. The keyboard was constructed not so much for ease of use or ease of learning, but for mechanical reasons connected with the way the pivoted type-bars swung. The layout of the keys was designed to prevent the most frequently used letters from jamming together at the printing point. This original arrangement, called the QWERTY layout after the first letters in the top row, was retained almost unaltered in subsequent machines. It still survives even in computer consoles, in spite of the fact that the original mechanical reasons for it have long since disappeared. Over the years, many improved keyboard layouts have been advocated, designed for greater ease of use, but none have so far succeeded in displacing the traditional QWERTY keyboard.

Tens of millions of people have used QWERTY typewriters since the 1870s. Morphic resonance might therefore be expected to facilitate markedly the learning of this skill and strongly stabilize the associated morphic fields. Typing has in fact intrigued experimental psychologists for decades, because the "rate at which typists (even average ones) perform exceeds by far the rate that laboratory tests quite common in psychology would lead a psychologist to predict."[19]

No doubt one reason why new, improved keyboard designs have failed to catch on is the difficulty of retraining typists and scrapping existing machines; but in addition, despite its inefficient design, the QWERTY layout may be easier to learn and use precisely because so many people have already become familiar with it.

There is empirical evidence that non-typists find the standard QWERTY layout easier to learn than a random layout;[20] and alphabetical ABCDE keyboards, designed for ease of learning by novices, have in some experiments proved harder[21] and in others at least no easier than the standard layout.[22] "Operators with little or no typing skill, for whom alphabetical arrays are often intended, were as fast or faster with the standard typewriter arrangement."[23]

In experiments specifically designed to test for the effects of morphic resonance, the rate at which novices learn to type on the QWERTY keyboard could be compared with other layouts that from the point of view of contemporary psychological theory should be of equivalent difficulty. A conclusive experiment might be difficult to perform in the Western world, because it would be difficult to establish that subjects had not already been exposed to QWERTY keyboards, even just by seeing them; it could be argued that any such exposure would tend to bias the results in favour of the standard layout. It would be important to find subjects who had never come across a standard QWERTY typewriter or computer keyboard. One way of doing this would be with students of English in, say, Russia, who are familiar with the Roman alphabet but who have never seen a Western typewriter before. The rate at which they learn to type on a standard QWERTY keyboard would be compared with the rate on a machine on which the keys were laid out in a different way. If the students learned quicker with the QWERTY keyboard, this would suggest that their learning was being facilitated by morphic resonance from typists in the West.

In the Western world, the reverse of this experiment could be carried out with students of Russian (or Greek or Hindi) who are familiar with the relevant alphabet but who have not previously been exposed to typewriters using this alphabet. In such an experiment, the rate of learning to type with, say, a standard Russian keyboard would be compared with the rate with a different layout, designed to be of comparable difficulty according to conventional theories. The hypothesis of formative causation would predict that the standard layout should be easier to learn, just because so many people in Russia have already learned it.

This experiment could be done quite easily using suitably programmed microcomputers whose keys were labelled with the appropriate patterns of Russian letters. A standardized learning procedure could also be programmed into the computer and the rate of learning recorded automatically.

These examples, only a few of the many conceivable ways in which the hypothesis of formative causation could be tested in the realm of human psychology, serve to illustrate that such experiments are not only feasible but can be carried out with facilities and equipment that are readily available in most university psychology departments and even in many secondary schools.

In this chapter we have explored the possibility that our learning of language and of physical and mental skills is facilitated by morphic resonance from many other people who have already learned them. Several experiments have been carried out to test for this effect, with results that are consistent with the hypothesis of formative causation. If further experiments provide a convincing weight of evidence in favour of this hypothesis, this new understanding could have far-ranging implications for education and training. For example, it might be possible to develop new methods of teaching that maximize the facilitation of learning by morphic resonance.

We now consider the role that morphic resonance may play in our own personal memories.

CHAPTER 11
Remembering and Forgetting

We can remember people, places, tunes, words, ideas, stories, events, and a host of other things. We usually take all this for granted and don't need to ask how our memories work.

The conventional theory is, of course, that everything we can remember is somehow stored inside our brains in the form of material patterns, the memory traces: there are such material patterns in our brains for every tune we know, for everyone we can recognize, for every word in our vocabulary, for every event we can recall—a myriad of memory traces for everything we are capable of remembering.

But this is just a speculative theory. No one has ever seen a memory trace; and scientists who have looked for such traces have so far failed to find them.

In this chapter I explore the alternative possibility that memories are *not* stored inside the brain. The spatio-temporal patterns we remember may not be inscribed in the brain in the form of material traces but may depend instead on morphic fields. The morphic fields through which our experience, behaviour, and mental activity were organized in the past can become present again by morphic resonance. We remember because of this resonance from ourselves in the past.

I first discuss the morphic fields of behaviour and of mental activity, and the general role of morphic resonance in memory. I then consider one of the essential preconditions for conscious memory: awareness. In general, we cannot remember something if we were not aware of it in the first place; and awareness arises against a background of unawareness, owing to habituation, which itself depends on morphic resonance. I go on to consider the role of morphic resonance in recognizing and recalling, and end with a discussion of how we come to forget things.

Behavioural and Mental Fields

According to the hypothesis of formative causation, the morphic fields that organize our behaviour are not confined to the brain, or even to the body, but extend beyond it into the environment, linking the body to the surroundings in which it acts. They co-ordinate sensation and action, bridge the sensory and motor regions of the brain, and co-ordinate a nested hierarchy of morphic fields, right down to those that organize the activity of particular nerve and muscle cells.

A similar conception was developed within the gestalt school of psychology in the 1920s and 1930s. This approach is currently out of fashion in the academic world, but it takes on a new significance in the light of the morphic field concept. Both approaches involve a conception of holistic patterns of organization which embrace the body and the environment. Gestalt psychologists often described these fields as "psychophysical fields."

For the gestalt psychologists, the behavioural environment was not to be understood in terms of objects alone, but in terms of the "dynamic properties" of the psychophysical fields. Kurt Koffka gave a simple illustration of this principle: Imagine yourself basking in the sun in a mountain meadow, relaxed and at peace with the world. Suddenly you hear a scream for help—your feelings and your environment immediately change:

> At first your field was, to all intents and purposes, homogeneous and you were in equilibrium with it. No action, no tension. As a matter of fact, in such a condition even the differentiation of the Ego and its environment tends to become blurred; I am part of the landscape and the landscape is part of me. And then, when the shrill and pregnant sound pierces the lulling stillness, everything is changed. Whereas all directions were dynamically equal before, now there is one direction that stands out, one direction into which you are being pulled. This direction is charged with force, the environment seems to contract, it is as though a groove had formed in a plane surface and you were being forced down the groove. At the same time there takes place a sharp differentiation between your Ego and the voice, and a high degree of tension arises in the whole field.[1]

Koffka pointed out that the first type of field, the homogeneous, is very rare; any action presupposes inhomogeneous fields, fields with lines of force. These fields organize behaviour towards ends or goals. Football players, for example, as they move towards the enemy goal line, "see the playing ground as a field of changing lines whose principal direction leads them towards the goal. . . . All the motor performances of the players (as shifting about on the field) are connected with the visual shifting." These responses are not a matter of logical thought; for a player in his tense state, "the visual situation produces the motor performances directly."[2]

The gestalt approach and the hypothesis of formative causation resemble each other in their conception of fields, but they differ in that the gestalt psychologists did not have the idea of morphic resonance. Rather, these psychologists adopted a conventional theory of memory traces. They believed that the fields could be remembered because of traces they left in the brain. As Koffka put it, "The field of the present process comprises the traces of previous processes."[3] By contrast, on the hypothesis of formative causation, it is not necessary for the fields to leave material traces in the brain, any more than the programs to which a radio set is tuned leave traces in the set. A field brings about material effects while the system is tuned in to it. But if the tuning is changed, then other fields come into play: the original field "disappears." It appears again when the body in relation to its environment re-enters a state similar to that in which the field was expressed before; the field once again becomes present by morphic resonance.

Behavioural fields organize our habitual activities, and usually do so without our being conscious of them. However, conscious mental activity, such as that involved in thinking out alternative courses of action, does not necessarily involve overt behaviour. It is more concerned with virtual or possible behaviour and activity. The fields that according to the present hypothesis are associated with this mental activity are therefore different from behavioural fields and can most appropriately be described as mental fields, rather than behavioural fields. These are again a kind of morphic field stabilized by morphic resonance from similar past patterns of activity, and will in the subsequent discussion usually be referred to as morphic fields. The distinction between morphogenetic, behavioural, and mental fields is of value when considering the kinds of organized activities with which these fields are associated; but it is not a hard and fast distinction. They are perhaps more like different regions of a spectrum of morphic fields, and merge into each other. For example, in the case of *Amoeba,* which moves by changing its shape, the associated fields could be thought of as intermediate between morphogenetic and behavioural fields. And in the case of a newly invented human skill, such as playing a new video game, the mental fields through which the game was conceived shade off into behavioural fields as the game is developed and the skill of playing it acquired and practised.

The connections between morphic fields and the activities of the brain are discussed in more detail in the following chapter. In this chapter I explore the idea that morphic resonance underlies the various aspects of our memories.

Memories and Morphic Resonance

Our original experiences of events, as well as our recalling of them, are influenced by our interests and motives. We generally remember what is significant and meaningful for us better than that which is not. Nothing has significance and meaning all by itself; things matter only in relation to the context and the subject. Systems of relationship and interaction have been given a wide variety of names: for example, F. C. Bartlett, one of the pioneers of memory research, referred to them as *schemata;*[4] Arthur Koestler thought of them in terms of perceptual and motor *hierarchies;*[5] and G. H. Bower has analysed what he calls "organizational factors in memory" in terms of the grouping together or classifying or *categorizing* of psychological elements on the basis of common properties, and the *relating* of such classes to one another in multiple ways.[6]

From the point of view of the hypothesis of formative causation, such schemata, hierarchies, or organizational factors can be regarded as morphic fields, organized in hierarchies and connected together in multiple ways through higher-level fields.

Our ability to identify and categorize things depends on patterns of relationship. For example, we can recognize a word whether it is spoken in a high- or low-pitched voice, with a regional or foreign accent, by an old person or a child, or if it is written by hand or typed. We recognize it through the *pattern* of sounds, the way the different elements or phonemes are related to each other in time, or the way the letters representing them are related as a sequence in space. Likewise we can recognize the form of a letter in a wide variety of typefaces and handwritings. We can recognize a tune when it is hummed or played on a piano, a violin, or a flute, in spite of the very different qualities of sound; and we can also recognize humming and piano, violin, and flute sounds by their characteristic qualities irrespective of the tune being played. Likewise we recognize plants, animals, and things—lilies, cats, and chairs—even though any particular one we come across may differ in detail from all the ones we have previously encountered.

On the present hypothesis these classes or categories can be thought of as linked to characteristic morphic fields, which organize our perceptual experiences, usually closely associated with the language through which we not only organize and describe our experience but communicate with others. These classes or categories of experience are part of our biological and cultural inheritance, and are stabilized by morphic resonance not just with our own past experience but with many other people's. Like all morphic fields, those underlying our perceptions, categories, and concepts are not rigidly defined in terms of exact positions and dimensions and frequencies, but are probability structures. This is why categorization can take place on the basis of *similarity,* and does not depend on exact identity.[7]

Any particular experience involves not only categorizing various elements of it, but relating these elements together. Again, these relationships can be thought of in terms of fields: by their very nature morphic fields interrelate and interconnect elements into integral wholes. They give meaning to the elements through their interconnections into such higher-level wholes. And a particular element of experience can, of course, have more than one meaning: it can be integrated into more than one higher-level field. Our conscious experience involves the formation of such patterns of interconnection, and memories depend on the reconstruction of such patterns of connection: what we consciously remember is not so much what happened in our bodies or in the external world, but rather the subjective experiences associated with what happened. These are organized by fields, and remembering them depends on self-resonance.

In short-term memory, elements of recent experience are preserved for a limited time, rather like echoes. This kind of memory may well be associated with reverberating patterns of electrical activity in the nervous system, maintained by self-resonance. If these elements are not related together by a higher-level field, they have no coherence. Their temporary coexistence soon fades away, and there is no cohesive pattern to be recalled. Long-term memory is different, and depends on the establishment of higher-level fields, which can then become present again by morphic resonance. This establishment of new fields depends on our awareness. And awareness is, as it were, the other side of the coin of habituation, to which we now turn our attention.

Habituation and Awareness

Our conscious memories are of events which took place in particular places at particular times, even if we cannot always "place" the memories geographically or chronologically. It is precisely because of the uniqueness of these past experiences that we can remember them consciously.

Our conscious experience takes place within a framework of repetitive habits: our own, other people's, and the world's in general. And, like all animals, we become habituated to patterns that are repetitive or continuous. In our own experience, habituation produces a sense of familiarity which enables us to take most aspects of ourselves and our environments for granted. But this involves an active kind of unawareness. Through the contrast with the familiar, of which we become unaware, we are aware of what is unfamiliar. The unfamiliar is what generally attracts our attention. Without attention, we are unable to establish the patterns of connection that allow us to remember.

Habituation can be understood in terms of self-resonance: the more similar the present patterns are to those from the past, the more specific the morphic resonance. The less the difference between the present and the past, the less we are *aware* of any difference and the less we notice about this aspect of our present experience.

Habituation is in fact fundamental to the way that our senses and perceptual systems work. The very functioning of the sensory system involves habituation: if the rhythmic electrical pattern aroused in the sense organs and the nervous system by a particular stimulus continues, this repeated pattern is subject to self-resonance and ceases to be noticed. We notice changes and differences in things, rather than what stays the same.

For example, we all know from our own experience that we cease to notice continued tactile stimuli, such as the contact of our bottoms with chairs and of our clothes with our skins. What we do notice are *changes* in touch or pressure: if someone touches us unexpectedly we are aware of it at once. We feel differences in surfaces or textures as we move our hands and fingers over them; again, we sense changes.

The same is true of the other senses. We soon cease to notice familiar smells, sounds, tastes, and sights. And habituation occurs over a wide range of time scales, from year to year, day to day, minute to minute, and even from second to second. Such short-term habituation in the visual system, for example, gives a sensory awareness of differences as the eyes scan over things; we notice boundaries more than continuous surfaces in between; and likewise we generally notice things that move within our field of vision rather than things that stay put.

Habituation over all time scales involves a kind of unconscious memory of the familiar, which is the background against which we can be aware of changes, movements, and differences. Our conscious memories depend on this awareness, for we cannot remember something if we were not aware of it in the first place.

We now consider the two principal aspects of our memories, recognizing and recalling, and the role that morphic resonance plays in them.

Recognizing

The sense of familiarity that results in a habituated unawareness under other circumstances can be experienced consciously in the act of recognition. Recognition involves an awareness that a present experience is also in some sense remembered: we *know* that we were in this place before, or met this person somewhere, or came across this fact or idea. But we may not be able to recall where or when. Recognition and recall are different kinds of memory processes: recognition depends on a similarity between our experience now and our experience before and involves an awareness of familiarity. Recall, by contrast, involves an active reconstruction of the past on the basis of remembered meanings or connections.

Normally we are able to recognize more easily than to recall. For example, we may not be able to recall the name of a common garden plant which we recognize, even though it may be "on the tip of our tongue" and even though we may recall what letter it begins with. But if someone reminds us of the name, we recognize it at once.

Many psychological experiments have demonstrated quantitatively the greater effectiveness of recognizing than recalling. In one such experiment subjects were asked to memorize 100 words, which were presented to them five times. On average they could recall only 38. But if they were asked to recognize the 100 words mixed with 100 unrelated filler words, almost all of them were recognized: the mean score was 96.[8] Even more striking differences are found in visual experiments. For example, in one experiment subjects were asked to look at and memorize a meaningless shape. If they were asked to reproduce it by drawing it, their ability to do so declined rapidly, within a matter of minutes. By contrast, weeks later their ability to select this shape from a range of similar shapes was almost perfect.[9] Most of us in fact have quite remarkable powers of visual recognition, which we usually take for granted. For example, people who were shown 2,650 colour slides of pictures for 10 seconds each were later tested with pairs of slides, one of which was a new picture, and asked to say which one they had seen before. They correctly identified over 90% even after several days. Subjects were able to recognize the pictures almost as well even when the original slides were shown for only a second each, and even when they were shown with right and left reversed.[10]

On the present hypothesis, recognition, like habituation, depends on morphic resonance with previous similar patterns of activity within the sensory organs and the nervous system: these patterns are similar for the reason that the sensory stimuli that give rise to them are similar to those that came before, if not identical. Recognition and habituation occur because many features of the body and of the environment remain more or less the same; objects endure, and patterns of activity repeat themselves.

Recalling

Recognizing involves primarily the *sensory* aspect of memory, and depends on the sensory organs and the sensory portion of the nervous system. Recalling involves primarily processes of active reconstruction, in other words the *motor* aspect of memory, and depends on the motor organs and the motor portion of the nervous system. This is clear enough in the case of our memories for physical skills, such as riding a bicycle or playing the piano, and it is also the case in the speaking or writing of language. All of these kinds of recall involve habitual patterns of activity that are more or less unconsciously organized. On the present hypothesis such patterns are organized by chreodes, and the chreodes are stabilized by morphic resonance from similar past patterns of activity.

Conscious recall, even if it does not show itself in any objectively observable physical activity, is also an active process. We call to mind past experience and factual knowledge when we are thinking, for example when we are trying to solve some practical problem; and these memories often contribute to a new pattern of organization which may solve the problem. Recalling also occurs in dreams and daydreams: here too it is part of an active, constructive process, and we are often surprised by the ways in which elements of our past experience are woven together. Often we recall things during conversations with other people, and often in response to particular circumstances and sensations, perhaps most strikingly in the case of evocative smells. And recalling goes on continually as part of our "inner lives": in our "stream of consciousness" or "internal dialogue." We "go over" things in our minds.

A mental recall of elements of past activities may well take place in animals that appear to think: for example in the chimpanzees, which, having played previously with boxes and sticks, were able to work out how to use them to reach a bunch of bananas suspended well above the ground (p. 173). Such mental activity presumably occurs in some sort of "mental space" and involves a combination of kinaesthetic, spatial, and visual memories.

The kind of recalling that is characteristically human is that which depends on language. We can tell other people about our experiences, in so far as we can communicate them in words. This verbal recall is active: it depends on speaking or writing, and it also depends on our ability to code our experience in words in the first place. Even when we are thinking silently, using language in our thinking, we are speaking in a virtual, as opposed to actual, manner. People who "talk to themselves" or "think out loud" actually utter their thoughts, and people who "think on paper" write them down. Our non-verbal auditory memory is also active: for example we can hum a tune we know, or recall it silently by humming it virtually, "under our breath."

The ability to recall a particular experience depends on the ways we connected up various aspects of our awareness at that time and place and the ways in which they were related to other experiences through morphic fields. To the extent that we use language to categorize and connect the elements of experience, we can use language to help reconstruct these past patterns. But we cannot recall connections that were not made in the first place.

Our short-term memory for words and phrases enables us to remember them long enough to grasp the connections between them and understand their meanings. We most often remember meanings—patterns of connection—rather than the actual words. It is relatively easy to summarize the gist of a recent conversation, but for most of us it is impossible to reproduce it verbatim. The same is true of written language: for example you may be able to recall facts and ideas expressed in the preceding chapters of this book, but you are probably unable to recall even a single sentence word for word.

Not only in relation to language but in general, our short-term memory provides the opportunity for categorized elements of our recent experience to be connected with each other, as well as with past experience that is consciously recalled. What is not connected together is forgotten. These connections, on the present hypothesis, depend on morphic fields. In the case of language, this process of connecting involves the verbal categorization of experience and the formation of connections through virtual or actual speech. The structures of language provide the basic framework for these connections, and are associated with nested hierarchies of chreodes (pp. 183–86).

In the case of spatial recall—for instance in remembering the layout of a particular house—the morphic fields that connect different things and places together are related to patterns of movement of the body, for example going through a door, along a corridor, climbing stairs, entering a room, and so on. These patterns of movement are organized by chreodes within morphic fields, and it is these that are recalled through morphic resonance. Such morphic fields are associated with bodily movements in relation to the environment, and they integrate these patterns of movement with relevant features of the environment perceived through the senses. The fields are spatio-temporal: spatial in the sense that they are extended in and around the body and embrace the environmental space, and temporal in that they are associated with patterns of activity that unfold over time.

The principles of memorizing and recalling have long been understood from a practical point of view in mnemonic systems. These provide techniques for establishing connections that enable items to be recalled more easily. Some depend on verbal connections and involve coding the information in rhymes, phrases, or sentences. For instance, "Richard Of York Gained Battles In Vain" is a well-known mnemonic device for the colours of the rainbow (Red, Orange, Yellow, Green, Blue, Indigo, Violet). Other systems are spatial and rely on visual imagery. For instance, in the "method of loci" one first memorizes a sequence of locations, for example the various rooms and cupboards of one's own house. Each item to be recalled is then visualized in one of these locations.

The basic principles of mnemonic systems were well known in classical times and were taught to students of rhetoric. In the Renaissance there was a revival of interest in the "Art of Memory," and several complex systems based on the method of loci were elaborated.[11] Modern mnemonic systems such as those advertised in popular magazines are the heirs of this long and rich tradition.[12]

Several people with exceptionally good memories have been studied to find out how they are able to recall so much. One such investigation is described by the Soviet neuropsychologist A. R. Luria in his classic monograph *The Mind of a Mnemonist*. His subject, S., when working as a junior newspaper reporter in Russia astonished his editor by his remarkable ability to write detailed reports without the aid of notes, an ability that he himself took for granted. He was sent by the editor to see Luria, who tested him with ever longer sequences of words and numbers—first thirty, then fifty, then seventy—and found that he could recall them perfectly with apparent ease in any order, even years later. He could memorize poems in foreign languages that he did not understand, as well as complex mathematical formulae. It turned out that he usually used his own version of the method of loci.

> When S. read through a long series of words, each word would elicit a graphic image. And since the series was fairly long, he had to find some way of distributing those images of his in a mental row or sequence. Most often (and this method persisted throughout his life), he would "distribute" them along some roadway or street he visualized in his mind. . . . This technique of converting a series of words into a series of graphic images explains why S. could so readily reproduce a series from start to finish or in reverse order; how he could rapidly name the word that preceded or followed one I'd selected from the

series. To do this, he would simply begin his walk, either from the beginning or the end of the street, find the image of the object I had named, and "take a look at" whatever happened to be situated on either side of it.[13]

Not all mnemonists use visual imagery; some rely on the construction of verbal and numerical associations.[14] But none remember in the passive way implied in the popular notion of "photographic memory"; both memorizing and recalling are active mental processes. What are recalled are the mental constructions by means of which the memorized items were linked together, not the items themselves in isolation.

Forgetting

If morphic resonance underlies the phenomena of memory, and if the effects of such resonance do not fall off with time, then why is anything forgotten?

At first sight, the trace theory would appear to offer a simple and straightforward explanation for forgetting in terms of the decay of traces, and the morphic resonance hypothesis would seem unable to explain why any memories are ever lost. However, this impression would be misleading. Even among orthodox memory theorists, it is not usually assumed that forgetting is generally explicable in terms of the decay of the hypothetical memory traces. If it were, then old people should remember the recent past best and the more remote past least. But in fact the reverse is often the case: they forget what happened recently and easily remember incidents from their childhood and youth. Moreover, all of us are familiar with the way in which incidents that have been forgotten for many years may suddenly come to mind, and so it would be difficult if not impossible to prove that anything had been permanently lost.

Various kinds of forgetting are widely recognized by memory theorists. These are generally interpreted in terms of hypothetical "retrieval mechanisms" and memory traces within the nervous system, but they turn out to be just as easily compatible with interpretations based on morphic resonance.

First of all, the majority of what we see, hear, and otherwise experience is forgotten more or less immediately as it passes out of our short-term memories. We pay no particular attention and form no new connections or associations between the disparate elements; consequently no characteristic connections or associations can be remembered. From the mechanistic point of view, this is because appropriate memory traces were not established in the first place; from the point of view of formative causation, it is because appropriate morphic fields were not established.

Second, forgetting depends on the context; we may remember things under some conditions and forget them in others. It is a matter of common experience that we can recall people's names or facts or words in foreign languages in their familiar contexts much better than in unfamiliar contexts. The context dependence of recall has often been demonstrated experimentally.[15] This kind of forgetting clearly cannot be explained in terms of the decay of memories, but rather serves to emphasize that recall involves patterns of interconnection.

Third, the phenomena of repression, to which Sigmund Freud drew attention, involve the inability to remember certain events, especially painful ones, which nevertheless continue to exert a powerful unconscious influence on behaviour. They are difficult, if not impossible, to recall consciously because of their disturbing significance. This kind of forgetting depends on patterns of interconnection that somehow prevent conscious recall of the events in question; no one supposes that they are forgotten because hypothetical memory traces have decayed.

Fourth, various kinds of memory loss occur in response to brain damage, and concussion often results in amnesia for events that occurred in a more or less extended period just before or after the accident. This kind of amnesia is, however, often reversible and the lost memories may well come back. The effects of brain damage are discussed in more detail in the next chapter; here it is sufficient to note that when brain damage results in a loss of memory, this does not prove that the lost memories were encoded in the damaged tissue. It could mean that this tissue was associated with memory retrieval rather than storage, using the conventional framework of explanation; or that the tissue was associated with the tuning in by morphic resonance to the person's own past states. This interpretation has already been discussed in connection with animal memory in chapter 9.

Last, much forgetting appears to occur because of the interference of subsequent similar patterns of experience and activity. Our experience is cumulative, and similar experiences tend to "run together" or to be confused in such a way that we cannot recall them separately. Such repetition strengthens habits, but at the same time works against conscious recall. For example, we cannot recall all the separate occasions on which we have driven a car, although these cumulative experiences underlie our driving skills. We also know from our own experience that if we visit an interesting place or meet an important person only once, we are likely to remember our impression in detail. But if we visit a place or meet a person many times, the first occasion is harder to remember; the details tend to be lost in a "blur," a kind of cumulative composite memory of the place or the person. In this context, the emphasis in mnemonic systems on forming striking and unusual images makes very good sense.

The way in which subsequent experiences reduce our ability to recall similar previous experiences is called retroactive interference in the psychological literature, and it has often been demonstrated experimentally.[16]

The "running together" or "blurring" or "confusion" which underlies this kind of forgetting fits well with an interpretation in terms of morphic resonance, which pools or fuses together influences from similar past patterns of activity. In this process the individual differences between similar past patterns of activity are not exactly lost, because they contribute to the overall probability structures of the morphic fields; but they can no longer be recalled separately. They are indeed confused or confounded with each other, in a way that recalls the root meaning of both these words. The Latin word *confundere* means to pour together or mix.

In this chapter we have seen how our experiences of remembering and forgetting can be interpreted in terms of morphic fields and morphic resonance. We now consider the relationship of our memories and our mental activity to our brains, and explore the new interpretation of this relationship that the hypothesis of formative causation provides.

CHAPTER 12

Minds, Brains, and Memories

No one knows how our conscious experience is related to our bodies and our brains. It is obvious that our subjective experience is influenced by what happens within and around our bodies. It is also obvious that the ways we think, speak, and behave depend on theories, beliefs, desires, hopes, fears, habits, memories, and intentions, none of which are physical objects; they are subjective, and yet they have objective, observable effects. How are the subjective and the objective realms related?

This question of the relation of the soul to the body, or of the mind to the brain, is one of the perennial problems of philosophy. In this chapter we begin by considering the two main schools of thought: materialism or physicalism on the one hand, and dualism or interactionism on the other. The hypothesis of formative causation sheds a new light on this long-standing debate and leads to a new interpretation of the fashionable analogy between the brain and a programmed computer. We then return to the problem of memory, and we consider how the effects of brain damage on loss of memory can be understood and also what the evocation of memories by the electrical stimulation of the brain might mean. We end by considering some of the implications of this idea.

Materialism versus Dualism

Materialists believe that the mind is in the brain. One form of materialism treats conscious mental activity as an epiphenomenon of the activity of the brain, rather like a shadow: the conscious mind is an effect of the physical activity of the brain, but it is not a cause; it has no function at all, and everything would go on just the same without it. Another form of material-ism asserts that conscious mental activity and brain processes are simply different aspects of the same reality. They can be spoken about in different ways—just as the evening star and the morning star are called by different names, to use a favourite example—but they are in fact identical.

An essential feature of materialism is that it assumes that the physical world is causally closed; in other words, physical processes cannot be subject to causal influences from the soul, self, consciousness, or spirit, or indeed from anything that is, at this stage of our understanding, undefinable in physical terms. What we think, say, and do is in principle fully explicable in terms of the physico-chemical processes in our brains, which are governed by the ordinary laws of nature. We have no genuine free will because there is nothing in us that is free or capable of altering what would in any case have happened for purely physical reasons. In so far as chance events within our bodies play a part in our decisions or creativity, they do not represent free will or choice; they are merely random.[1]

The main alternative to materialism since the time of Descartes has been some form of dualism or interactionism, according to which the mind—or ego, soul, psyche, spirit, or conscious self—somehow interacts with the body through the brain. The view can also be phrased more dynamically by saying that conscious mental activities interact with the physical activity of the body through processes in the brain. This interaction has been thought of by means of a variety of analogies: for example the mind is like the driver driving the car, like a pianist playing a piano, with the brain as a kind of keyboard, or like the software of a computer interacting with the hardware of the brain. This last analogy can be taken further if the conscious self is compared to the programmer, the source of the software through which he interacts with the computer hardware (Fig. 12.1).

Although materialism is the official philosophy of modern science, interactionism is vigorously defended by a number of philosophers, including Karl Popper,[2] and is surprisingly widely supported by scientists, including quantum physicists such as Werner Heisenberg and Wolfgang Pauli,[3] and neurophysiologists such as Wilder Penfield,[4] John Eccles,[5] and Roger Sperry.[6]

However, the familiar debate between materialists and dualists has taken on a new twist in recent years. Many people who regard themselves as materialists or physicalists have come to think of the mind-brain relation-ship in terms of the computer metaphor, with the mind like the software and the brain like the hardware. But the very same analogy is also used by interactionists. This information or program paradigm is, in fact, much closer in spirit to dualism than it is to traditional monistic materialism.

The hypothesis of formative causation introduces the idea of morphic

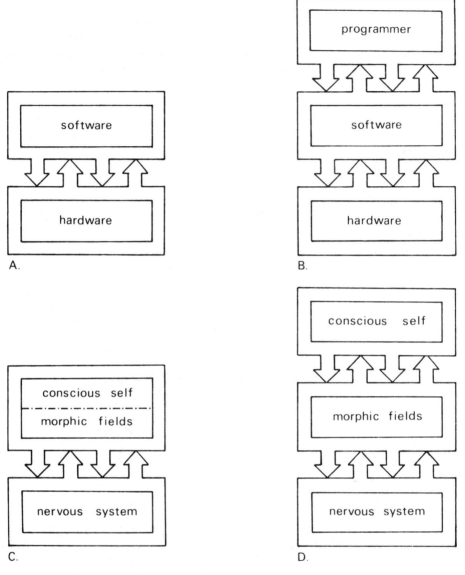

Figure 12.1 *Top*: The computer analogy for mind-brain interaction. A: The mind is compared to the software. B: The software corresponds to the programs of the unconscious and conscious mind, and the conscious self to the programmer. *Bottom*: C: The "physicalist" interpretation of the conscious self as a subjective aspect of the morphic fields interacting with the nervous system. D: The "interactionist" interpretation of the conscious self as interacting with the morphic fields, which interact with the nervous system.

fields, which interact with the nervous system and play a role similar to the programs or software in the computer metaphor. These fields provide a new context for the traditional physicalist-interactionist debate, and can be interpreted in the light of *either* philosophical theory. Figure 12.1 illustrates this diagrammatically. In the spirit of physicalism, the conscious mind can be regarded as a subjective aspect of the morphic fields that organize the activity of the brain; these fields can be experienced as it were from within. The conscious mind is not something over and above these behavioural and mental fields; it somehow exists within them. Or, in the spirit of interactionism, the conscious self can be supposed to interact with these fields, perhaps by containing them and including them, and may also serve as the creative ground through which new fields arise.

These two interpretations parallel the two interpretations of the computer metaphor in which the conscious mind is regarded either as an aspect of the software, or as analogous to the conscious programmer who interacts with the computer *through* the software.

Programs of the Brain

Chreodes within morphic fields are like programs in that they are structures of organization and are purposive: they are directed towards goals. The need for some such concept is made very clear by the way in which the computer metaphor has come to dominate modern thinking about the organization of mental activity. Much of this thinking can be translated into the terminology of formative causation simply by substituting the phrase *morphic field* for *program*. J. Z. Young, for example, after many years of research on the nervous system has proposed that the lives of human beings, like those of animals, are governed by sets of programs.

> Some of these programs may be called "practical" or physiological and they ensure that we breathe, eat, drink, and sleep. Others are social, and regulate our speaking and other forms of communication, our agreeing, and our loving and hating. We also have long-term programs, those that ensure continuing not of ourselves but of the race, programs for sexual activity and mating, programs for growth, adolescence, and indeed, for senescence and dying. Perhaps the most important programs of all are those used for the activities that we call mental, such as thinking, imagining, dreaming, believing, and worshipping.[7]

Young quotes the definition of *program* from *Webster's Third New International Dictionary* as "a plan of procedure; a schedule or system under which action may be taken toward a desired goal." He emphasizes that these programs of the brain are "plans of action" which are chosen in advance to meet particular kinds of situations.

Clearly the concept of such programs goes far beyond any reductionistic analysis to the physics of nerve impulses or the molecular biology of the nerve cells. Indeed, a holistic approach to the organization of the brain's activities seems practically unavoidable, and phrases such as *integrated patterns* and *organized systems* abound in the literature on the functioning of the brain. Even Francis Crick, the doyen of molecular biology, has been forced to the conclusion that the activity of various brain mechanisms must depend on "some kind of overall control system."[8]

Advocates of general systems theory have emphasized that the integrative activities of the brain need to be understood in terms of the dynamics of self-organizing systems,[9] and a start has been made on the mathematical modelling of such systems.[10] Here again, the computer analogy predominates:

> The brain is a communication mechanism which is used and directed by the self-organization of information. It has no more to do with this information than the computer with the information it processes. Although the comparison between brain and computer should not be carried too far since, to some extent, they represent very different principles, it may be useful to also distinguish between "hardware" and "software" in the brain. The network of neurons, then, represents the "hardware," and its possibly multilevel self-organization dynamics the "software."[11]

If this is regarded as a form of materialism, it is not the traditional monistic kind, but the modern dualistic kind which accepts the primacy of information over matter and energy, the acceptance of which Norbert Weiner and others have argued is essential for the survival of materialism in the modern world (p. 88). This modern dualistic approach has recently been summarized as follows:

> What mentality depends on is not a particular physical substrate, but the functional organization of the processes that it makes possible. There is still no need to invoke mystical properties in explaining the mind, but this approach can be informed by the theory of computability.[12]

And so we come back to computers, and to the ambiguous programming metaphor.

In recent years attempts to model the workings of the brain have influenced, and in turn been influenced by, the field of artificial intelligence.[13] The hope is that advances in computing techniques will enable better models to be made of "information processing" within the nervous system; and conversely that better models of the nervous system will lead to new insights that further the development of artificial intelligence. But doubts remain, not least among those engaged in such research.

> How does the analogy of the computer account for cognition? Most neurobiologists, it must be acknowledged, are suspicious of the analogy, but have little to put in its place. But the computer people are also at a loss to know just how far to take the analogy. . . . In the past few years, the artificial-intelligence community has slipped into the habit of asserting that the only proof that perception and cognition have been understood is that they (or somebody) can construct a machine that will replicate the process. The strategy is sensible enough: if a machine that will replicate the process of human vision can be built, a demonstration of its power would persuade all kinds of people that the problem of vision had been tackled seriously. Sceptics will, however, complain that simulation is not the same as understanding.[14]

The problem here is comparable to that faced by computer models of morphogenesis: just what do the models correspond to? No one imagines that a developing organism or a human brain actually *is* an electronic computer made of silicon chips and other inorganic components. The plausibility of these models depends not on any physical resemblance of computers to organisms, but on the clear-cut distinction between software and hardware. The organizing, goal-directed programs are concerned with form, pattern, interrelationship, and information. They are not reducible to the interactions of the electrons, atoms, and molecules that make up material structures. These goal-directed programs are in fact like morphic fields.[15] The principal difference between the two concepts is that the programs are supposed to be "written in genes and brains"[16] and stored as memory traces, just as the memory of computers depends on storage devices such as magnetic discs, whereas the morphic fields are not written in brains, but become present by morphic resonance. The program theory of the organization of the brain and the hypothesis of formative causation thus lead to very different interpretations of the nature of memory, to which we now turn.

Brains and Memories

The conventional idea that memory must be explicable in terms of physical traces within the nervous system is, as already stated, an assumption rather than an empirical fact. The assumption has been questioned by a number of philosophers at least since the time of Plotinus in the third century A.D.[17] The most stimulating critique still remains Henri Bergson's *Matter and Memory* (1911). But a number of arguments have been advanced more recently that appear to raise fundamental *logical* problems for any trace theory of memory.[18]

One of these problems arises in connection with the retrieval of memories from the hypothetical memory stores in which, according to the mechanistic theory, the memories are present in a coded form. When these memories need to be consulted or reactivated, they are called up by a retrieval system. For the retrieval system to be able to identify the stored memories it is looking for, it must be able to recognize them. But to do this it must *itself* have some kind of memory. There is thus a vicious regress: if the retrieval system is itself endowed with a memory store, then this in turn requires a retrieval system with memory—and so on.[19]

In spite of such arguments, the lack of empirical evidence for memory traces, and the difficulties faced by mechanistic models of memory storage within a dynamic nervous system (pp. 165–68), the idea of traces has been remarkably persistent. One reason for its durability has been the apparent lack of any alternative; another is that it appears to be supported by two well-known lines of evidence: that brain damage can lead to loss of memory, and that electrical stimulation of certain parts of the brain can evoke memories. We now consider this evidence in more detail.

Brain Damage and Loss of Memory

Brain damage can result in the loss of memory in two distinct ways, known as retrograde and anterograde amnesia. In retrograde amnesia, or "backwards" loss of memory, there is a loss of the ability to remember things that happened before the damage occurred. In anterograde amnesia, there is a loss of the ability to remember things that happen after the brain has been damaged.

From the mechanistic point of view, retrograde amnesias may be due either to the destruction of memory traces or to a destruction of the ability to retrieve the memories from the memory store (or to a combination of both). By contrast, from the point of view of the hypothesis of formative causation, the potential for past patterns of activity to influence the present by morphic resonance cannot be destroyed; rather, brain damage can affect the ability of the brain to tune in to its past patterns of activity.

Anterograde amnesias, from a mechanistic point of view, involve the loss of the ability to form new memory traces; whereas from the point of view of formative causation, they involve the loss of the ability to establish new morphic fields.

As we shall now see, the known facts can be interpreted plausibly from both points of view. The purpose of the present discussion is to show that the effects of brain damage on the loss of memory provide no persuasive evidence in favour of the materialist theory, as they are usually assumed to do. The hypothesis of formative causation fits the facts just as well, if not better.

The best known example of retrograde amnesia is the loss of memory of events that preceded concussion. In concussion, as a result of a sudden blow on the head, a person loses consciousness and becomes paralysed. The loss of consciousness may last for only a few moments or for many days, depending on the severity of the impact. As a person recovers and regains the ability to speak, he may seem normal in most respects, but may be unable to recall events that immediately preceded the accident or occurred weeks, months, or years before. Typically, as recovery proceeds, the first of the forgotten events to be recalled are those that occurred longest ago, and the memory of more recent events then progressively returns.

In such cases, the trauma obviously affects the ability to recall past experiences, but the amnesia cannot possibly be due to the destruction of memory traces, for the lost memories return. From the conventional point of view, it depends on the recovery of the ability to retrieve these memories from the memory stores; while from the point of view of formative causation, it depends on the recovery of the ability to tune in to them by morphic resonance.

However, the events immediately preceding the blow on the head may never be recalled: there may be a permanent blank period. For example, a motorist may remember approaching the crossroads where the accident occurred, but nothing more. A similar "momentary retrograde amnesia" also occurs as a result of electroconvulsive therapy, administered to mental patients by passing a burst of electric current through their heads. Such patients usually cannot remember what happened immediately before the administration of the shock.[20]

One possible explanation might appear to be in terms of the repression of events closely associated with an unpleasant experience. But other kinds of head injury which do not result in loss of consciousness, for example certain injuries due to bullet wounds or crushing, do not usually result in the events that preceded the injury being forgotten. The generally accepted explanation for such amnesias is that they represent a failure of long-term memories to be established. Events and information in short-term memory are forgotten because a loss of consciousness prevents them from being connected up into patterns of relationship which can be remembered (pp. 207–8).

The failure to make such connections, and hence to turn short-term memories into long-term memories, often persists for some time after a concussed patient has recovered consciousness: this anterograde amnesia is also sometimes described as "memorizing defect." People in this condition rapidly forget events almost as soon as they occur. They may, for instance, forget a meal they have just eaten or news they have just heard.

From a conventional point of view the failure of such patients to establish long-term memories, for whatever primary reason, is due to a failure to lay down memory traces. From the point of view of the hypothesis of formative causation, it is due to a failure to establish new morphic fields.

Various characteristic memory defects occur as a result of damage to the central cortex caused by strokes, accidental injury, or surgery. Some, such as massive lesions of the frontal lobes, have general effects on the ability to concentrate, and hence affect the formation of fresh memories. Others have quite specific effects on abilities to recognize and recall.[21] The ability to recognize faces, for instance, may be lost as a result of a lesion in the secondary visual cortex in the right hemisphere. A sufferer may fail to recognize the faces of even his wife and children, although he still knows them by their voices and in other ways.

This inability to recognize faces is called prosopagnosia (from the Greek *prosopon,* face, and *agnosis,* not knowing) and is one of many kinds of loss of power to recognize the import of sensory stimuli. Neurologists have described agnosias for colours, sounds, animate objects, music, words, and so on. These are sometimes described in terms such as mind-blindness or word-deafness.

Some neurologists consider that agnosias in general are best explained in terms of defects in the higher levels of the hierarchical system through which features detected by the sense organs are combined into patterns, recognized, and named. Others have suggested that some agnosias are better explained in terms of disconnections between intact brain areas such as the language and visual regions of the cortex, preventing objects being named and thus accounting for the apparent inability to recognize them. Significantly, none of these mechanistic interpretations of the agnosias attributes them to the destruction of memory traces.

The same is true of other kinds of disorder, such as the aphasias (disorders of language use) due to lesions in various parts of the cortex in the left hemisphere; and the apraxias, the loss of previously acquired abilities to manipulate objects in a co-ordinated way. These are generally attributed to disturbances of organized patterns of activity in the brain rather than to a loss of memory traces.[22]

On the present hypothesis, these abilities are lost because the brain damage affects parts of the brain with which the morphic fields are normally associated. If an appropriate pattern of brain activity is no longer present, the fields cannot be tuned in to or bring about their organizing effects.

This interpretation makes it much easier to understand the fact that lost abilities often return; patients often recover partially or completely from brain damage even though the damaged regions of the brain do not regenerate. The appropriate patterns of activity come into operation somewhere else in the brain. This is almost impossible to understand if programs are "hard-wired" into the nervous system; but fields can move their regions of activity and reorganize themselves in a way that fixed material structures cannot. Such recoveries are reminiscent of the regenerative abilities of plants and animals, and they pose the same kind of problem for mechanistic explanation.

In general, after traumatic head injury,

> memories and skills return at a rapid rate during the first six months, with recovery sustained at a lower rate for up to 24 months. Defects in sensory, motor, and cognitive functions caused by brain injury due to penetrating wounds are characterized by an enormous resiliency of function in the great majority of cases, ultimately leading to little or no detectable defect.[23]

One of the leading researchers on the long-term effects of brain damage, Hans Teuber, after years of studying the recovery of wounded veterans of World War II and the Korean and Vietnam wars, concluded that "this far-reaching restitution of function remains, in my view, unexplained."[24]

We are far from understanding how the brain is organized, how memory works, how brain damage leads to amnesia, or how people can recover from brain injuries. The mechanistic interpretations of these phenomena are still vague and speculative, despite decades of intensive research. The hypothesis of formative causation offers a new approach to these problems, which may turn out to be more fruitful; but at present the question is open.

The evidence regarding evocation of memories by electrical stimulation of the brain is just as ambiguous as the information about memory loss resulting from brain damage; it can be interpreted in terms of morphic resonance, or in terms of hypothetical memory traces, which remain as elusive as ever.

The Electrical Evocation of Memories

In the course of operations on conscious patients with various neurological disorders, Wilder Penfield and his colleagues tested the effects of mild electrical stimulation of various regions of the brain. As the electrode touched parts of the motor cortex, the appropriate limb movements would occur. Stimulation of the primary auditory or visual cortex evoked auditory or visual hallucinations: flashes of light, buzzing noises, and so on. Stimulation of the secondary visual cortex gave rise to complex, recognizable visual hallucinations of flowers, animals, familiar people, and so on. And in epileptics, when some regions of the temporal cortex were touched, some patients recalled apparently specific memory sequences, for example an evening at a concert or a telephone conversation. The patients often alluded to the dream-like quality of these experiences.[25]

The electrical evocation of these memories could mean that they were stored in the stimulated tissue, as Penfield initially assumed; or it could mean stimulation of that region activated other parts of the brain that were involved in remembering the episode.[26] But it could also mean that the stimulation resulted in a pattern of activity that tuned in to the memory by morphic resonance.

Significantly, Penfield himself, as a result of further reflection on these and other findings, abandoned his original interpretation:

> In 1951 I had proposed that certain parts of the temporal cortex should be called "memory cortex," and suggested that the neuronal record was located there in the cortex near points at which the stimulating electrode may call forth an experiential response. This was a mistake. . . . The record is *not in the cortex.*[27]

Penfield, like Lashley and Pribram (pp. 162–64), gave up the idea of localized memory traces within the cortex in favour of the theory that they were distributed in various other parts of the brain instead, or as well. The advantage of this hypothesis is that it accounts for the recurrent failure of attempts to find these traces; the disadvantage is that it is untestable in practice. In the light of formative causation, the elusiveness of the memory traces has a very simple explanation: they do not exist. Rather, memory depends on morphic resonance from the patterns of activity of the brain itself in the past. We tune in to ourselves in the past; we do not carry all our memories around inside our brains. But what if we tune in to other people as well?

Tuning In to Other People

On the hypothesis of formative causation, the reason we have our *own* memories is that we are more similar to ourselves in the past than we are to anyone else; we are subject to a highly specific self-resonance from our own previous states. But we are also similar to members of our own family, to members of social groups to which we belong, to people who share our language and culture, and indeed to some extent we are similar to all other human beings, past and present.

If we are influenced by morphic resonance from particular individuals to whom we are in some way linked or connected, then it is conceivable that we might pick up images, thoughts, impressions, or feelings from them, either during waking life or while dreaming, in a way that would go beyond the means of communication recognized by contemporary science. Such resonant connections would be possible even if the people involved were thousands of miles apart. Is there any evidence that such a process actually happens? Perhaps there is: for such a process may be similar to, if not identical with, the mysterious phenomenon of telepathy. There is a wealth of anecdotal evidence for the occurrence of telepathy,[28] many people claim to have experienced it themselves,[29] and it has been detected in many parapsychological experiments.[30] This evidence is, of course, much disputed, largely because from the conventional scientific point of view telepathy, like the other alleged phenomena of parapsychology, is theoretically impossible. By contrast, in the context of morphic resonance, it is theoretically possible.

Morphic resonance might also provide a new interpretation for a relatively rare but well-documented phenomenon: memory of past lives. Some young children spontaneously claim to remember a previous life, and sometimes give details about the life and death of the previous person whom they claim to be. Careful research has shown that some of the details they give could not have been known to them by normal means. Dozens of case studies of this type have now been documented in detail.[31] (Descriptions of previous lives have also been given by adults under hypnosis, but many seem to contain a large element of fantasy and the evidence for "paranormal" memory is much less impressive than in the spontaneous cases in young children.)

Those who accept the evidence for memories of previous lives usually explain it in terms of reincarnation or rebirth. However, the hypothesis of formative causation provides a different perspective: in such cases a person may for some reason tune in by morphic resonance to a person who lived in the past. This might help to account for the transfer of memories without our having to suppose that the present person *is* the other person whose memories he or she can pick up.

However, the principal way in which we are influenced by morphic resonance from other people may be through a kind of pooled memory. We have already discussed the collective influence of other people's habits on the learning of languages and the acquisition of physical and mental skills, and considered ways in which this possibility can be, and has been, tested by experiment (chapter 10). The idea that a collective memory underlies our mental activity follows as a natural consequence from the hypothesis of formative causation. A very similar idea already exists in the concept of the collective unconscious worked out by Carl Jung and other depth psychologists.

Collective memories are like habits in the sense that the repetition of similar patterns of activity effaces the particularity of each individual instance of the pattern; all similar past patterns of activity contribute to the morphic field by morphic resonance and are, as it were, merged together. The result is a composite or average of these previous similar patterns, which we can think of by analogy with composite photographs (Fig. 6.4). Jung called such habitual patterns archetypes and thought that they were built up by collective repetition:

> There are as many archetypes as there are typical situations in life. Endless repetition has engraved these experiences into our psychic constitution. . . . When a situation occurs which corresponds to a given archetype, that archetype becomes activated.[32]

Before returning to a discussion of Jung's ideas in chapter 14 in the context of the social and cultural aspects of human mental life, we consider the role of formative causation in the organization of animal societies.

CHAPTER 13

The Morphic Fields of Animal Societies

From the point of view of the hypothesis of formative causation, social groups are ordered by social morphic fields, fields which embrace and contain the individual organisms within the social unit or holon. These fields, like morphic fields at all other levels of complexity, are shaped and stabilized by morphic resonance.

In this chapter we consider the organization of animal societies and in the following two chapters the organization of human societies and cultures. This discussion explores the role of morphic fields in the co-ordination of social behaviour and of morphic resonance in social and cultural inheritance.

Animal Societies as Organisms

Societies of termites, ants, wasps, and bees may contain thousands or even millions of individual insects. They can build large and elaborate nests (Fig. 13.1), exhibit a complex division of labour, and reproduce themselves. They have often been compared to organisms or described as superorganisms.

Not surprisingly, there has been a long-standing debate as to whether such societies really are new kinds of organisms or systems at a level above that of individual animals; or on the other hand whether they are complex aggregates that are fully explicable in terms of the properties and behaviour of the individuals that comprise them. Should they be thought of holistically as systems or organisms at a new level of complexity with irreducible properties? Or can they be understood in a reductionist manner as nothing but the sum of their parts and the interactions between them?

Figure 13.1 Mounds made by compass termites in Australia. Their broad sides (top) face east and west; their narrow sides (bottom) face exactly north and south. They thus expose a minimum surface area to sunlight in the middle of the day, avoiding excessive heating. (After von Frisch, 1975)

At present, animal societies are generally studied by biologists in a reductionist spirit. But this mechanistic method has only relatively recently replaced the holistic approach. Edward O. Wilson, the founder of sociobiology, has described the decline of the superorganism concept as follows:

> During some forty years, from 1911 to about 1950, this concept was a dominant theme in the literature on social insects. Then, at the seeming peak of its maturity it faded, and today it is seldom explicitly discussed. Its decline exemplifies the way inspirational, holistic ideas in biology often give rise to experimental, reductionist approaches which supplant them. For the present generation, which is so devoted to the reductionist philosophy, the superorganism concept provided a very appealing mirage. It drew us to a point on the horizon. But, as we worked closer, the mirage dissolved—for the moment at least—leaving us in the midst of unfamiliar terrain, the exploration of which came to demand our undivided attention.[1]

However, as in the case of the reductionist approach to morphogenesis, behaviour, memory, and psychology, this approach to animal societies has not so far resulted in a mechanistic understanding of them. Rather, Wilson continues, "there exists among experimentalists a shared faith that characterizes the reductionist spirit in biology generally, that in time all the piecemeal analyses will permit the full reconstruction of the system in vitro." But, he freely admits, "at the present time we cannot come close" to this accomplishment.[2] So here, as in other areas of biology, the question remains wide open. The reductionist faith has been fruitful in stimulating many detailed investigations, but there is no evidence that it will ever provide convincing explanations for the holistic properties of organisms at any level of complexity.

According to the hypothesis of formative causation, the organization of social systems depends on nested hierarchies of morphic fields, with the overall field of the society organizing the individual animals within it through their morphic fields, which in turn organize their component organs, and so on down to the cellular and subcellular levels.

Some kinds of animal societies involve such a close integration of the individuals into a higher-level whole that there seems to be general agreement, even among sociobiologists, that they are best regarded as unitary organisms. A wide variety of colonial invertebrates consist of individuals so harmoniously linked together that at first sight they are easily mistaken for a straightforward single organism. This type of organization reaches its extreme in colonies of the order Siphonophora, which vaguely resemble jellyfish and live in the open ocean, where they use their stinging tentacles to capture fish and other small prey. One well-known example is the Portuguese man-of-war; another is *Nanomia* which consists of many specialized individual organisms (Fig. 13.2). At the top is an individual modified into a gas-filled float. Below it are organisms that act like little bellows, squirting out jets of water which propel the colony; by altering the shape of their openings they are able to alter the direction of the jets. Through their co-ordinated action the *Nanomia* colony is able to dart about vigorously, moving at any angle and in any plane, even executing loop-the-loop curves. Lower on the stem there are other organisms which are specialized for the ingestion and digestion of nutrients for the rest of the colony. Long branched tentacles arise from them and are used to capture prey. There are also bracts, consisting of inert, scalelike organisms that fit over the stem and help protect it from physical damage. Finally, there are sexual organisms, which produce gametes which through fertilization can give rise to new colonies.

These specialized individuals within the colony are effectively like organs in an organism, and some are even connected together and co-ordinated through nerves. Such forms of life seem to be both colonies *and* organisms.[3] Other colonial invertebrates, such as the corals, can likewise be regarded as both at the same time.

Societies of Insects

The social insects, like colonial invertebrates, also exhibit a marked specialization of individuals. The queen is generally larger than the other members of the society and is specialized for egg laying, in some termite colonies producing up to 30,000 eggs a day.[4] Among the sterile workers in many ant and termite societies there are distinct castes, including soldiers with formidable mouthparts. Even in bee societies, in which the workers look alike, there is a remarkable division of labour. In the hives of honeybees, for example, some clean the cells and feed the queen and larvae, some construct and seal the cells of the honeycomb, some guard the hive, and others go out foraging. Any given worker can play all these roles, and typically does so sequentially, starting as a cleaner and ending as a forager.

The members of insect societies communicate with each other through the exchange of food, by means of various chemical substances, by touch, and in a variety of other ways, the most remarkable being the famous dance of honeybees by means of which returned foragers indicate to others where food can be found.[5]

These societies have striking self-organizing properties. Honeybees, for example, maintain the temperature of their nest with remarkable constancy; from spring to autumn the interior temperature is almost always between

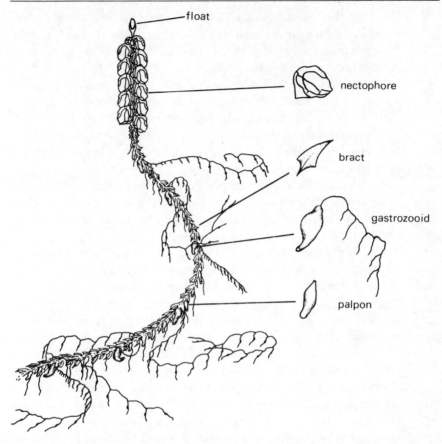

float

nectophore

bract

gastrozooid

palpon

Figure 13.2 A colony of the siphonophoran *Nanomia cara*. The floats that buoy up the colony, the nectophores that propel it, the gastrozooids that capture and digest prey, and other member organisms such as the bracts are modified to such an extreme that they are comparable to organs of a single animal. (After G. O. Mackie, 1964, Royal Society Proceedings B159, 366–91)

34.5 and 35.5 degrees centigrade. This degree of control is facilitated by the way they seal up all crevices and gaps, leaving only a single entrance hole. The output of heat from the insects themselves is the source of warmth, and in response to cold weather they huddle together more or less closely in clusters. In warm weather the workers cool the hive by fanning with their wings; and if this alone does not suffice, some of them collect water and carry it into the nest, distributing it on the brood cells; others spread this into a film, from which evaporation is rapid, while a third group fans the moist air away from the brood cells and out of the nest.[6]

On the present hypothesis, over and above the behavioural fields of individual insects are the morphic fields of the society as a whole, which co-ordinate the activities of the individuals. These fields are spatially extended and embrace the entire colony; the individuals live within them. It is through these supra-individual fields that the colony comes into being and maintains its structure and organization, in spite of the continual turnover of individual workers, whose life-span is generally much shorter than that of the colony as a whole. The self-organizing properties of these fields are what enable the colonies to adjust to accidents, damage, and environmental fluctuations and to repair their nests.

The need for some such concept can be illustrated by considering the way in which termites construct their nests, which can reach enormous sizes and in some species are extremely complex, even incorporating what can only be described as an air-conditioning system.

The African fungus-growing termite *Macrotermes natalensis* forms vast colonies that last for years and at maturity contain about two million insects at any one time. The nest develops from a small underground chamber made by the royal couple, and can grow to a height of over ten feet above the ground. At the base of the mound is the nest proper, with the royal cell in its centre. In its many chambers, which are connected by numerous passages, are masses of finely chewed wood, on which the termites cultivate the fungus that they eat. Above this there is a large air space, and this is enclosed by the outer casing of the mound, on the outside of which are ridges or buttresses. Channels as thick as a man's arm radiate into many small ducts within the buttresses. The air in the fungus chambers is heated by the fermentation process and by the termites themselves; this hot air rises and is forced into the duct system of the ridges, the walls of which are so porous that they allow gas exchange to take place: carbon dioxide escapes and oxygen penetrates from the outside. From these "lungs," the cooled and regenerated air now flows back down another system of wide ducts into the cellar, whence it returns to the nest.[7]

These structures are built by workers from pellets of soil which are glued together with excrement or saliva. But how do they know where to put these materials? In the words of E. O. Wilson:

> It is all but impossible to conceive how one colony member can oversee more than a minute fraction of the work or envision in its entirety the plan of such a finished product. Some of these nests require many worker lifetimes to complete, and each new addition must somehow be brought into a proper relationship with the previous parts. The existence of such nests leads inevitably to the conclusion that the workers interact in a very orderly and predictable manner. But how

can the workers communicate so effectively over such long periods of time? Also, who has the blueprint of the nest?[8]

Detailed observations of the building activity have shown that the nest structure that has been completed, more than direct communication among the insects, influences what work will be done. For example, in the building of arches, workers first construct columns, and then if another column is being built sufficiently close by, they bend the column towards the other one (Fig. 13.3) until the tilted growing ends of the two columns meet.[9] No one knows how they do this. The workers cannot see the other column: they are blind. There is no evidence that they run back and forth at the base of the columns measuring the distance. Moreover, "it is improbable that in the midst of all the confused scampering in the vicinity, they can recognize distinct sounds from the column by conduction through the substrate."[10] By a process of elimination, it is generally assumed that they must be able to locate the other column by somehow smelling it.[11]

In short, very little is actually understood about the way in which the termites construct these prodigious structures. Moreover, the conventional

Figure 13.3 The construction of an arch by workers of the termite species *Macrotermes natalensis*. Each column is built up from pellets of soil and excrement, carried by the insects in their mandibles. When the column reaches a certain height, the termites, which are blind, begin to extend it at an angle towards a neighbouring column. (From *Animal Architecture* by Karl von Frisch, illustrated by Turid Holldobler. Copyright © 1974 by Turid Holldobler. Reprinted by permission of Harcourt Brace Jovanovich, Inc.)

idea that instinctive abilities are somehow "programmed" or "hard-wired" in the nervous system might lead us to expect that termites that build such complex nests have larger and more complex nervous systems than species that build much simpler nests. But in fact they do not.[12]

The hypothesis of formative causation provides an alternative approach, and suggests that the structures of the nests are organized by morphic fields embracing the nests as wholes, with a nested hierarchy of fields within them associated with the various elements of the overall structure. These fields are not inside the individual termites; rather the individual insects are inside the social fields.

If this is so, the organizing activity of the fields would extend beyond the range over which individual insects can communicate with each other by smell or by mechanical means. This idea is experimentally testable, and indeed there is already suggestive evidence that some such effects occur.

Over sixty years ago, the South African naturalist Eugene Marais made a series of observations on the way in which the workers of the species *Eutermes* repaired large breaches that he made in their mounds. He was struck by the way the workers set to work on the breach from every side and yet worked in a co-ordinated way so that the new parts all joined correctly, even though termites working on different sides of the breach did not come into contact with each other and, being blind, could not see each other.

He then carried out a simple but remarkable experiment. He took a large steel plate several feet wider and higher than the termitary and drove it right through the centre of the breach in such a way that it divided the mound and indeed the entire termitary into two separate parts.

> The builders on one side of the breach know nothing of those on the other side. In spite of this the termites build a similar arch or tower on each side of the plate. When eventually you withdraw the plate, the two halves match perfectly after the dividing cut has been repaired. We cannot escape the ultimate conclusion that somewhere there exists a preconceived plan which the termites merely execute.[13]

The same thing happened when the steel plate was driven in first and then a breach was made on either side of it.

Apparently this fascinating experiment has never been repeated; it would obviously be well worth doing so, preferably using material that would provide better acoustic insulation than a steel plate.

Marais thought that the queen was like the "brain" of the colony and was somehow connected with the entire colony directly, over and above the chemical and other influences that were physically carried to other members of the colony by the workers who tended her. On the present hypothesis, such a linkage could be thought of in terms of the extended morphic field embracing both the queen and all the other members of the colony. Marais claimed to have demonstrated the existence of such non-material connections by means of simple experiments such as the following:

> While the termites are carrying out their work of restoration on either side of the steel plate, dig a furrow enabling you to reach the queen's cell, disturbing the nest as little as possible. Expose the queen and destroy her. Immediately the whole community ceases work on either side of the plate.[14]

Again, no further work seems to have been done along these lines; but clearly it would be of interest to find out just how immediate this effect is; for an effect mediated by the morphic field of the colony could be immediate, whereas an effect that depended only on normal sensory communication could not. In such experiments it would not be necessary to kill the queen; merely removing her from the colony would probably suffice.

Probably it is with termites or other social insects that the most decisive experiments could be done to distinguish between the field approach to animal societies and the conventional mechanistic approach.

Schools, Flocks, and Herds

In vertebrates too the co-ordination of individuals within a group is sometimes so close that it is almost impossible not to think of them as a kind of composite organism.

Many species of fish form schools or shoals.

> At a distance a fish school resembles a large organism. Its members, numbering anywhere from two or three into the millions, swim in tight formations, wheeling and reversing in near unison. Either dominance systems do not exist or are so weak as to have little or no influence on the dynamics of the school as a whole. When the school turns to the right or left, individuals formerly on the flank assume the lead.[15]

Schools exhibit characteristic patterns of behaviour, particularly in response to potential predators. When under attack, a school may respond by leaving a gaping hole or vacuole around the predator (Fig. 13.4). More often the school splits in half and the two halves turn outwards, eventually swimming back around the predator and rejoining. This is known as the fountain effect, and leaves the predator ahead of the school. Each time the predator turns, the same thing happens.

Figure 13.4 The formation of an empty space within a school of fish around a predator. (From *The Oxford Companion to Animal Behaviour,* edited by D. McFarland, © Oxford University Press 1981. Reproduced by permission.)

The most spectacular of the schools' defences is the "flash expansion," so called because on film it looks like a bomb bursting as each fish simultaneously darts away from the centre of the school as the group is attacked. The entire expansion may occur in as little as one-fiftieth of a second, and the fish may accelerate to a speed of ten to twenty body lengths per second within that time. Yet the fish do not collide. "Not only does each fish know in advance where it will swim if attacked, but it must also know where each of its neighbours will swim."[16]

How schooling behaviour is co-ordinated remains a mystery. Vision no doubt plays a part; but some species continue to swim in schools at night. Moreover, in experiments in which fish were blinded by being fitted with opaque contact lenses, they were still capable of joining and maintaining their position indefinitely within a school of normal fish. Perhaps they could judge the position of their neighbours by special pressure-sensitive organs, known as the lateral lines, which run along their length. But this idea has been tested by severing the lateral lines at the gill covers. Such fish still school normally.[17]

Even if the means by which they are aware of each other's positions were understood, this would still not account for such rapid co-ordinated responses as the flash expansion: a fish could not sense *in advance* where its neighbours were going to move.

If, however, the school is organized by a morphic field that embraces all the fish within it, the properties of this field could underlie the behaviour of the school as a whole and help account for the co-ordinated behaviour of the individual fish. Of course, a detailed understanding of the operation of the field would have to be determined by appropriate experiments; so little is known at present that it can only be conceived of in rather vague and general terms.

Flocks of birds, like schools of fish, show such a remarkable co-ordination of their individual members that they too have often been compared to a single organism. The naturalist Edmund Selous, for example, wrote as follows of the movement of a vast flock of starlings:

> Each mass of them turned, wheeled, reversed the order of their flight, changed in one shimmer from brown to grey, from dark to light, as though all the individuals composing them had been component parts of an individual organism.[18]

He also observed the way in which flocks of peewits, gulls, and other birds took off all at once, often for no apparent external reason.

> A flight of ox-birds [dunlins]—some 150 to 200 I should say—went down on a mudflat here. After a little time, and when I had the glasses full on them, taking in the whole line, they rose all together instantaneously, without any visible extraneous cause. Soon they came down again, and now a swan flies across the line, not more than a foot or two above them. If all went up now, who could be surprised? But this was not the case. A certain number did, when the swan cut the line, but only a few inches from the ground, and settled again almost at once while the rest stood where they were. Some minutes later, under no discernible provocation, all rose in flight again on the instant.[19]

Selous studied the behaviour of such flocks over a period of thirty years, and became convinced that it would admit of no normal sensory explanation: "I ask how, without some process of thought transference so rapid as to amount practically to simultaneous collective thinking, are these things to be explained?"[20]

Recently, the banking movements of large flocks of dunlins have been studied by taking slow-motion films and studying the way in which the movement of the flock was initiated. These revealed that the movement was not exactly simultaneous, but rather started either from a single individual or from two or three birds together. This initiation could occur anywhere within the flock, and manoeuvres always propagated through the flock as a wave radiating from the initiation site. These waves moved very rapidly, taking on average 15 milliseconds (15 thousandths of a second) to pass from neighbour to neighbour.

In the laboratory, tests were carried out with captive dunlins to find out how rapidly they could react to a sudden stimulus. The average startle reaction time to a sudden light flash was 38 milliseconds. This means that it is very unlikely that they can bank in response to what their neighbours do, since this banking response occurs much quicker than the measured startle reaction time.

However, when the films were examined very carefully, it turned out that at the very beginning of the flock's movement the neighbours of the initiating bird reacted more slowly than the rate at which the reaction wave spread through the rest of the flock. On average the immediate neighbours took 67 milliseconds to react.

Wayne Potts, who carried out this fascinating study, has proposed what he calls the chorus-line hypothesis to explain it. He bases this on experiments carried out in the 1950s with human chorus lines. The dancers rehearsed particular manoeuvres; then these were initiated by a particular person without warning, and the rate at which they propagated along the line was estimated from films. This was on average 107 milliseconds from person to person, nearly twice as fast as an average human visual reaction time of 194 milliseconds. Potts suggests that this was accomplished by the individuals seeing the approaching manoeuvre wave and estimating its arrival time in advance. He takes his findings with the dunlins to support this hypothesis: the slower reaction time of the immediate neighbours of the initiating birds occurred because they could not see and advancing wave, for this wave had not yet developed.[21]

At first sight, this appears to provide a straightforward mechanistic explanation for the banking phenomenon and to remove the need for "mystical" factors such as thought transference or, Potts might wish to add, morphic fields. But does it? It is worth examining Potts's hypothesis in detail because it is the most plausible—indeed almost the only—conventional explanation for the way in which such flocks behave as wholes.

First of all, because these waves can propagate in any direction through the flock, Potts has to assume that the birds sense and notice such waves almost immediately, even if they are coming from directly behind them. This would require them to have practically continuous 360-degree visual attention, which does not seem a very plausible assumption.

However, for the purpose of argument let us assume with Potts that this is the case. A second problem now appears. The birds do not respond to the approaching manoeuvre wave in a non-specific reflex manner like the startle reaction to a sudden flash of light. They change their pattern of flight in a precise way; the angle, speed, and duration of their turning are precisely integrated with those of the rest of the flock; thus, densely packed as they are, none of the birds collide with each other. Although there are a limited number of patterns of flock manoeuvre, such as banking, they are not stereotyped in their quantitative details. They are more flexible than a well-rehearsed chorus line going through standard routines. This means that not only would the birds have to sense an advancing wave, they would also have to sense from it exactly how they should turn. The reaction time for such a response could well be longer than for the non-specific startle reaction. If so, this would mean that the birds would have to sense the manoeuvre wave, including its direction, angle, and speed, even farther in advance than Potts assumes, thus creating an even stricter requirement for unblinking 360-degree awareness of the flock all around them.

This would also mean that the birds would have to perceive and respond to the manoeuvre wave as a gestalt, grasping the movement of the flock as a whole and responding precisely to it in accordance with their position within it. But this continuum of the flock as a whole and the movement of patterns through it sounds very like an example of a field phenomenon. It involves, on the present hypothesis, the morphic field of the flock.

The conventional alternative to this conclusion would involve assuming that the manoeuvre wave was apprehended in the birds' perceptual or "data-processing" space, and responded to through the programmed activities of the individual birds' nervous systems. But this merely begs the question, because the nature of such programs or organizing principles in the nervous system is completely unknown. On the present hypothesis these organizing principles are themselves morphic fields.

Potts's chorus-line hypothesis begs a further question. He assumes that the very rapid propagation of manoeuvres along a well-rehearsed human chorus line is itself explicable mechanistically in terms of known physical principles. There is no evidence for this. From the present point of view, the chorus-line routines may themselves depend on morphic fields which are stabilized by morphic resonance from previous rehearsals and performances.

Thus even if we accept the chorus-line hypothesis, it cannot in itself provide an explanation for the banking behaviour of flocks: the nature of the manoeuvre wave and the birds' response to it remain unexplained mechanistically and would more appropriately be accounted for in terms of morphic fields.

Just as many species of fish form schools and birds flocks, so many mammalian species form herds or packs. These groups also move in a co-ordinated way, sometimes at great speed, and the animals do not collide. Here too the organization of the group can be thought of in terms of morphic fields.[22] Again, this is not an *alternative* to communication between members of the herd through senses such as sight, sound, and smell. Morphic fields do not supplant the need for sensory communication; rather they provide the structured context within which the animals' communications and responses occur.

The Organization of Animal Societies

Schools, flocks, and herds provide spectacular examples of the co-ordination of activities of individual animals within a larger whole; but in fact all animal societies by their very nature involve patterns of organization and co-ordination. Social animals respond and relate to each other within the context or framework of these structures.

In many species the social structure is relatively simple and may be only temporary—for instance, when males and females come together during the reproductive period and co-operate in mating and in provision for their young, then separate again to lead a more or less solitary life. At the other end of the spectrum are a wide variety of complex and enduring social structures such as those of termites and chimpanzees.

Ethologists have described in detail many patterns of social organization, for example dominance hierarchies, as in the pecking order of chickens, and complex co-operative activities, such as the hunting behaviour of a pack of wolves.[23]

Everyone agrees that the patterns of organization of such societies are to a large degree inherited. The conventional assumption is that they are to a large extent genetically programmed and that the social order somehow "arises" or "emerges" from the interactions between the individual animals. But this merely restates the problem in different words. How do patterns of social organization arise or emerge? On the present hypothesis, they are attributable to morphic fields, and these fields are stabilized by morphic resonance from similar past societies: the patterns of social organization are not inherited in the genes.

Once again, the concept of such social morphic fields is not an alternative to the many known forms of interaction and communication between animals within the group. But a physical or chemical signal, gesture, or call from one animal to another has a meaning only within a context—indeed, anything that is meaningful or significant is only so by virtue of its relationship to other things. The morphic fields are what underlie such patterns of relationship.

Likewise, morphic fields are not *alternatives* to physiological influences on behaviour such as hormones. An increased amount of sex hormones in the blood-stream of birds, for example, can result in their entering a breeding phase. But the characteristic behaviour of birds in courtship, nest-building, incubating the eggs, and rearing their young is not attributable merely to the chemistry of the sex hormones: different species have very different patterns of behaviour but the same sex hormones. Rather, such hormones bring about specific physiological and biochemical changes within the birds, which tune them in to the morphic fields of breeding behaviour, including the social fields that pattern the complementary activities of the males and females.

Cultural Inheritance

Just as an individual animal behaves in a way that is characteristic of its species and within this framework shows various habits and peculiarities of its own, so one animal society resembles others in the same species more or less closely, but at the same time has its own customs, habits, or traditions. Many of these behaviour patterns are related to the particular territory or micro-environment that the group inhabits and are adopted by new members of the group, especially by young animals growing up within it. There is in effect a kind of social memory. Spectacular examples of such traditional behaviour are provided by many migratory animals. Herds of reindeer, for instance, follow traditional migration pathways and return annually to the same calving grounds; various species of ducks, geese, and swans migrate in flocks of mixed ages along traditional routes year after year; and some of the breeding grounds of colonial birds are known to have been used for centuries.[24]

Such patterns of group behaviour have an autonomy that depends on the group itself; they are not individually inherited. Young animals from one group raised within another of the same species generally adopt the patterns of the group within which they grow up. These habits or traditions develop through the participation of many individual members over many generations. They are forms of cultural inheritance.

There is general agreement among biologists that cultural inheritance cannot be explained genetically, but depends instead on another kind of transmission which takes place within the context of the group.

Perhaps the simplest forms of cultural inheritance can be explained in terms of individual imitative learning. For instance, in species of birds that learn their songs by listening to those of nearby adults, local dialects often develop. But in so far as the cultural traditions depend on the behaviour of the group as a whole, the tradition is transmitted by the group itself.

From the point of view of the hypothesis of formative causation, the forms of social behaviour are shaped by the morphic fields of the group. By morphic resonance these fields will be influenced by the behaviour under similar circumstances of all similar groups in the past, right back through the history of the species. But because a given group in general resembles itself in the past more closely than it resembles other groups, it will be most strongly and specifically influenced by morphic resonance from its own past patterns of activity. This self-resonance is the means by which the traditions of the group are transmitted. The morphic fields of the group contain a kind of group memory.

In this chapter, we have seen how the idea of social morphic fields enables animal societies to be conceived of as social morphic units, or social holons or superorganisms, and how this idea provides a way of understanding the co-ordination of the behaviour of individual organisms within the social unit: the colony, school, flock, herd, pack, group, or pair. The inheritance of these social fields takes place by morphic resonance from previous similar social units; and self-resonance from a group's own past stabilizes its particular characteristics and traditions. We now consider how these principles apply to human societies and cultures.

CHAPTER 14

The Fields of Human Societies and Cultures

Human societies and cultures have characteristic patterns. In the traditional societies of the past, social and cultural structures often remained quite stable over many generations, despite the fact that the individual people within them were continually changing. Even modern societies have distinct, enduring patterns: for example, the American way of life is characteristically different from the Polish or the Japanese. And within modern societies there are of course many distinct social, cultural, and religious entities: families, businesses, town councils, trade unions, police forces, factories, churches, string quartets, clubs, schools, political parties, and so on. All of these have their characteristic patterns of organization, written or unwritten rules, customs, and traditions.

The existence of patterns of social and cultural organization is recognized to varying degrees by everyone. We could not function as members of society without some knowledge of the prevailing manners, expectations, status hierarchies, and so on. In this chapter, we explore the idea that such patterns are organized by social and cultural morphic fields.

This approach involves more than the mere introduction of a new terminology, for at least two reasons. First, it enables us to see patterns of social and cultural organization in a much broader context than usual; for social and cultural morphic fields are of the same general nature as the morphogenetic fields of protein molecules or willow trees or chick embryos, the behavioural fields of spiders or blue tits, the social fields of termites or flocks of birds, and the mental fields involved in doing arithmetic or making plans. Human social and cultural patterns depend on formative causation, which is expressed through morphic fields in systems at all levels of complexity.

Second, the morphic fields of societies and cultures, like morphic fields of other kinds, are stabilized by morphic resonance from previous similar systems. This principle sheds new light on the *inheritance* of social and cultural patterns, which is still very poorly understood.

We begin with a discussion of the organization of human societies and cultures, and the ways in which this organization is interpreted in the conventional theories of the social sciences.

Human Societies as Organisms

Despite their great diversity, all human societies have certain fundamental features in common. All involve the incorporation of individuals into social groups; all have language; all have structures of kinship and social organization; all have myths and rituals which are in some way related to the origin of the social group and its continuation; all have customs, traditions, and manners; all impose upon the people within them a variety of expectations, obligations, rules, and laws; all have systems of morality; and all function as more or less cohesive, self-organizing wholes.

Moreover, all societies and social groups involve an awareness of the group as a unit. People not only belong to families, tribes, clans, communities, nations, teams, schools, regiments, colleges, companies, corporations, clubs, or associations, but they *know* that they are members of the group and have some conception of it as an entity. They are likewise aware of the existence of other such social entities to which they do not belong.

The idea that societies are wholes that are greater than the sum of their individual parts seems to be taken for granted almost universally. All of us have grown up with it. The parallel between societies and organisms is so pervasive that it is built into conventional phrases such as the *body* politic, the *arm* of the law, and the *head* of state. Economies too are thought of as if they are living organisms: they develop and grow, create demands, consume resources, can be healthy or sick, and so on. Political discourse is replete with phrases that take for granted the reality of collective entities such as parties, pressure groups, social classes, trade unions, companies, corporations, governing bodies. Such vaguely defined concepts as the will of the people, the national interest, spheres of influence, and the defence of the realm are not mere abstractions: they play a major role in shaping political actions and have come to have enormous effects on the world.

Organic views of society are traditional everywhere, and still predominate even in the West. The only major challenge to them has come from the philosophy of individualism, which began to play an important part in political philosophy in the seventeenth century. Its development paralleled the rise of mechanistic science and the atomistic philosophy of nature. Individualism represents an atomistic conception of society. The community is not a higher form of unity to which the individual is subordinate; rather the individual is the primary reality, and societies are collections of individuals. However, carried to its logical extreme, individualism leads to a doctrine of pure anarchy, and few people are inclined to go this far. In political thought individualism is usually taken to mean only that the state should not interfere more than necessary with individual liberty. This is the central tenet of the liberal political tradition and its modern right-wing outgrowths. The supremacy of the state in the maintenance of the law and order, in the imposition of taxes, in foreign relations, in the waging of war, and in many other ways is accepted more or less unquestioningly. In practice, collectivist ideologies such as those of socialism and individualist ideologies such as those of the New Right differ only in degree. All are fundamentally collectivist. They all recognize social wholes, such as political parties, judicial systems, armies, and nation-states, which are greater than the sum of their parts.

From the point of view of the hypothesis of formative causation, all such social entities are organized by morphic fields. As in the case of other organized systems at all levels of complexity, from molecules to ecosystems, these morphic fields are organized in nested hierarchies of fields within fields (Fig. 5.9).

Cultural Inheritance

Culture comes from the Latin root *colere,* to till or cultivate; in English the word still retains this primary meaning in the context of agriculture. Just as agriculture involves the imposition of a new order on the earth, which in its natural state is wild and uncultivated, so human culture is by implication not natural. It does not arise spontaneously in growing children; we are all inculturated or cultivated as we grow up. In this sense culture is opposed to nature.

But in another sense culture is natural; no human beings exist without culture, and cultures themselves can be compared to living organisms. They have forms, which are inherited and reproduced again and again; they are to various degrees self-organizing; and they change and evolve. The same ambiguity is inherent in agriculture itself; in one sense it is artificial, but in another sense the crops that are grown in the fields are natural: they have a life of their own; they develop in accordance with the natural rhythms of day and night, the seasons and the weather; and both crops themselves and systems of agriculture change and evolve.

There is near universal agreement that the inheritance of culture cannot be explained genetically.[1] It is quite evident that as babies grow up they learn the language of their natural or adoptive parents and assimilate the prevailing culture. Further, within a given society customs and traditions are passed on from generation to generation, and however this transmission occurs it can hardly be genetic. Even sociobiologists, the most extreme advocates of neo-Darwinism, do not claim that cultural forms are genetically programmed. E. O. Wilson, for example, confines the role of genetic evolution to the innate human capacity—indeed the overwhelming tendency—to develop one culture or another. "To the extent that the specific details of culture are non-genetic, they can be decoupled from the biological system and arrayed beside it as an auxiliary system."[2] Richard Dawkins has taken this approach further in proposing the concept of *memes,* which he defines as "units of cultural inheritance."[3] He compares them to selfish genes:

> Examples of memes are tunes, ideas, catch-phrases, clothes fashions, ways of making pots or of building arches. Just as genes propagate themselves in the gene pool by leaping from body to body via sperm and eggs, so memes propagate themselves in the meme pool by leaping from brain to brain via a process which, in the broad sense, can be called imitation. . . . As my colleague N. K. Humphrey neatly summed up, "memes should be regarded as living structures, not just metaphorically but technically."[4]

Dawkins appears to regard memes as atomistic units of cultural inheritance, just as he regards genes as atomistic units of biological inheritance; and this aspect of his proposal has been widely attacked by social scientists and anthropologists, most of whom think of cultures organismically, as wholes with coherent patterns of interconnection between their various elements. Nevertheless, the meme concept is helpful in focussing attention on the analogies between biological and cultural inheritance, and on the distinctness of the two processes as well.

Morphic fields have some of the characteristics that Dawkins attributes to memes: they are "living structures," propagated within societies by a process which in a broad sense can be called imitation. But cultural morphic fields are not atomic units of culture that can be shuffled around and permutated at random; like all other types of morphic fields, they are structured in nested hierarchies of fields.

The personal and mental life of every human being is shaped by culture, not least through languages and the cultural heritages that they embody: think, for example, of the differences between people brought up as Germans and as Italians. And every human society has structures and patterns that are inseparable from the cultural heritage of that society. On the present hypothesis, as children grow up they come under the influence of various social morphic fields, and tune in to many of the chreodes of the culture, the learning of which is facilitated by morphic resonance: for example American boys learn to play in baseball teams, and English boys in cricket teams. The social roles that people take up—the roles of schoolchildren, secretaries, goalkeepers, mothers, bosses, workers, and so on—are shaped by morphic fields stabilized by morphic resonance with those who have played these roles before. Likewise, the patterns of relationship among the various social roles— for example between workers and bosses—are shaped by the morphic fields of the social unit, maintained by resonance from the group's own past and from other more or less similar groups.

Theories of Social and Cultural Organization

In the nineteenth century the primary preoccupation of social theorists was social change and development. The century began in the wake of the French and American revolutions and as the industrial revolution in England was gathering momentum. Social changes were unmistakeable realities, and it was in this context that sociology began. Its founders, such as Saint-Simon and Auguste Comte, thought of society as a developing organism that could be understood in the positivistic spirit of science. And not only could society be understood in terms of sociological laws, but this knowledge could be used to control human behaviour, and in particular to help bring about the development of socialism. Against this background Karl Marx formulated his theory of social change through conflict between classes and attempted to discern the laws that developing societies would follow as they progressed towards the final communistic state, a classless society in which historical tensions disappeared. Since class conflict was the "motor of history," the achievement of this final state would be the "end of ordinary history."

Developmental theories of society were not the monopoly of the socialists and communists. Thoroughly capitalistic theories flourished as well, especially in Britain and America, particularly under the influence of Herbert Spencer. His primary interest was in social evolution, and he did much to popularize the evolutionary concept in general, leading rather than following Darwin in the use of the word *evolution* (p. 45). But although Spencer emphasized the idea of society as an organism, he interpreted it in a paradoxically individualist manner. Society is an organism "whose corporate life must be subservient to the parts instead of the lives of the parts being subservient to the corporate whole."[5]

Darwin and his followers emphasized the importance of competition between individual organisms in the struggle for survival. The principle of the survival of the fittest combined with an individualistic theory of society provided a seemingly scientific justification for capitalism and meant that inequalities in wealth, position, and power were inevitable. This principle was not, however, confined to *individuals* within a given society, but extended to entire social *groups*. The competition and conflict between these groups was assumed to have raised the evolutionary level of society in general. This idea spawned a variety of speculative theories of social evolution collectively known as social Darwinism.[6] Such theories had considerable political influence, and were commonly invoked in support of imperialism in general and the British Empire in particular. In the United States they provided a convenient explanation for the dominance of the "advanced" white races and their expansion into the territory of the "primitive" Red Indians; in Australia for taking over the land of the "backward" aborigines; and so on. The following account of social evolution from the 1911 edition of the *Encyclopaedia Britannica* summarizes the general principles:

> The first organized societies must have been developed, like any other advantage, under the sternest conditions of natural selection. In the flux and change of life the members of these groups of men which in favourable conditions first showed any tendency to social organization became possessed of a great advantage over their fellows, and these societies grew up simply because they possessed elements of strength which led to the disappearance before them of other groups of men with which they came into competition. Such societies continued to flourish, until they in turn had to give way before other associations of men of higher social efficiency. In the social process at this stage all the customs, habits, institutions and beliefs contributing to produce a higher organic efficiency of society would be naturally selected, developed and perpetuated.[7]

Different authors filled in the details of this general scheme as they saw fit; and here, as elsewhere, Darwinism lent itself to almost untrammelled speculation.

Functionalism and Structuralism

There was a widespread reaction against this kind of armchair theorizing in the first few decades of this century, and many sociologists and anthropologists emphasized the need for the empirical study of societies as they actually *are,* regardless of how they came to be that way. The most popular theoretical framework for these studies was called functionalism, and it remained predominant in various forms until about the 1960s. The primary metaphor was physiological: just as structures such as the heart, the liver, and the kidneys function in relation to the needs of the organism as a whole and to its maintenance in a more or less steady state, so also do social institutions and activities have functions that are related to the maintenance of the society as a whole as it exists in its environment.

Closely akin to functionalism is systems theory, which provided the dominant model in sociology in the 1950s and 1960s.[8] It emphasized the principles of interaction, feedback, and homeostasis, familiar on the one hand to physiologists and on the other hand to control engineers. Systems theory is much influenced by cybernetics, the theory of communication and control, and has been applied in the study of political processes, industrialization, complex organizations, and so on. It provides the basis for computer models, which have come to be widely used in commercial, governmental, and military organizations.

The structuralist school, which has grown up since the second world war, has much in common with functionalism and shares its assumption that societies are organic wholes. Rather than trying to explain all social and cultural structures in terms of their social functions, structuralists attempt to discern the unobservable structures that underlie observable phenomena such as myths, systems of kinship, classifications of animals and plants, and patterns of exchange of goods. In several respects, structuralism has superseded functionalism, which can be seen not so much as an opposing theory but as a "rudimentary version" of structuralism.[9]

The structuralist approach has been widely influential not only in anthropology and sociology, but also in linguistics, especially through the work of Noam Chomsky (pp. 184–85), in the study of art and literature, and as an approach to biological form.[10] The mathematical models of morphogenetic fields of René Thom (pp. 101–2), and Goodwin and Webster (pp. 67–69) were put forward in a general structuralist context.

But what *are* these underlying structures? They sometimes sound very like Platonic Ideas or Forms. Some structuralists do indeed seem to belong to the Platonic or idealist tradition; others deny that they are idealists and seek to reduce these structures to physico-chemical mechanisms. Lévi-Strauss, for example, takes the latter course, and refers the structures of culture and society to hypothetical mechanisms in the brain. His own intellectual development was much influenced during the 1940s by the pioneering work on cybernetics, computing, and information theory; he has proposed that the "algebra of the brain" can be represented as a rectangular matrix of at least two (but perhaps several) dimensions which can be read up and down or side to side like a crossword puzzle.[11] The binary oppositions, represented by + and −, are comparable to the binary codes on which computers work. And so we come back once again to the computer metaphor for the human mind.

Both structuralism and functionalism face a grave difficulty in so far as they imply that societies are harmoniously integrated organisms whose institutions serve to maintain a more or less steady state. Many societies and social institutions are far from harmonious and are not in equilibrium: they change. Think for example of the changes in Russia, in Brazil, in Kenya, or indeed anywhere else over the last century. Neither functionalism nor structuralism seems to offer an adequate explanation of such changes; this is perhaps their greatest weakness, and a major reason for a decline in their influence. Explanations of social change in terms of conflict, competition, opposition, and strain seem more plausible than functionalist theories of a social steady state or structuralist theories of changeless patterns in the human mind; and these are what Marxists and Darwinians have been emphasizing in one way or another all along. Meanwhile the empirical study of social change, for example in response to urbanization or rural development, has become the focus of many contemporary sociologists and anthropologists.

An interpretation of social and cultural patterns in terms of morphic fields provides a way of retaining the important insights of functionalism and structuralism while at the same time moving beyond the Platonist-reductionist dualism which they have so far failed to supersede. Functionalism stresses the functional interrelations among the parts of a society, and structuralism the underlying patterns or structures. Both seem eminently compatible with the idea of morphic fields.[12] Such fields structure human language, thought, customs, culture, and society and they organize the interrelations of the component parts. They are stabilized by self-resonance from a society's own past and by morphic resonance from previous similar societies. Since morphic fields are probability structures, social and cultural regularities would be expected to be statistical in nature rather than precisely determinate, and this is indeed in accordance with the facts.

But what about social and cultural change? Morphic fields in general have a stabilizing and conservative effect; they cannot in themselves account for the initiation of change. Such changes no doubt depend on a variety of factors, including contact or conflict between different societies, classes, or cultural systems; on changes in the environment; on the development of new technologies; and so on. Here, as elsewhere, the origin of new fields depends on circumstances and on creative processes that cannot be explained in terms of repetition (see chapter 18). But once new patterns of activity have arisen, the spread and adoption of these innovations may well be facilitated by morphic resonance. And often-repeated patterns of social change—in the process of urbanization, for example—may be shaped by chreodes and stabilized by morphic resonance.

Group Minds

Intangible social influences are a matter of common experience. There are many phrases in everyday language that refer to them: the *power* of tradition, social *pressure*, the *force* of conformity, and so on. All of us have experienced the feelings of shame that are associated with social disapproval and the positive feelings engendered by social approval; and we are familiar with the invisible influences referred to by terms such as social solidarity, loyalty, morale, and team spirit.

Emile Durkheim thought of such organizing influences as aspects of the *conscience collective*. The French word *conscience* embraces the meanings of both *consciousness* and *conscience* in English. He defined this as "the set of beliefs and sentiments common to the average members of a single society which form a determinate system that has a life of its own." It has its "own distinctive properties, conditions of existence, and mode of development." It transcends the lives of individuals: "they pass on and it remains."[13]

Sigmund Freud, too, was driven to the conclusion that some such concept was necessary:

> I have taken as the basis of my whole position the existence of a collective mind, in which mental processes occur just as they do in the mind of the individual. . . . Without the assumption of a collective mind, which makes it possible to neglect the interruptions of mental acts caused by the extinction of the individual, social psychology in general cannot exist. Unless psychical processes were continued from one generation to another, if each generation were obliged to acquire its attitude to life anew, there would be no progress in this field and next to no development. This gives rise to two further questions: how much can we attribute to psychic continuity in the sequence of generations? And what are the ways and means employed by one generation in order to hand on its mental states to the next one? I shall not pretend that these problems are sufficiently explained or that direct communication and tradition—which are the first things to occur to one—are enough to account for the process.[14]

Freud concluded that an important part of this collective mental inheritance was transmitted unconsciously.

William McDougall (who carried out the experiments on the inheritance of learned behaviour in rats described in chapter 9) was an influential social psychologist who likewise came to the conclusion that societies have an autonomy that is best conceived of in terms of a group mind:

A society, when it enjoys a long life and becomes highly organized, acquires a structure and qualities which are largely independent of the qualities of the individuals who enter into its composition and take part for a brief time in its life. It becomes an organized system of forces which has a life of its own, tendencies of its own, a power of moulding all its component individuals, and a power of perpetuating itself as a self-identical system, subject only to slow and gradual change. . . . We may fairly define a mind as an organized system of mental or purposive forces; and, in the sense so defined, very highly organized human society may properly be said to possess a collective mind. For the collective actions which constitute the history of any such society are conditioned by an organization which can only be described in terms of the mind, and which yet is not comprised within the mind of any individual; the society is rather constituted by the system of relations obtaining between the individual minds which are its units of composition.[15]

Ideas such as these had a widespread influence in the first few decades of this century, but have become rather unrespectable among intellectuals in more recent years. This is partly because of the increasingly reductionist climate of the academic world, and perhaps also because of the frightening manifestations of the collective psyche in Nazi Germany and in other nationalist movements. The idea of invisible organizing principles over and above the individuals within a society has, of course, remained, but they are referred to by more neutral terms such as patterns of relationship,[16] social structures, and social consensus. These are, however, just as elusive as the group mind, and raise the same kinds of problems: attempts to reduce them to mechanisms within the brains of individual people seem inadequate and unconvincing, while interpretations in terms of changeless Platonic Forms seem incompatible with changing historical reality. The hypothesis of formative causation enables these structures, patterns, and consensuses to be embraced within the idea of morphic fields, together with the notions of the group mind and the *conscience collective*.

Collective Behaviour

Collective behaviour is a term used by sociologists to refer to "the ways in which people behave together in crowds, panics, fads, fashions, crazes, cults, followings, reform and revolutionary social movements, and other similar groupings."[17] It has been defined as "the behaviour of individuals under the influence of an impulse that is common and collective, in other words, that is the result of social interaction."[18] Many studies have been made of the spread of rumours, jokes, fads and crazes, hysterical contagion, the behaviour of rioting mobs, and so on; but there are no generally agreed theories that can account for these phenomena.[19]

As we have seen, the behaviour of schools, flocks, herds, and packs of social animals suggests the idea that fields embrace all the individuals within them (pp. 231–37). The idea of such fields of influence may also shed much light on human collective behaviour. Crowds, for example, have often been compared to composite organisms, with their own laws and properties. One useful classification of crowds by Elias Canetti distinguishes various types with quite distinct properties, which from the present point of view can be taken to represent different types of crowd field. One basic type is the open crowd:

> The crowd, suddenly there where there was nothing before, is a mysterious and universal phenomenon. . . . As soon as it exists at all, it wants to consist of *more* people: the urge to grow is the first and supreme attribute of the crowd. . . . The natural crowd is the *open* crowd; there are no limits whatever to its growth; it does not recognize houses, doors or locks and those who shut themselves in are suspect. . . . The open crowd exists so long as it grows; it disintegrates as soon as it stops growing.[20]

Canetti contrasts this extreme type of the spontaneous crowd with the closed crowd:

> The closed crowd renounces growth and puts the stress on permanence. The first thing to be noticed about it is that it has a boundary. . . . The boundary prevents disorderly increase, but it also makes it more difficult for the crowd to disperse and so postpones its dissolution. In this way the crowd sacrifices its chance of growth, but gains in staying power. It is protected from outside influences which would become hostile and dangerous and it sets its hope on *repetition*.[21]

But within crowds of both basic types there is equality: "It is for the sake of equality that people become a crowd and they tend to overlook anything which might detract from it." Moreover, the crowd has a goal or direction. "A goal outside the individual members and common to all of them drives underground all the private differing goals which are fatal to the crowd as such. Direction is essential for the continuing existence of the crowd. . . . A crowd exists so long as it has an unattained goal."[22]

Crowds are temporary, and precisely for this reason can reveal to us some of the features of collective social organization that are so easily taken for granted in more permanent groups. *Teams* are another kind of temporary group of which most of us have had direct experience. Here too, although a team is more structured and disciplined than a crowd, the individual is subordinate to the collective behaviour directed towards a common goal—in many games quite literally the scoring of goals.

> When a collection of individuals first jells as a team, truly begins to react as a five-headed or eleven-headed unit rather than as an aggregate of five or eleven individuals, you can almost hear the click: a new kind of reality comes into existence. . . . A basketball team, for example, can click in and out of this reality many times during the same game; and each player, as well as the coach and fans, can detect the difference. . . . For those who have participated in a team that has known the click of communality, the experience is unforgettable.[23]

When successful sportsmen are questioned about their experiences as members of teams, some speak of a "sixth sense" which enables them to be in the right place at the right time; others speak of empathy and intuition. In general, "an incredible power of communication often develops between members of a team where one can anticipate the moves of the other."[24]

Such phenomena can be, and often are, interpreted in terms of group minds. An interpretation in terms of morphic fields provides an alternative that incorporates the group mind concept and in addition provides a natural explanation for the building up of group habits by morphic resonance from the group itself in the past and from other groups that resembled it. Think, for example, of the Manchester United football team, or the Boston Symphony Orchestra, or the local Methodist church: each has its own characteristic traditions and ethos and, at the same time, generic resemblances to other football teams, orchestras, or Methodist churches.

The Collective Unconscious

The collective unconscious of Carl Jung has much in common with the concept of the group mind, and what Jung called the archetypes resemble what Durkheim called "collective representations."[25] Jung wrote as follows:

> The collective unconscious is a part of the psyche which can be negatively distinguished from a personal unconscious by the fact that it does not, like the latter, owe its existence to personal experience and consequently is not a personal acquisition. While the personal unconscious is made up eventually of contents which have at one time been con-

scious but which have disappeared from consciousness through having been forgotten or repressed, the contents of the collective unconscious have never been in consciousness, and therefore have never been individually acquired, but owe their existence exclusively to heredity. Whereas the personal unconscious consists for the most part of complexes, the content of the collective unconscious is made up essentially of archetypes.[26]

One of the reasons why Jung adopted this idea was that he found recurrent patterns in dreams and myths which suggested the existence of unconscious archetypes, which he interpreted as a kind of inherited collective memory. He was unable to explain how such inheritance could occur, and his idea is clearly incompatible with the conventional mechanistic assumption that heredity depends on information coded in DNA molecules. Even if it were to be assumed that the myths of, say, a Yoruba tribe could somehow become coded in their genes and their archetypal structure be inherited by subsequent members of the tribe, this would not explain how a Swiss person could have a dream that seemed to arise from the same archetype. Jung's idea of the collective unconscious simply does not make sense in the context of the mechanistic theory of life; consequently it is not taken seriously within the current scientific orthodoxy. However, it makes very good sense in the light of the hypothesis of formative causation.

By morphic resonance, structures of thought and experience that were common to many people in the past contribute to morphic fields. These fields contain as it were the average forms of previous experience defined in terms of probability. This idea corresponds to Jung's conception of archetypes as "innate psychic structures."

> There is no human experience, nor would experience be possible at all, without the intervention of a subjective aptitude. What is this subjective aptitude? Ultimately it consists in an innate psychic structure. . . . Thus the whole nature of man presupposes woman both physically and spiritually. His system is tuned in to prepare for a quite definite world where there is water, light, air, salt, carbohydrates, etc. The form of the world into which he is born is already in born in him as a virtual image. Likewise parents, wife, children, birth, and death are inborn in him as virtual images, as psychic aptitudes. These *a priori* categories have by nature a collective character; they are images of parents, wife and children in general. . . . They are in a sense the deposits of all our ancestral experiences.[27]

Although Jung thought the collective unconscious was common to all humanity, he did not regard it as entirely undifferentiated. "No doubt, on an earlier and deeper level of psychic development . . . all human races had a common collective psyche. But with the beginning of racial differentiation essential differences are developed in the collective psyche as well."[28]

Marie-Louise von Franz has taken this idea further (Fig. 14.1). Below the level of the personal unconscious lies a "group unconscious" of families, clans, tribes, on so on. Below this is a "common unconscious" of wide national units. "We can see, for instance, that Australian or South American Indian mythologies form such a wider 'family' of relatively similar religious motifs which, however, they do not share with all of mankind." Below this

Figure 14.1 Diagram showing the structure of the collective unconscious as interpreted by von Franz. A, ego consciousness; B, personal unconscious; C, group unconscious; D, unconscious of large national units; E, unconscious common to all humanity, containing universal archetypal structures. (After von Franz, 1985)

lies "the sum of those universal psychic archetypal structures that we share with the whole of mankind."[29]

Such a conception is in general agreement with the idea of morphic resonance, the specificity of which depends on similarity: members of particular social groups are in general more similar to past members of the same groups than to social groups of entirely different races and cultures; but underlying all human groups are certain general similarities through which all participate in a common human heritage.

CHAPTER 15
Myths, Rituals, and the Influence of Tradition

According to the hypothesis of formative causation, new social and cultural morphic fields arise in the course of human history, then through repetition become increasingly habitual. They organize particular social and cultural patterns.

The structuralist approach, as we saw in the preceding chapter, involves an attempt to discern these underlying social and cultural patterns, and it has much in common with an interpretation in terms of morphic fields. However, structuralists have not so far been able to escape the Platonist-reductionist duality inherent in the mechanistic world view. Some structuralists treat these patterns as if they are Platonic Forms transcending time and space, and hence are incapable of evolutionary change; others, like Lévi-Strauss, attempt to reduce them to hypothetical mechanisms in human brains (p. 245).

This reductionist approach involves an attempt to bridge the gulf between the "soft" sciences of sociology and anthropology and the "hard" sciences of biology and chemistry, in which the mechanistic paradigm is still predominant. However, the mechanistic approach to the functioning of the brain is much softer than it might at first appear. Very little is actually known about the organization of physico-chemical processes in the brain or about the nature of memory (chapters 9–12). Moreover, mechanistic speculations about the "programming" of the brain involve dualistic metaphors in which the "hardware" of the brain is organized by "software" whose physical nature remains entirely obscure (chapter 12).

An interpretation of social and cultural structures in terms of morphic fields provides a different way of bridging the gulf between the "soft" and the "hard" sciences. Social and cultural fields are of a nature similar to the morphic fields that organize biological and chemical systems, although they are not, of course, *reducible* to these biological and chemical fields. Like the morphic fields of systems at all levels of complexity, social and cultural fields are stabilized by morphic resonance from similar systems in the past, including self-resonance from the systems' own past. Thus the idea of formative causation transcends the dilemma of conventional structuralism, with its Platonic or reductionist alternatives; and it offers a potentially more fruitful approach to the understanding of cultural inheritance and the evolution of cultural habits.

In this chapter, we consider the nature of myths, rituals, traditions, and initiations in the context of morphic resonance. The interpretation of these phenomena in terms of formative causation does not so much contradict conventional structuralist interpretations as go beyond them. It could even be regarded as a kind of evolutionary structuralism.

Myths and Origins

Myths are stories of origins. They concern the doings of gods, heroes, and superhuman beings and account for the way things are as they are. "They are both explanations and examples: examples in the sense that they are repeatable, and serve as models and justifications for all human actions."[1] In traditional societies there is no sense of a progressive development: what happens now repeats what happened before, and this repetition always refers to the first time it happened, in the mythic time of origins. This time was in the past, but it is also in some sense present now, because the original patterns are continually repeated.

The following description is by an anthropologist who spent much of his life among the Northern Aranda aborigines of Australia.

> The *gurra* ancestor hunts, kills, and eats bandicoots; and his sons are always engaged upon the same quest. The witchetty grub men of Lukara spend every day of their lives in digging up grubs from the roots of acacia trees. . . . The *ragia* (wild plum tree) ancestor lives on the ragia berries which he is continually collecting into a large wooden vessel. The crayfish ancestor is always building fresh weirs across the course of the moving flood of water which he is pursuing; and he is forever engaged in spearing fish. If the myths gathered in the Northern Aranda area are treated collectively, a full and very detailed account will be found of all the occupations which are still practised in Central Australia. In his myths we see the native at his daily task of hunting,

fishing, gathering vegetable food, cooking, and fashioning his imple-
ments. All occupations originated with the totemic ancestors; and here
too the native follows tradition blindly: he clings to the primitive
weapons used by his forefathers, and no thought of improving them
ever enters his mind.[2]

This fidelity to the past conceived of as a timeless model is alien to our
modern way of thinking. We see the past in terms of stages in a progressive
historical process. But in traditional societies all over the world the mythic
attitude prevailed. Every technique, rule, and custom was justified by the
simple argument that "the ancestors taught it to us." In the words of Lévi-
Strauss:

> Mythical history thus presents the paradox of being both disjointed
> from and conjoined with the present. It is disjointed from it because
> the original ancestors were of a nature different from contemporary
> men: they were creators and these are imitators. It is conjoined with
> it because nothing has been going on since the appearance of the
> ancestors except events whose recurrence periodically effaces their par-
> ticularity.[3]

This sounds very like a description of morphic resonance, through
which patterns of activity are repeated again and again, stabilized by this
resonance from all similar past patterns, right back to the time each morphic
field first came into being.

A common modern attitude is to regard the myths of traditional
societies as fanciful stories that are not only untrue but prevent progress. By
contrast with the myths of primitive peoples, it is often assumed that modern
scientific accounts of the origin of the universe, the evolution of life, and the
development of civilization are objective and true. But this attitude is sim-
plistic. The disciplines of science and of history are themselves influenced by
the prevailing culture and shaped by the dominant paradigms. They involve
implicit assumptions that are often deeply habitual. Scientific theories are like
myths in that they are mental constructs, ways of making sense of the world;
they are also like myths in that they have a cultural dimension. Scientific
paradigms are shared by members of scientific communities, and indeed play
a major role in defining the activities of these communities. And in the light
of formative causation, both myths and scientific paradigms are shaped by
morphic fields and maintained by morphic resonance. We return to a discussion of paradigms towards the end of this chapter.

Scientific theories themselves have origins, and they are often associated
with stories that sound very like myths. Thus, for example, according to

Descartes himself his philosophy was inspired by an encounter with the Angel of Truth in a dream; and Newton's theory of gravitation, the grandest theory of classical physics, is said by popular legend to have come to him under an apple tree when a fruit of the tree fell on his head. There are few great innovators whose life stories are devoid of legendary features; and some, like Einstein, are widely seen as being endowed with the spirit of genius; others, like Marx, Darwin, and Freud, are often compared to Old Testament prophets.

The main difference between modern theories of progress and traditional myths is that the theories of progress do not so much refer back to prototypic models in the past, but refer forward to future goals, often envisioned as states of peace, prosperity, brotherhood, and wisdom. But these notions of progress have developed within a culture shaped by the Judaeo-Christian tradition, and the most distinctive feature of this tradition is its myth of history: the idea of historical progress towards an end that in some sense recreates the primal paradise before the Fall. This model of history is itself a morphic field, strongly stabilized by morphic resonance. Western civilization has developed and is still developing within this field; Western science has grown up within it (chapter 3). To what extent are modern scientific theories of the origin of the universe and of evolution new versions of this traditional model of history?

Superficially, there may seem to be no connection between this mythic view of history and the development of science and technology. Science and the accompanying growth of rational understanding are, after all, commonly supposed to have liberated Modern Man from the archaic systems of belief perpetuated by religion. From this point of view, science is altogether different from primitive mythical thought: through a heroic struggle against the forces of priestly prejudice, great men such as Galileo and Darwin have led humanity out of the darkness of superstition into the light of rational knowledge. But this familiar story sounds very like a myth itself.

The growth of scientific knowledge is often assumed to have revealed convincingly that all traditional myths are false: they are made-up stories which at best have a certain poetic value. In particular, the biblical story of the creation in the first chapter of the book of Genesis cannot be taken seriously in the light of modern theories of cosmology and evolution. According to this story, in the beginning "the earth was without form, and void, and darkness was upon the face of the deep." God first of all created light; then the firmament of heaven; then the earth and the seas; then plants; then the sun and the moon; then the creatures of the sea and the birds of the air; then reptiles and mammals on land; and finally the first man and woman.

According to modern science, what first appeared in the Big Bang was light. Then as the universe grew the galaxies and stars were formed; then the solar system came into being; then as the earth cooled, the seas and the dry land were formed; then life arose in the primeval broth; then plants began to evolve; then animals, first in water and then on dry land; then birds and mammals evolved from reptiles; and finally *Homo sapiens* arose from apelike ancestors.

This sequence differs from the ancient story in the book of Genesis in several respects, perhaps the most notable being the creation of the sun and moon after the earth and the vegetation upon it. In the scientific account, of course, the sun is supposed to have been formed before this planet, or at least around the same time. Opinions still differ as to the origin of the moon: some astronomers hold that it came into being together with the earth and the planets; others maintain that it originated later and may even have split off from the earth.[4] Another difference is that Genesis places the origin of birds before the origin of reptiles, whereas evolutionary theory derives birds from the reptiles.

Nevertheless, the broad outlines of the Genesis myth and the contemporary scientific account are not dissimilar; they have a strong family resemblance. The scientific account is of course far more detailed, and attributes creativity to chance rather than to God. But both, by their very nature as accounts of origins, refer to events that happened before there were people to witness them and can therefore only be imagined, calculated, inferred, or modelled. They can never be statements of observable or observed facts.

The creation theories of science have grown up within the Judaeo-Christian cultural matrix, with its paradigmatic conception of a beginning, a Fall, a historical progress towards the end of history, and an end that in some sense re-establishes the beginning. The theory of the Big Bang and the modern doctrine of universal evolution bear a striking resemblance to this fundamental myth of our culture. (And it may not be a mere coincidence that since we now live in fear of ending our civilization in a cataclysmic nuclear war, we have come to have a creation story that begins with a vast explosion.)

Rituals

A sociological definition of rituals is that they are "formal actions following a set pattern which expresses through symbols a public or shared meaning." Symbols are "any gesture, artifact, sign, or concept which stands for, signifies, or expresses something else."[5]

All cultures have rituals: the Jewish feast of the Passover, the Christian mass, and marriage services are familiar religious examples; the annual presentation of the national budget, the state opening of parliament in Britain, and the inauguration of the President of the United States are political rituals; national customs such as Guy Fawkes night in England or the Thanksgiving dinner in America have a ritual quality; and everyday life contains many more or less unconscious ritual elements, such as the conventional forms of greeting and saying goodbye. The word *goodbye* itself, for example, is a form of blessing whose usage persists even though its original meaning, "God be with you," is largely forgotten.

Many rituals are associated with stories of origins that tell of the original act that the ritual commemorates: the original passover in Egypt on that dreadful night when the first-born sons of the Egyptians were slain; the Last Supper of Jesus with his disciples on the eve of his sacrifice on the cross; the foiling of the gunpowder plot of Guy Fawkes; the thanksgiving dinner of the Pilgrim Fathers after their first harvest in the New World. Other rituals, such as those of birth, marriage, and death, concern the passage from one state of being to another. But all, through their very repetition, in some sense connect the present with the past. We have a need or appetite for the past which rituals help to satisfy.

The relationship of rituals to time has been summarized by Lévi-Strauss as follows: "Thanks to ritual, the 'disjointed' past of myth is expressed, on the one hand, through biological and seasonal periodicity and, on the other, through the 'conjoined' past, which unites from generation to generation the living and the dead."[6]

One example that he has used to illustrate these principles concerns the rituals of certain Australian tribes, which fall into three categories: rites of control, historical rites, and mourning rites. The rites of control are concerned with the regulation of natural and spiritual phenomena by fixing the quantity of spirit or spirit-substance allowed to emanate from places established by the ancestors within the tribal territory. (The annual presentation of the national budget could be seen as a secular reflection of a rite of control.)

> The commemorative or *historical rites* recreate the sacred and beneficial atmosphere of mythical times—the "dream age," as the Australians call it—mirroring its protagonists and their great deeds. The *mourning rites* correspond to an inverse procedure: instead of charging living men with the personification of their remote ancestors, these rites assure the conversion of men who are no longer living men into ancestors. It can thus be seen that the function of the system of ritual is to overcome and integrate three oppositions: that of diachrony and synchrony [i.e., of change across time and of simultaneity]; that of the periodic and

non-periodic features which either may exhibit; and, finally within diachrony, that of reversible and irreversible time, for, although present and past are theoretically distinct, the historical rites bring the past into the present and the rites of mourning the present into the past, and the two processes are not equivalent; mythical heroes can truly be said to return, for their only reality lies in their personification; but human beings die for good.[7]

The effectiveness of rituals is believed in all cultures to depend on their conformity to the patterns handed down by the ancestors: by their very nature ritual forms are highly conservative. The gestures and actions should be done in the correct way; and ritual forms of language are conserved even when the language is no longer in everyday use: thus the liturgy of the Coptic church is in the otherwise extinct language of ancient Egypt; until recently the Roman liturgy was in Latin; the brahminic rituals of India are in Sanskrit; and so on. But why is the effectiveness of rituals so universally believed to depend on their close similarity to the way they have been done before? Why should this similarity of ritual forms in the present to those in the past be regarded as essential to establishing a connection with the ancestors?

The idea of morphic resonance suggests a natural answer. Through morphic resonance, rituals really can bring the past into the present. The greater the similarity between the way the ritual is done now and the way it was done before, the stronger the resonant connection between the past and present performers of the ritual.

Initiations

Culturally transmitted specific forms of behaviour, language, and thought do not by definition arise spontaneously in human beings as they develop. They are acquired by processes of imitation. All involve *initiation* in a broad sense of the word. This is true of the learning of languages, songs, dances, social customs and manners, physical and mental skills, crafts, professions, and so on. Much imitative learning, usually from parents, elders, and teachers, is picked up informally, and is taken for granted. According to the hypothesis of formative causation all such learning is facilitated by morphic resonance, both from those who are directly imitated and from all those who have done the same things before (chapter 10). We now consider social and religious initiations, which take place through rituals that both mark and bring about the transition of a person from one social role to another or from one status to another.

Initiation rituals are concerned with the crossing of boundaries such as those between boyhood and manhood or the unmarried and the married state. From an anthropological point of view, they can be seen to lie in the broader category of rites of passage, which includes the rituals associated with birth and death and also with the crossing of boundaries in space and time, for example from one territory to another or from one year to another.

Typically, rites of passage have three phases. In the first, the initial state is stripped away: the state of childhood in rites of maturity, the responsibilities of life in many funeral customs, and so on. These are rites of separation: the individual is separated from his or her initial state and left in transition. This threshold state is characterized by danger and ambiguity, symbolized for example by being blindfolded or removed into the bush or forest far away from normal life or by having to undergo various unpleasant trials. A further ritual of integration ends this phase and emphasizes the individual's integration into his or her new state. Such rituals show many similarities across cultures: washing, head-shaving, circumcision and other bodily mutilations, the crossing of streams and other obstacles all indicate separation; while anointing, eating, and dressing in new clothes are common integrative actions.[8]

Initiation rituals serve to effect the passage of individuals across social or religious boundaries, and at the same time they serve to define these boundaries and make them manifest. Thus, for example, the Gisu of Uganda say that they initiate boys to make them into men in order that they shall not remain uninitiated boys. There is an inherent circularity in this procedure in the sense that the initiation serves to define the very categories that it presupposes. Such rituals are not simply a way of marking biological maturity, since they are carried out with boys at different stages of maturity; rather, they are concerned with the crossing of boundaries that are culturally defined.

Various traditional types of initiation survive in modern societies, in marriage rituals for example. And many other kinds of ceremony reflect some of the features of initiation rites: the passing of tests and the awarding of certificates in schools; the gaining of university degrees and the graduation ceremonies at which they are conferred; inductions into professional bodies; the commissioning of army officers; and so on.

A frequent theme in many initiation rites is the death of a person's previous social or religious identity and the birth of the new. The person is "born again" in a new religious or social role.

Social roles are associated with norms, which involve both patterns of expectation and often-repeated and commonly occurring patterns of behaviour. The socially expected norm does not necessarily correspond to actual behaviour or simply reflect the most frequent pattern; nevertheless, a relationship between the norm and the behaviour of a person in this social role is maintained by sanctions against deviation from the norm. In the language of sociology, norms are acquired by socialization and by internalization. The latter concept expresses "the process by which an individual learns and accepts as binding the social values and norms of conduct relevant to his or her social group or wider society."[9] The ritual initiations of individuals mark transitions into roles that are already established and governed by such norms; the person takes on a role with its associated norms and is shaped by it.

This taking on of a new role, often symbolized by putting on new clothes, can be thought of in terms of entering into a new morphic field, which on the one hand is socially recognized in terms of a norm and on the other hand comes to structure the behaviour and mental activity of the individual. It is in this sense that it is internalized. The patterns of behaviour of individuals within social roles, and also the patterns of social expectation, are stabilized and maintained by morphic resonance from previous members of the society.

Similarly, in religious initiations, such as confirmation and ordination within the Christian church, a person enters a new way of being, a new norm. His or her development within this morphic field can be thought of as following a chreode, a canalized pathway of change (Fig. 6.2), and indeed in many traditions the metaphor of a Path or Way is commonly employed. The prototypic Paths are often regarded as having been established by the founders of the tradition, for example the Buddha, Jesus, and Mohammed, and the initiate begins to follow this Path, which has already been traversed by many people before. For example, Jesus Christ proclaimed that "I am the Way"; and in the Christian tradition it is believed that Christ is in some sense present in the lives of those who follow him, and that they are helped to follow his path by all those who have followed it before them, referred to in the Apostles' Creed as the "communion of saints."

The influence of past followers of a path is explicitly recognized in many—perhaps all—religious traditions. On the present hypothesis the initiate is tuned in by morphic resonance to those who have followed this chreode before. One example of this principle is provided by the use of mantras in various oriental traditions. These are sacred words or phrases which are transmitted from guru to disciple during initiation rites and in the course of spiritual training:

> The mantra has power and meaning only for the initiated, i.e., for one who has gone through a particular kind of experience connected with the mantra. . . . However, this experience can only be acquired under

guidance of a competent Guru (being the embodiment of a living tradition) and by constant practice. If after such preparation the mantra is used, all the necessary associations and the accumulated forces of previous experiences are aroused in the initiate and produce the atmosphere and power for which the mantra was intended.[10]

Traditions, Schools, Styles, and Influences

Histories of religions, the arts, ideas, and cultural movements abound in concepts such as heritage, tradition, and influence. Within broad categories such as Islam, Christianity, the West, the East, the Middle Ages, the Renaissance, Classicism, and Romanticism, historians describe the appearance and development of schools, sects, styles, movements, and so on, and trace patterns of interconnection and influence between them.

This is too vast an area to be discussed in any detail in this book, but it is worth observing in general terms that an interpretation in terms of nested hierarchies of morphic fields might make good sense of many of these phenomena, and the idea of morphic resonance might point towards a new understanding of heritages, traditions, and influences.

Religions can be grouped in families, such as Judaism, Christianity, and Islam, the religions of the Book, as Moslems call them; the family of religions of Indian origin, including Hinduism, Buddhism, and Jainism; the families of religions of Australian aborigines, North American Indians; and so forth. The religions in each family share certain fundamental beliefs and attitudes, and could be considered to participate in a broad morphic field, within which are the fields of specific religions such as Islam. Within these fields are schools or sects, which may be further subdivided into orders and denominations, each with its own traditions, beliefs, and practices associated with its characteristic morphic fields. There are fields within fields within fields. In Christianity, for example, the entire Church is conceived of as an organic whole, the mystical body of Christ, "which is the blessed company of all faithful people."[11] Within this are the Eastern Orthodox, Roman Catholic, and Protestant traditions; and each of these is further differentiated, as for example in the Franciscan and Jesuit orders in the Roman Catholic Church. Characteristically all these orders and subdivisions, while acknowledging the church as a whole, have their own stories of origin centred around their founders— for example St. Francis of Assisi and St. Ignatius Loyola—and so do the various Protestant sects founded by Luther, Calvin, John Wesley, and others. Each order, sect, and denomination has its own traditions and its own ways of initiating and incorporating new members. And then each individual religious community and local church has its own collective life and local traditions. As new members grow up within them or are converted to them, they enter more or less fully into the spirit of the tradition. From the present point of view, they tune in by morphic resonance to the fields and chreodes of the tradition.

A similar pattern is apparent in broad cultural movements such as the Renaissance and the various schools of art and thought that developed within them. For example, schools of painting, including the Florentine, Venetian, and Flemish, were characterized by styles, artistic forms, and atmospheres and had a spirit which enables their productions to be recognized by anyone with sufficient experience. The same can be said of schools of architecture, sculpture, literature, and music. Here is an example from the history of music:

> The French violin school, born in the first years of the eighteenth century, originated from the spirit created by Corelli's sonatas. The French musicians received these works with enthusiasm, but they had already evolved an instrumental style strong enough to receive the new style as an incentive rather than an overwhelming influence. In imitation of the Italian example French composers took to the writing of sonatas, but at first they remained faithful to the spirit of the suite, their early sonatas being rather loosely connected dance pieces with an occasional aria thrown in. . . . The unifying feature of all these forms was the ground design of two animated movements separated by a quiet and restful movement of convincing aesthetic effect.[12]

Examples of this type could be multiplied indefinitely: there are dozens of them in almost every book on the history of the arts; and many comparable examples can be found in the history of ideas.[13] Schools of art and of thought are made up of people who have been incorporated into them, often through a process of apprenticeship or training, and who enter into their spirit.[14] The influences of different schools on each other involve an *in-fluence*—literally, a flowing in—of forms, styles, and spirit. Such transfers between traditions as well as transmission within a tradition can be thought of in terms of morphic resonance.

This hypothesis also suggests that styles and forms of art represent morphic fields which are expressed in the individual paintings, sonnets, sonatas, et cetera. Just as the morphic fields of an animal species are expressed in individual animals, and just as these individuals contribute cumulatively to the morphic fields of the species, so the individual works of art produced within a given school have a cumulative influence on the morphic fields of the school. These fields, like the social and cultural fields we have already considered, work through the behavioural and mental fields of individual members, and are in turn influenced by these individuals' thoughts and actions; nevertheless, they are fields at a supra-individual level and have a life or spirit or atmosphere of their own.

The notion of morphic resonance helps us to understand the maintenance of forms and styles, the continuity of traditions, and the transmission of influence; but once again it cannot account for creativity, for the origin of new fields.

The Fields of Science

The natural sciences all acknowledge certain common principles and recognize great founding fathers such as Galileo, Descartes, and Newton. They are divided into several broad fields, including physics, chemistry, geology, and biology. These in turn have developed under the influence of great historical figures, for example Darwin in the case of biology. They are subdivided into fields such as organic chemistry and botany, and these in turn embrace a range of more specialized disciplines: botany, for instance, includes taxonomy, mycology, plant anatomy, plant physiology, plant pathology, plant genetics, and so on. These in turn are subdivided into specialized subdisciplines: crop physiology, for example, is just one branch of plant physiology. Each of these disciplines and subdisciplines has its own history and its own great men, whose portraits often look down upon those who are trained and work within it. Each discipline has its own textbooks, journals, newsletters, professional societies, and conferences. Science is practised by professional communities, which regulate themselves and train those who enter into them. The members of these communities share interests and attitudes, and recognize others within the same field on the basis of their shared training and experience.

In the present context, the fields of science can indeed be seen as fields—morphic fields. On the one hand these embrace the members of the professional community and are social fields that co-ordinate and maintain the solidarity and cohesion of the group: they are a kind of *conscience collective*. On the other hand they order the way in which the subject matter is perceived and categorized, the ways in which problems are tackled, and in general provide the framework for thought and practice within the discipline.

Such morphic fields correspond closely to what Thomas Kuhn has called paradigms. "A paradigm is what the members of a scientific community share, *and,* conversely, a scientific community consists of men who share a paradigm."[15] Kuhn argues that normal science is a cumulative and progressive activity that consists of solving puzzles within the context of a shared paradigm; but that scientific revolutions, which are extraordinary and relatively infrequent, involve the establishment of a new paradigm or framework. Typically this does not, at least at first, make sense to practitioners brought up within the old paradigm; a period of controversy ensues, which ends only when existing professionals have either been converted to the new paradigm or have died off and been replaced by a new generation familiar with it. This now provides the consensus for a further period of normal science.

Kuhn uses the word *paradigm* in two main senses:[16]

> On the one hand, it stands for the entire constellation of beliefs, values, techniques, and so on shared by the members of a given community. On the other it denotes one sort of element in that constellation, the concrete puzzle-solutions which, employed as models or examples, can replace explicit rules as a basis for the solution of the remaining puzzles of normal science.[17]

The first sense of the term *paradigm* is sociological, and Kuhn suggests as an alternative the term *disciplinary matrix;* for the second sense, of shared example, he suggests the alternative word *exemplar.*[18]

He illuminates both senses of the word by a consideration of the "educational initiation that prepares and licenses the student for professional practice."[19] In part, this involves learning something of the development of the field through text-book accounts:

> Characteristically, textbooks of science contain just a bit of history, either in an introductory chapter, or, more often, in scattered references to the great heroes of an earlier age. From such references both students and professionals come to feel like participants in a long-standing historical tradition. Yet the textbook-derived tradition in which scientists come to sense their participation is one that, in fact, never existed. . . . Partly by selection and partly by distortion, the scientists of earlier ages are implicitly represented as having worked upon the same set of fixed problems and in accordance with the same set of fixed canons that the most recent revolution in scientific theory and method has made to seem scientific.[20]

For this reason the text-books and the historical traditions they imply have to be rewritten after each scientific revolution. And once rewritten in this way, science again comes to seem largely cumulative.

Kuhn goes on to point out that scientists are not the only group that tends to see its own past developing linearly towards its present vantage point:

the temptation to write history backwards is very widespread. But scientists are particularly prone to this, partly because the results of scientific research show no immediate and obvious dependence on their historical context and partly because in periods of normal science the contemporary position seems so secure. Giving more historical detail about either the present or the past can serve only to reveal the role of human idiosyncrasy, error, and confusion:

> Why dignify what science's best and most persistent efforts have made it possible to discard? The depreciation of historical fact is deeply, and probably functionally, ingrained in the ideology of the scientific profession, the same profession that places the highest of all values upon factual details of other sorts.[21]

But absorbing this quasi-mythological perspective is only part of the professional initiation. Much of it consists of learning by doing, involving practice problem-solving with pen and paper and with instruments in the laboratory. As the student progresses from his freshman classes to his doctoral dissertation and beyond, the problems assigned to him become more complex and less completely precedented. But they continue to be modelled closely on previous achievements. These models are paradigms in the sense of exemplars. The student does not learn by exclusively verbal means; he also acquires a tacit knowledge that comes only through practice. One result of this experience is that a student acquires the ability to perceive similarities between new problems and the familiar exemplars: "He views the situations that confront him as a scientist in the same gestalt as other members of his specialists' group. For him they are no longer the same situations he had encountered when his training began. He has meanwhile assimilated a time-tested and group-licensed way of seeing."[22]

This acquired way of seeing is not confined to the perception of problems, but applies also to literal perception through the senses. For example, a novice looking at photographs of cloud chambers sees only a chaotic assembly of droplets; but a trained particle physicist sees the tracks of electrons, alpha particles, and so on. Likewise a novice looking through a microscope at a slide of plant tissue sees only a confusion of colours, lines, and blobs; but a plant anatomist sees cells of specific types, and nuclei, chloroplasts, and other structures within them.

Kuhn's account of scientific development in terms of tradition-bound periods punctuated by revolutionary breaks has many parallels in the history of the arts, ideas, political thought, and other areas of human activity. His metaparadigm has already been widely applied. Here for example is a passage from a recent book on the history of art:

We now know that scientific progress requires more than merely "adding to" existing knowledge and the systematic building up of achievement. We also know, since the shift into Modernism, that progress is not made, as was once thought, by the accumulation of knowledge within existing categories: it is made by leaps into new categories or systems. Art is not a descriptive statement about the way the world is, it is a recommendation that the world ought to be looked at in a given way.[23]

Indeed, such parallels are almost inevitable, given that Kuhn originally borrowed from cultural historians their well-established perception of periodization in terms of revolutionary breaks in style, taste, and institutional structure.[24] He has himself drawn attention to the similarities between the uses of paradigms as exemplars for problem-solving in science and styles in art. "I suspect, for example, that some of the notorious difficulties surrounding the notion of style in the arts may vanish if paintings can be seen to be modelled on one another rather than produced in conformity to some abstract canons of style."[25]

An interpretation of paradigms in terms of morphic fields does not merely involve the substitution of one term for another, but helps to place Kuhn's insights in the wider context of formative causation, both within human cultures and throughout the entire realm of nature. The stabilization of these fields by morphic resonance helps to account for the continuity and conservatism of scientific traditions. As new members are initiated into the professional communities of scientists, through morphic resonance they come under the cumulative influence of other members of the community, going right back to the founders of the tradition, and assimilate the traditional habits.

Once again the appearance of new morphic fields, new paradigms, cannot be explained entirely in terms of what has gone before. New fields start off as insights, intuitive leaps, guesses, hypotheses, or conjectures. They are like mental mutations. New associations or patterns of connection come into being suddenly by a kind of "gestalt switch." Scientists often speak of "scales falling from their eyes" or of a "lightning flash" that "illuminates" a previously obscure problem, enabling its components to be seen in a new way that for the first time enables it to be solved. Sometimes the relevant illumination comes in sleep. Here, for example, is the famous description by the chemist Friedrich von Kekulé of the dream through which he discovered the structure of the benzene ring:

> I turned my chair to the fire and dozed. . . . Again the atoms were gambolling before my eyes. This time the smaller groups kept modestly

in the background. My mental eye, rendered more acute by repeated visions of this kind, could now distinguish larger structures, of manifold conformation; long rows, sometimes more closely fitted together; all twining and twisting in snakelike motion. But look! What was that? One of the snakes had seized hold of its own tail, and the form whirled mockingly before my eyes. As if by a flash of lightning I awoke.[26]

This is how the mathematician Henri Poincaré described the origin of one of his fundamental discoveries, the theory of Fuchsian functions:

> For fifteen days I strove to prove that there could not be any functions like those I have since called Fuchsian functions. . . . One evening, contrary to my custom, I drank black coffee and could not sleep. Ideas rose in crowds; I felt them collide until pairs interlocked, so to speak, making a stable combination. By the next morning I had established the existence of a class of Fuchsian functions. I had only to write out the results, which took but a few hours.[27]

Another great mathematician, Karl Gauss, described how he had finally proved a theorem on which he had worked unsuccessfully for four years:

> At last two days ago I succeeded, not by dint of painful effort but so to speak by the grace of God. As a sudden flash of light, the enigma was solved. . . . For my part I am unable to name the nature of the thread which connected what I previously knew with that which made my success possible.[28]

The naturalist Alfred Russel Wallace discovered the principle of natural selection, independently of Darwin, through a sudden illumination while he was suffering from a severe attack of malarial fever in the Dutch East Indies.[29] And so on. As Kuhn has expressed it: "No ordinary sense of the term 'interpretation' fits these flashes of intuition through which a new paradigm is born."[30]

But to describe such creative intuitions is not to explain them, of course. We come back to the mystery of origins.

CHAPTER 16

The Evolution of Life

Evolutionary Faith

The idea of evolution underlies much of modern political, economic, and social theory; it pervades geology and biology, and has recently become the basis of our entire cosmology. Indeed it shapes the way we think about practically everything. But it is more than a predominant way of thinking: it also has a deep intuitive appeal.

One reason for this is that it is a theory of original unity: it explains the diversity of the world in terms of a common source. On the grandest scale, everything in the universe has its ultimate origin in the primordial explosion. The sun arose from the same galactic cloud as other stars in our galaxy; the earth from the same spinning disc of matter as our brother and sister planets. All forms of life are thought to have come from a common ancestral form, perhaps even from a single primal cell. We are therefore related to all living things, and ultimately to everything in the universe.

One of the great themes of traditional creation myths is the division of a primal unity into many parts, the emergence of the Many from the One. Clearly modern evolutionary theories fulfil this mythic role.

Another aspect of the idea of evolution that has a numinous appeal is its affirmation of a continuing creativity in the universe, in life, and in humanity. The creative process did not occur only in the distant past, in the mythic time of origins; it has been going on ever since, and is still going on now. Our fascination with innovations and human creativity is one way in which we experience evolution as a living idea, or more than an idea: a faith, which is not just a matter of belief, but also involves confidence, reliance, and trust. Like other faiths, the evolutionary faith has a self-fulfilling quality, and enables innovations to be made at an ever-accelerating rate, at least in the scientific, technological, and economic realms.

Even those who reject a simple faith in material progress, or who fear that dreadful worldwide disasters will result from the advances of technology, do not generally reject the basic idea of evolutionary development. Rather, they stress that as well as material progress there is an urgent need for progress in the political, social, moral, or spiritual realms, or in all these areas together.

Whether or not the evolutionary faith is recognized as essentially religious or ideological in nature, it arouses what seem very like religious passions in its defenders; and like traditional religious faiths, evolution is interpreted very differently by different sects and schools of thought.

Within biology, as in other fields, debates on the nature of evolution have usually been conducted in a sectarian spirit, and still are: neo-Darwinians versus neo-Lamarckians, gradualists versus saltationists, sociobiologists versus Marxists, and so on. The passions aroused can be intense.[1] Truth itself seems to be at stake, and the opponents (whoever they are) seem to be propagating profoundly false theories, generally with dangerous political, social, or religious implications.

The various interpretations of evolution are indeed closely linked to social, political, or religious systems. Thus, for example, a school of biology in Japan emphasizes the importance of co-operation by groups of organisms in the evolutionary process, whereas neo-Darwinians emphasize competition between individuals. The two sides accuse one another of merely reflecting the social assumptions of their own culture.[2] Likewise, sociobiologists emphasize the importance of selfish genes in competition with each other; their Marxist opponents see in this a reflection of right-wing political doctrines;[3] and Marxists themselves of course have their own collectivist ideology. Materialists believe that the entire evolutionary process is purposeless and results from the interplay of blind chance and necessity;[4] pantheists emphasize the spontaneity and creative power of nature;[5] while theists believe that nature itself in some sense arises from the divine being and that the evolutionary process has a spiritual purpose.[6]

The various schools of thought commonly criticize each other on the grounds that they start from preconceived assumptions. And so they do. But who does not?

Darwin himself certainly based his ideas on assumptions that have been questioned again and again. We now turn to a re-examination of these assumptions, partly because this helps to clarify most of the subsequent evolutionary controversies and partly because this historical context makes the evolutionary implications of the hypothesis of formative causation easier to grasp. We then consider the evolution of morphic fields, and the possible role of morphic resonance in the phenomena of atavism and parallel evolution.

The Ambiguity of Darwin

Underlying Darwin's evolutionary vision was a strong sense of the autonomy, spontaneity, and creativity of nature. He could not but think of nature as alive. But in order to affirm the creativity of nature on earth, he had to deny its dependence on the transcendent God of contemporary Protestant theology. And in his denial of any creative role to God as thus conceived, he adopted the doctrine of materialism.[7] He accordingly attempted to expel as much mystery as possible from the working of nature, reducing everything to the operation of blind laws and blind chance. He took these laws to include some of the social and economic principles that were influential in Victorian England, including Malthus' theory of population and an emphasis on individual competition and self-advancement; he also took for granted a kind of common-sense utilitarianism.

Darwin's attitude to nature was inevitably ambiguous. His sense that nature is alive seems to have been one of his primary intuitions; but then he continually denied it, or at least relegated it to unconsciousness. As a materialist, he assumed on theoretical grounds that nature is dead. He was well aware of this ambiguity:

> The term "natural selection" is in some respects a bad one, as it seems to imply conscious choice; but this will soon be disregarded after a little familiarity. . . . For brevity's sake I sometimes speak of natural selection as an intelligent power. . . . I have, also, often personified the word Nature; for I have found it difficult to avoid this ambiguity; but I mean by nature only the aggregate action and product of many natural laws—and by laws only the ascertained sequence of events.[8]

Darwin's primary model of the evolutionary process was the development through human activity of the many breeds and varieties of domesticated animals and cultivated plants. Drawing on the experience of animal and plant breeders, he concluded that three fundamental principles were involved in evolutionary transformation: the spontaneous variability of living organisms and the tendency for offspring to resemble their parents; the effects of the environment and of habit; and selection. We will discuss these in turn, and then consider how they can be re-evaluated in the light of formative causation.

Spontaneous Variation

Darwin gave many examples of spontaneous variation: the loss or gain of entire structures such as vertebrae, petals, nipples, and even whole limbs; sudden and dramatic changes in patterns of growth and development; and many other kinds of variation, for example in colour, pattern, and behaviour. He showed how such variants or sports had on many occasions been selected by breeders and formed the basis of a new breed or variety: one example was the dwarf ancon sheep (Fig. 16.1); another was the spontaneous appearance of nectarines from peaches.[9]

Biologists who subsequently continued this line of enquiry provided many more examples of spontaneous jumps or discontinuities; new types arose from the old directly, without passing through a series of intermediate forms.[10] From the beginning of this century, these have generally been referred to as mutations, from the Latin root *mutare,* to change. The study of the inheritance of such discontinuous variations has been the basis of the science of genetics from Mendel onwards. In the course of the present century it has become clear that at the genetic level, mutations can involve changes ranging from the loss or gain of entire chromosomes, through large-scale changes in chromosome structure, right down to changes in single base-pairs within DNA molecules.

The word *mutation* is now commonly used to refer to such genetic changes, especially to changes within DNA molecules; and this seems to imply that the spontaneous variation of organisms can be reduced to the spontaneous variation of genes. But a mutant organism is one in which such changes are *expressed,* and this expression takes place within the context of its co-ordinated development and behaviour. For example, genetic mutations in fruit flies that lead to the development of four wings instead of two (Fig. 5.6) alter the morphogenesis of an entire segment of the fly; but as we have seen, the mutant gene does not contain or program this development (pp. 89–90). Recall the television analogy: the mutation of a condenser in the tuning circuit may cause it to tune in to a different channel; but the program on this channel is not encoded in the mutant condenser. Darwin himself was very aware of what he called the "co-ordinating power" common to all organic beings, which according to the hypothesis of formative causation is due to morphic fields. Darwin attributed this organizing capacity to the *nisus formativus,* the formative impulse postulated by early nineteenth-century vitalists. For example,

we may infer that, when any part or organ is either greatly increased in size or wholly suppressed through variation and continued selection,

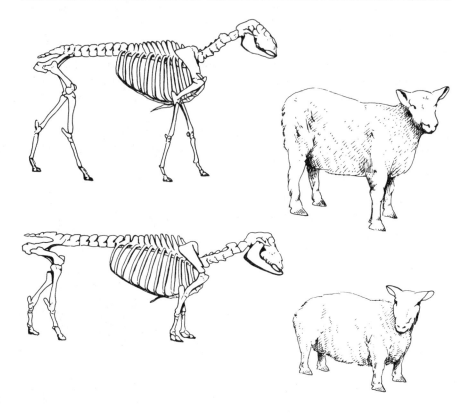

Figure 16.1 A normal sheep (above) compared with the dwarf "ancon" or "otter" breed, together with their corresponding skeletons. (After Stanley, 1979) This breed originated with a sudden mutation, rather than passing through gradual intermediate forms. Darwin described it as follows: "In 1791, a ram-lamb was born in Massachusetts, having short crooked legs and a long back, like a turnspit-dog. From this one lamb the *otter* or *ancon* semi-monstrous breed was raised; as these sheep could not leap over the fences it was thought they would be valuable. . . . [T]he ancons have been observed to keep together, separating themselves from the rest of the flock when put into enclosures with other sheep" (Darwin, 1875).

the co-ordinating power of the organization will continually tend to bring again all the parts into harmony with each other.[11]

Ever since Darwin, a number of influential biologists have seen in such large-scale mutations the most probable way in which new types of organisms have arisen.[12] The geneticist Richard Goldschmidt, for example, expressed a rather extreme form of this view:

Species and the higher categories originate in single macro-evolutionary steps as completely new genetic systems. The genetical process which is involved consists of a re-patterning of the chromosomes, which results in a new genetic system. . . . This new genetic system . . . produces a change in development which is termed a systemic mutation. . . . The facts of development, especially those furnished by experimental embryology, show that the potentialities, the mechanics of development, permit huge changes to take place in a single step.[13]

Goldschmidt called such radically mutant organisms "hopeful monsters." Such an emphasis on large-scale changes reduces the creative role of natural selection in the evolutionary process and is controversial precisely for this reason. It locates the main source of evolutionary creativity within living organization itself: new kinds of organisms simply appear spontaneously. Darwin knew that this was true of domesticated plants and animals and that it had been the source of many established breeds and varieties. But he denied that it played an important part in natural evolution, preferring to emphasize the creative power of natural selection instead. Although spontaneous variation remained an essential feature of his theory as the source of evolutionary novelty, he tried to minimize its role; and the simplest way to do this was to concentrate attention on small variations rather than large ones. The smaller they were, Darwin felt, the less mysterious they seemed, and the more scientific his theory became (p. 48).

The Effects of Habit

In Darwin's day, it was widely assumed that acquired characteristics could be inherited. Darwin shared this belief and cited many examples to support it.[14] In this respect he was a Lamarckian, not so much because he was influenced by Lamarck as because he and Lamarck both accepted the inheritance of acquired characteristics as a matter of common sense. It was a view that had always been so widely accepted that it was simply taken for granted.[15] As we saw in chapter 8, according to the hypothesis of formative causation, the inheritance of acquired habits of form and behaviour is due to the inheritance of morphic fields by morphic resonance.

Lamarck placed a strong emphasis on the role of behaviour in evolution: animals' development of new habits in response to needs led to the use or disuse of organs, which were accordingly either strengthened or weakened. Over a period of generations, this process led to structural changes which became increasingly hereditary (p. 140). Lamarck's most famous example was the giraffe:

It is interesting to observe the result of habit in the peculiar shape and size of the giraffe: the animal, the largest of the mammals, is known to live in the interior of Africa in places where the soil is nearly always arid and barren, so that it is obliged to browse on the leaves of trees and to make constant efforts to reach them. From this habit long maintained in all its race, it has resulted that the animal's fore-legs have become longer than its hind legs, and that its neck is lengthened to such a degree that the giraffe, without standing up on its hind legs, attains a height of six metres.[16]

In this respect too Darwin agreed with Lamarck, and he provided various illustrations of the hereditary effects of the habits of life. For example, domestic fowls, ducks, and geese have almost lost both the habit of flying and the power of flight. He carefully compared the skeletons of domesticated breeds with the wild parent species and showed that there had been a general decrease in weight and size of wing bones and an increase in leg bones relative to the rest of the skeleton, which he thought was probably the indirect result of the action of the muscles on the bones: "There can be no doubt that with our anciently domesticated animals, certain bones have increased in size and weight owing to increased or decreased use."[17] He supposed that similar principles had obtained under natural conditions: ostriches, for example, may have lost the power of flight through disuse and gained stronger legs through an increased use of them over successive generations.[18]

Darwin was in fact very conscious of the power of habit, which was for him almost another name for nature. "Nature, by making habit omnipotent, and its effects hereditary, has fitted the Fuegian to the climate and productions of his miserable country," he wrote succinctly.[19] Francis Huxley has summarized Darwin's attitude as follows:

A structure to him meant a habit, and a habit implied not only an internal need but outer forces to which, for good or evil, the organism had to become habituated. . . . In one sense, therefore, he might well have called his book *The Origin of Habits* rather than *The Origin of Species*. Like many others, he was never quite sure what a species was.[20]

Many biologists have followed Darwin in his emphasis on the cumulative effects of habit, and the ostrich remains a favourite example. These birds are born with horny calluses on their rumps, breast, and pubis, just where these will press upon the ground when they sit (Fig. 16.2). It is easy to suppose that their ancestors developed these calluses through their habit of sitting; over many generations the tendency to develop them became more and more pronounced, and eventually they developed even in embryos.

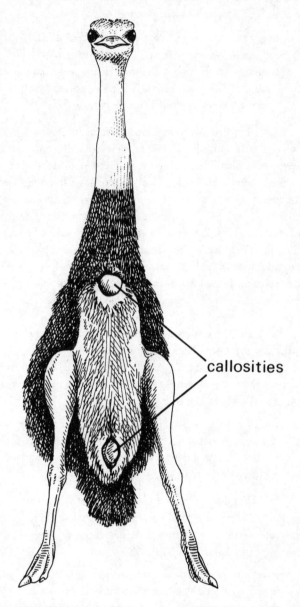

callosities

Figure 16.2 The underside of an ostrich's body, showing the callosities on which it rests while sitting. (Cf. Duerden, 1929)

Likewise, wart-hogs have hereditary calluses on their knees, corresponding with their habit of kneeling while they root in the ground.[21] So do camels, again in perfect agreement with their habit of kneeling (p. 140). We ourselves show such effects, as Darwin pointed out:

> Every one knows that hard work thickens the epidermis on the hands; and when we hear that with infants, long before birth, the epidermis is thicker on the palms and soles of the feet than on any other part of the body . . . we are naturally inclined to attribute this to the inherited effects of long-continued use or pressure.[22]

Such Lamarckian ideas have an immediate intuitive appeal and seem to accord well with common sense. The problem is that no one has been able to propose a plausible mechanism whereby acquired characteristics could be inherited materially. Darwin had a try: in his hypothesis of pangenesis he proposed that all the units of the body throw off tiny gemmules of "formative matter," which are dispersed throughout the body, multiply, and aggregate in the buds of plants and in germ cells, through which they are transmitted to the offspring.[23] This theory never attracted much support, and now seems highly implausible. More modern attempts to conceive of Lamarckian inheritance in terms of the transfer of specifically altered genetic material from various parts of the body to the germ cells have likewise met with no success.

Mendelian geneticists, following Weismann, have always denied axiomatically that such inheritance can occur (pp. 78–81). But then how can the inherited calluses of the ostrich, for example, be explained? The natural selection of chance mutations is the ready-made neo-Darwinian answer; but as C. H. Waddington observed:

> Can we really be satisfied with a theory that suggests that, purely by chance, a hereditary change has turned up which produces callosities in just the right places, and that the sitting habit of the ostrich has nothing to do with it?[24]

Waddington, in his own experiments with fruit flies, showed that acquired characteristics could indeed be inherited. He explained this effect in terms of chreodes or pathways of development (Fig. 6.2), and called the process genetic assimilation (Fig. 8.3):

> After a time we shall find that the path leading to the adapted condition is better defined than the main path, and also that it has become easier for development to choose that path. The threshold between the adapted alternative and the original main track will have been lowered.

If this lowering goes far enough, the alternative will become the main track, and genetic assimilation will be complete.[25]

He attempted to account for this effect in terms of the selection and accumulation of mutant genes within the population, and thus provided a seemingly orthodox neo-Darwinian explanation for it, one that has been quite widely adopted by modern evolutionary theorists. But as we have seen, recent reinvestigations have failed to support his explanation: the effect occurs even in the absence of genetic selection in favour of four-winged flies and seems much more Lamarckian than Waddington supposed (pp. 141–46).

As we saw in chapter 8, the inheritance of acquired characteristics is problematic from a mechanistic point of view. Both neo-Darwinians and neo-Lamarckians make the conventional mechanistic assumption that heredity is explicable in terms of chemical genes. But despite the attractiveness of Lamarckism from an evolutionary point of view, and despite the direct evidence for the inheritance of acquired characteristics, there is no evidence that specific genetic modifications actually happen or even that they are possible. The idea of formative causation provides a way of transcending this long-continued controversy. Acquired characteristics can be inherited, but not because of modifications of the DNA. Rather, they depend on modifications of morphic fields, which are inherited non-genetically by morphic resonance. Through repetition, new patterns of development and behaviour become increasingly habitual. Organisms do indeed inherit habits of behaviour and of bodily development, as both Lamarck and Darwin supposed.

Natural Selection

That natural selection plays some part in the evolutionary process seems beyond dispute: countless species and indeed whole ecosystems have become extinct, while others have not. Natural processes, including competition between organisms, climatic and ecological changes, and even global catastrophes leading to mass extinctions, have in some sense had a selective effect. Natural selection weeds out organisms and species that are not adequately fitted to their environment, for whatever reason.

Darwin, however, gave natural selection a more positive and creative role:

It may metaphorically be said that natural selection is daily and hourly scrutinizing, throughout the world, the slightest variations; rejecting those that are bad, preserving those that are good; silently and insensibly working, *whenever and wherever opportunity offers,* at the improvement of each organic being in relation to its organic and inorganic conditions of life.[26]

In so far as it relates to the development of locally adapted varieties, races, and subspecies, this idea is plausible; and it has never been widely disputed, at least as a partial explanation of evolutionary adaptation: Lamarckians, like Darwin himself, have emphasized that the hereditary effects of the habits of life also play an important role. The main problem that Darwin and Darwinians have always faced is to account for the origin of species themselves, or of genera, families, and the higher orders of living organization. The idea that such large-scale evolutionary processes all took place gradually over very long periods of time has been challenged again and again. How could complex structures such as eyes, wings, and feathers have evolved gradually before they became functional wholes? Why do plants and animals fall into distinct types, such as ferns, conifers, insects, and birds, rather than lying on a continuous spectrum of living forms?

The fossil record has always seemed more consistent with the idea that new forms of life have arisen suddenly, or at least rapidly. But Darwin and the Darwinians have always argued that these discontinuities in the fossil record exist because of its patchiness and imperfections. This argument has been disputed from Darwin's day onwards, and is currently under attack from a number of leading paleontologists:

> Today the fossil record—a rich store of information that was long untapped—is forcing us to revise this conventional view of evolution. As it turns out, myriads of species have inhabited the earth for millions of years without evolving noticeably. On the other hand, major evolutionary transitions have been wrought during episodes of rapid change, when new species have quickly budded off from old ones. In short, evolution moves by fits and starts.[27]

But the firm adherence of Darwin and his followers to the doctrine of gradual change has little to do with empirical evidence: behind this controversy lies a major question of dogma. Darwin's principal objective was to replace the idea of the design of nature by God with spontaneous natural processes. The conventional theology of his day sought to account for the intricate adaptations of plants and animals in terms of divine intelligence; Darwin postulated natural selection instead. And to prevent the reintroduction of God to account for sudden creative jumps, he had to deny that such jumps were of any significance in the evolution of life, or at least to minimize their importance. He equated this denial with science itself (p. 48). Most of his followers have shared this attitude, summarized by Richard Dawkins as follows:

In the context of the fight against creationism, gradualism is more or less synonymous with evolution itself. If you throw out gradualness you throw out the very thing that makes evolution more plausible than creation. Creation is a special case of saltation—the *saltus* is the large jump from nothing to fully formed modern life. When you think of what Darwin was fighting against, is it any wonder that he continually returned to the theme of slow, gradual, step-by-step change?[28]

The replacement of divine design by natural selection has led Darwinians into an intellectual habit which, ironically, resembles that of old-fashioned theodicy, the attempt to justify the ways of God to men. According to this kind of theology, God as a perfect and omniscient being must have created the best of all possible worlds, hence everything that happens must have some providential reason, even if this is not at first apparent. Darwinians, confronted with any given feature of a species, generally assume that it must have some purpose or adaptive value, and then speculate about the selective pressures that must have given rise to it. Such speculations are usually untested and untestable; they are in fact rather like fables: how the rhino got its horn, how the peacock got its tail, and so on. One of the appeals of Darwinism is that it permits a limitless supply of stories to be spun. But varied and ingenious though they often are, they all have the same moral of competitive success, and all take place in a monotonously utilitarian world.

Darwin, with his customary and disarming honesty, came to recognize that he had exaggerated the role of natural selection:

> I was not, however, able to annul the influence of my former belief, then almost universal, that each species had been purposely created; and this led to my tacit assumption that every detail of structure, excluding rudiments, was of some special, though unrecognized, service. Anyone with this assumption in his mind would naturally extend too far the action of natural selection, either during past or present times.[29]

Some of those who have accepted the idea of sudden jumps have indeed seen in them the expression of the creative power of nature, if not of God himself.[30] But others think of them as entirely a matter of chance, hence similar in nature to the small random mutations on which neo-Darwinian theory is based. In this case, there is a difference only of degree between the evolutionary role of large rather than small random changes. Yet others stress that mutations occur in the context of developmental and behavioural patterns which constrain the possible changes that can occur. A mutant horse may have extra toes, for example, but there are no mutant horses with wings, feathers, or flowers.

One of Darwin's staunchest supporters, T. H. Huxley, warned him at an early stage against his insistence on gradual change:

> Mr. Darwin's position might, we believe, have been even stronger than it is, if he had not embarrassed himself with the aphorism *Natura non facit saltum* which turns up so often in his pages. We believe that nature does make jumps now and then, and a recognition of this fact is of no small importance.[31]

But Darwin did not heed this warning, and was plunged into continual controversy as a result. As he rather ruefully noted after years of dispute: "There are, however, some who still think that species have suddenly given birth, through quite unexplained means, to new and totally different forms."[32] And this remains as true today as when he wrote it over a century ago.

As we have seen, the main reason why Darwin and his neo-Darwinian followers have insisted so strongly on gradual changes stems from the attempt to exclude as much mystery as possible from the evolutionary process, and above all to leave no openings for the creative activity of God. The disadvantages of the dogma of gradualism are that it conflicts with the fossil record, which suggests that large evolutionary changes have in fact occurred quite suddenly, and with the well-established fact that sudden mutations with large-scale effects do indeed occur—for example in the ancon sheep that Darwin described (Fig. 16.1) and in mutant fruit flies (Figs. 5.6 and 8.2).

The notion that biological evolution involves *both* sudden jumps *and* gradual changes seems to fit the facts far better than a dogmatic insistence on gradualism alone. Moreover, it is consistent with what we know about evolution in other spheres, including the evolution of science itself. New theories and paradigms come into being through sudden, intuitive leaps; but then within an established field of science, within the framework of a generally accepted paradigm which soon becomes habitual, research progresses in a relatively gradual and cumulative manner (pp. 265–69).

The theory of evolution by natural selection itself is no exception. As we saw in the preceding chapter, it came to Wallace as a sudden illumination, during an attack of malarial fever in the tropics; and it came to Darwin himself in a similarly sudden manner. First, in 1837, he underwent a conversion to evolutionism, abandoning his former belief in the constancy of species, and "suddenly everything appeared in a new light."[33] Then, on 28 September 1838, came the crucial moment of illumination. In Darwin's own words:

> Fifteen months after I had begun my systematic enquiry, I happened to read for amusement Malthus on population, and being well prepared

to appreciate the struggle for existence which everywhere goes on from long-continued observation of the habits of animals and plants, it at once struck me that under these circumstances favourable variations would tend to be preserved and unfavourable ones to be destroyed. The result of this would be the formation of new species. Here, then, I had at last got a theory by which to work.[34]

After this essential insight, Darwin's theory matured gradually over many years before he published *The Origin of Species* in 1859. Thus Darwin's own intellectual evolution, like the evolutionary process in general, seems to have involved both sudden jumps and gradual adaptations and changes.

The Evolution of Morphic Fields

Evolution occurs at all levels of organization, from atoms to galaxies. At each level of organization, the organized systems—insulin molecules, fruit flies, the instinctive patterns of nest-building behaviour in wasps, flocks of birds, tribal societies, scientific theories—are wholes: gestalts, morphic units, holons. The hypothesis of formative causation inevitably implies that the evolutionary process is closely connected with the evolution of morphic fields. There are four major consequences of this view.

First, the appearance of new patterns of organization—of new kinds of crystals, for example, or new classes of organisms such as the mammals, or new scientific theories—is associated with the appearance of new morphic fields. Possible creative sources of new fields are discussed in the final chapter; for the purpose of the present discussion the important point is that the appearance of new fields inevitably involves a jump or discontinuity. These fields are wholes, and precisely because of their irreducible integrity they have to appear suddenly. Wholes at all levels of complexity, like the quanta of quantum physics, either exist or do not; by their very nature they cannot come into being gradually.

Of course new morphic fields involve continuity with what went before, as well as discontinuity. All new fields embrace lower-level morphic units that existed prior to their appearance; these lower-level holons are the parts that are brought into relationship with each other in the new synthesis. For example, new kinds of molecules include within them atoms, such as carbon and oxygen, which evolved many billions of years ago; when cells with nuclei first arose, they probably incorporated pre-existing microbial cells within themselves;[35] many elements of the ancestral reptile form were included in the body plan of the first birds; new instincts include behavioural elements that have already been practised for countless generations; new theories include already existing ideas within them, as for instance Darwin's theory of evolution by natural selection was a new conceptual synthesis that incorporated the pre-existing ideas of evolutionary transformation and of the struggle for existence. In general, new patterns include old ones within themselves; nevertheless they are new and come into being suddenly; they have a wholeness and integrity that do not admit of gradual appearance.

Second, morphic fields are subject to natural selection. The fields of new patterns of organization that are not viable will not be stabilized by morphic resonance. Only those patterns that are capable of surviving are able to occur again and again; and the more frequently they occur, the stronger will their morphic fields become, owing to the cumulative effects of morphic resonance, hence the more probable the recurrence of the patterns. Natural selection favours some habits more than others; and the more successful a habit is, the more strongly it is stabilized by morphic resonance. In the realm of biology, this process is involved in the evolution of dominance and in the tendency for the most common patterns of form and behaviour to predominate—in the language of genetics this is known as the dominance of the wild type (chapter 8).

Third, this hypothesis allows for an inheritance of acquired characteristics that does not depend on selective modification of genes, as neo-Lamarckian theories do, but is instead based on inheritable modifications of the morphic fields in response to the habits of life. The inheritance of morphic fields by morphic resonance from previous similar organisms would also allow new patterns of development or behaviour to spread more quickly than conventional genetic inheritance would allow. One possible example is the spread of the habit of opening milk bottles by tits in Europe (pp. 177–80).

Fourth, morphic fields undergo differentiation or specialization in the sense that some versions of the general patterns they organize become more probable than others. Indeed, much of the evolutionary process seems to involve the appearance of variations on basic morphic themes. Domesticated animals and plants provide a clear illustration of this principle: think, for example, of the variations on the dog theme represented by breeds such as bulldogs and dachshunds.

Many paleontologists have deduced from the fossil record that when new evolutionary lines begin—when new basic body-plans appear—there is often an "intense radiation of types, an 'explosive phase' in the early part of their phylogeny, and that only a limited number of the branches continue to develop, and with decreasing speed."[36] One example is the adaptive radiation of the mammals after the sudden extinction of the dinosaurs over 60 million years ago. Most of the orders of mammals came into existence within about 12 million years: carnivores, whales, dolphins, rodents, marsupials, ant-eaters, horses, camels, elephants, bats, and many others. Indeed most of the basic mammalian forms that appeared then still exist.[37]

In such evolutionary branchings at whatever level—orders, families, genera, or species—it is perhaps most likely that the various alternative versions of a common basic field came into being by saltation. Many of the new variants may have arisen as a result of chromosomal changes and genetic mutations; some may have been stabilized as a result of historical accidents, for example because they occurred in small, isolated populations; some may have evolved in adaptive response to the conditions of life; some may have evolved gradually in the classical Darwinian manner. But however they arose in the first place, if they proved viable and reproduced successfully, their characteristic versions of the basic morphic fields would have been progressively reinforced by morphic resonance.

None of these considerations denies the role of natural selection at the genetic level; organisms whose genetic constitution is associated with more viable patterns of organization will be favoured by natural selection, and changes in gene frequencies within populations will consequently occur, as neo-Darwinians suppose. But the present hypothesis means that evolution involves more than a change in gene frequencies: it involves the natural selection and stabilization of patterns of organization brought about by morphic fields. These fields themselves evolve. Their expression is affected by the conditions and habits of life, as well as by genetic mutations.

This hypothesis is in surprisingly good agreement with much of Darwin's thought, including his strong sense of the power of habit. It differs in that it allows for both sudden and gradual changes and leaves open the question of creativity.

Extinction and Atavism

Countless species have become extinct for many reasons, including climatic changes and global catastrophes. Complex ecosystems have died out. Entire human languages and cultures have disappeared, and many skills and elements of culture have been lost. What has happened to their morphic fields?

According to the hypothesis of formative causation these fields in some sense still exist, although they cannot be expressed because there is nothing to tune in to them. Even the fields of the dinosaurs are potentially present here and now; but there are no appropriate tuning systems, such as living dinosaur eggs, that can pick them up by morphic resonance.

If for any reason—for example a genetic mutation or an unusual environmental stress—any living system comes into resonance with the fields of an ancestral or extinct type, then these fields could be expressed again, and archaic structures could suddenly reappear.[38] Such a phenomenon is in fact well known, and is usually referred to as reversion, atavism, or throwing back.

Darwin drew attention to many examples of throwing back known to plant breeders: "With most of our cultivated vegetables there is some tendency to reversion to what is known to be, or may be presumed to be, their aboriginal state."[39] A similar tendency is well known in domesticated animals, and has often been observed when they have run wild. Feral pigs, for example, become more bristly, tend to redevelop tusks, and the coloured stripes of young wild pigs reappear in their young. As Darwin commented: "In this case, as in many others, we can only say that any change in the habits of life apparently favours a tendency, inherent or latent in the species, to return to their primitive state."[40]

Such phenomena are closely related to what geneticists call the "dominance of the wild type" (pp. 146–49); in terms of morphic resonance, the ancestral wild-type fields have been around much longer and are more strongly stabilized than those of the domesticated forms, and hence tend to predominate unless prevented from doing so by human activity and selection.

Darwin believed that atavism underlay many of the mysteries of spontaneous variation, and he concluded his discussion of the subject by reflecting that the germ "is crowded with invisible characters . . . separated by hundreds or even thousands of generations from the present time: and these characters, like those written on paper with invisible ink, lie ready to be evolved whenever the organization is disturbed by certain known or unknown conditions."[41]

A striking human example of atavism is the occasional birth of babies with tails; others include the appearance of hind legs in whales and wings in flightless insects.[42] One of the best-studied cases is the appearance of extra toes in horses. Modern horses represent the limit of an evolutionary trend for the reduction of toes: they have only one. The hoof is its toenail. Their putative ancestors over twenty million years ago had three or four, and yet further back their progenitors had the original mammalian complement of five. Modern horses develop only vestiges of the second and fourth toes as short splints of bone mounted high above the hoof.

Abnormal horses with extra digits appear from time to time and have been the object of much interest since at least the time of Julius Caesar. A close examination of such animals has shown that in most cases the additional toe is a duplicate of the functional third digit, but some hark back to their remote ancestors by developing one or both of their side splints into toes complete with hooves (Fig. 16.3).[43]

In the reptilian ancestors of the birds, the two leg bones between the knees and the ankles, the tibia and fibula, were equal in length, and the ankle region below included a series of small bones. In most modern birds, including hens, the fibula is reduced to a splint and the embryonic ankle bones are engulfed by the growing tibia and fuse with it (Fig. 16.4). In some simple and ingenious experiments with chick embryos the archaic pattern has been

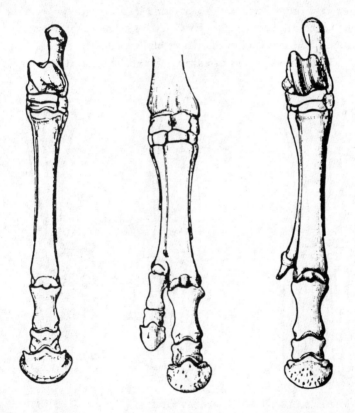

Figure 16.3 Left, the toe of a normal horse, with the vestiges of the second and fourth digits as small splints on either side (cf. Fig. 4.4). Centre, a mutant in which an extra toe has been formed by duplication of the third digit; the side splints are still present. Right, an atavistic mutant in which one of the side splints has developed into an extra toe. (After Marsh, 1892)

found to reappear. This happened, for example, when a small mica plate was inserted between the tibia and fibula at an early stage of embryonic development. The fibula grew to its full extent, and the ankle bones remained separate.[44]

Perhaps the most bizarre example of all is the development of teeth from embryonic chick tissue. *Archaeopteryx,* which is commonly supposed to be the first fossil bird, possessed teeth, but no fossil birds for the last 60 million years have produced them. Epithelium was taken from the embryonic gill arches of chicks (the structures that give rise to jaws in vertebrates) and cultured in the laboratory together with embryonic mouse tissue capable of forming the bone and dentin of teeth, but not the enamel. These combined tissues gave rise to teeth with chick enamel.[45] Moreover the teeth did not look like those of mice. Stephen Jay Gould has suggested that these may have resembled "the actual form of a latent bird's tooth."[46]

On the basis of the hypothesis of formative causation, comparable

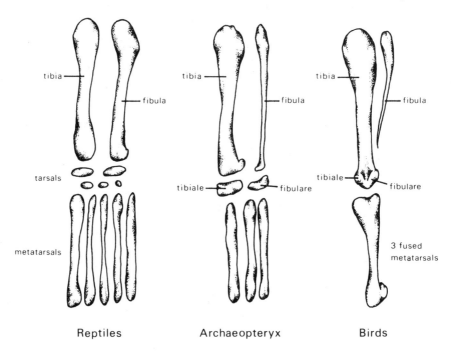

Reptiles   Archaeopteryx   Birds

Figure 16.4 A comparison of the lower parts of the hind limbs of reptiles, *Archaeopteryx,* and modern birds. Note the reduction in size of the fibula and in the number of tarsal and metatarsal bones in *Archaeopteryx,* and their further reductions and fusions in birds. (After Hall, 1984)

phenomena of atavism would also be expected in the social and cultural realms. Certainly this possibility has often been entertained by those who have throughout the ages feared a breakdown of civilization and a reversion of the social order to a primitive or barbarous state. But there may be many less spectacular ways in which extinct ancestral patterns have reappeared in the course of social and cultural evolution: some as deliberate revivals, and others quite unconsciously. This is obviously a fertile field for speculation.

Evolutionary Plagiarism

The phenomena of atavism suggest that morphic fields can leap from past to present species across gaps of time and space. But there is no reason why fields should leap only from ancestors to their own descendants: they may also jump sideways, as it were, from one group or type of organism to another, even if these organisms live on different continents. Such a transmission could occur by morphic resonance if, through genetic mutation or through the influence of the environment, organisms of one species tune in to the morphic fields of another. This process would permit a kind of morphic plagiarism, even in the absence of any material connection in space or time between the organisms that copy and those that are copied. Morphic fields are not subject to a principle of evolutionary copyright.

In the human realm, there are many examples of parallel social and cultural patterns that seem to have originated independently in different parts of the world. Parallel inventions and discoveries in science and technology are relatively common: a classical case is the independent invention of the differential calculus by Newton and Leibniz.

In some cases, such parallels may be explicable in terms of "diffusion," or in other words transfers by normal means of human communication. Others probably arose in circumstances when individuals or social groups were confronted with similar problems and hit on similar solutions. And no doubt solutions that worked were often similarly favoured by a process of natural selection.

The idea of morphic resonance is complementary to all of these well-known explanations. Diffusion would be aided by morphic resonance, which facilitates the learning process (chapter 10). Confrontation of similar problems is likely to favour a tuning in to solutions already arrived at by others elsewhere. And, owing to morphic resonance, successful patterns of activity, through repetition, show an increasing tendency to reappear. Thus social and cultural morphic fields would be expected to move from one group to another, both by diffusion and by leaps, by a kind of action at a distance.

Morphic plagiarism through the picking up of morphic fields by one group of organisms from another via morphic resonance may have occurred frequently in the course of biological evolution. It may well underlie the phenomenon of parallel evolution, in which similar patterns appear in more or less closely related plant and animal species. In some cases, striking similarities are also found among organisms that are only very distantly related, and these are generally referred to as evolutionary convergences.

Richard Dawkins has pointed out that on statistical grounds, following standard neo-Darwinian assumptions:

> It is vanishingly improbable that the same evolutionary pathway should ever be followed twice. And it would seem similarly improbable, for the same statistical reasons, that two lines of evolution should converge on the same endpoint from different starting points. It is all the more striking . . . that numerous examples can be found in real nature, in which independent lines of evolution appear to have converged, from very different starting points, on what looks very like the same endpoint.[47]

In the plant kingdom, the most familiar examples of parallel evolution are in the forms of leaves, where very similar patterns have appeared again and again in separate genera and families. So striking are the resemblances that many species and varieties are actually named on the basis of these leaf forms, which are borrowed, as it were, from other kinds of plants: *salicifolius* means willow-leaved; *ilicifolius,* holly-leaved; and so on.

In some cases, parallel evolution occurs to a remarkable extent in particular geographical areas. One example, to which F. W. Went has drawn attention, is of shrubs in New Zealand with intricate, interlaced branches and small orb-shaped leaves.

> The frequency with which one encounters this divaricate habit in New Zealand (whereas it is hardly known anywhere else in the world) is indicated by the fact that in New Zealand there are a total of about 50 species of shrubs, divided over 21 families, which are divaricate with interlaced, tortuous branches and reduced leaves. Many of these shrubs look so much alike, without flowers, that one cannot guess even the family, since various species . . . have exactly the same habit.[48]

Strangely enough, in most cases this habit of growth appears only in young shrubs, as a juvenile phase, and then gives way to the growth habit typical of the genus; in other species it appears only in the mature phase, and not in the juvenile.

Went has carefully considered the standard explanation that this habit is simply an adaptation to the environment favoured by natural selection, and has shown that it is hardly consistent with the facts. First, the idea that it protects the plants against browsing animals is implausible because New Zealand is the only extensive geographical area without large native herbivores. Second, the habit appears only at certain stages of growth and has not completely replaced the more usual one. Third, only some species show this habit, and other closely related forms that survive equally well in New Zealand do not. Finally, since this habit "occurs in so many shrubs from different habitats, it does not seem to be an adaptation to the environment."[49]

This puzzling phenomenon is not confined to New Zealand: many other examples can be found elsewhere in the world. So striking are these parallelisms that Went considers an explanation in terms of chance mutations highly implausible and has come to the conclusion that they indicate some kind of "non-sexual character transfer." To account for this he has gone so far as to suggest that entire segments of chromosomes may somehow have been transferred bodily from one genus or family to another.[50] However, this hypothetical mechanism could work only over short distances, and many other examples of parallelism occur in widely separated places, in both the plant and animal kingdoms.

In butterflies, for example, many close similarities are found in the patterns of wing colouration, both within and between families (Fig. 16.5). Some of these are familiar as standard examples of mimicry, which is generally supposed to be favoured by natural selection if a species of similar appearance living in the same environment is protected from its enemies by an unpleasant taste. Predators tend to avoid both the unpalatable insects and those that mimic them. But many other cases of parallelism can have no such explanation, especially when the species that look alike occur in quite different places.[51]

Perhaps the most spectacular examples are provided by the two main

| Melinæa imitata | Helinconius telchinia | Dismorphia praxinoe |
| (Ithomiinæ) | (Heliconiinæ) | (Pieridæ) |

Figure 16.5 Three species of South American butterflies which closely mimic each other. They belong to quite distinct families. Their colours are the same: black, white, and brilliant orange (stippled areas). (After Hardy, 1965)

branches of the mammals, the placentals and marsupials, which are thought to have diverged from a common protomammalian ancestor over 60 million years ago. The marsupials of Australia have evolved in isolation from placental mammals elsewhere, yet have given rise to a whole range of similar forms: pouched versions of ant-eaters, moles, flying squirrels, cats, wolves, and so on (Figs. 16.6 and 16.7). Moreover, much the same phenomenon occurred in South America, where marsupials independently gave rise to a range of parallel forms.[52] And among placental mammals themselves, there are many striking examples of parallel evolution, such as the porcupines of South America and the Old World. These are so alike that it has even been suggested that they crossed the Atlantic on rafts of vegetation.[53]

Even more mysterious is the convergent evolution of similar structures in organisms that are otherwise extremely different. The eyes of vertebrates, for example, have many features in common with the eyes of cephalopods, such as the octopus. When visiting an aquarium it is a strange experience to look an octopus in the eye; this extraordinary similarity in an animal so completely different is uncanny. Yet there are, of course, also differences; most notably, the retina of vertebrates is inverted—the nerves leading from the photocells point forward towards the light—whereas the cephalopod retina is not.

The standard neo-Darwinian explanation of such parallelisms and convergences is twofold: first, that they have evolved on the basis of random mutations that survived because of similar selection pressures; second, that such convergences on similar end-points occur because of similar structural constraints: there may be only a very limited number of ways of designing an eye, for example. As Dawkins has expressed it, such convergent resemblances

> provide most impressive demonstrations of the power of natural selection to put together good designs. . . . The basic rationale is that, if a design is good enough to evolve once, the same design *principle* is good enough to evolve twice, from different starting points, in different parts of the animal kingdom.[54]

But what are these "good designs," and what are the "design principles"—the "principles" of porcupines, for instance? They remain unexplained in mechanistic terms (chapters 5 and 6). From the present point of view, they are inherent in morphic fields. This hypothesis does not so much contradict the standard explanation as go beyond it. Natural selection still plays an important role; but it is no longer the great creative power, the designer and sustainer of all forms of life, the ultimate explanatory principle which replaces the God of nineteenth-century natural theology. The designs

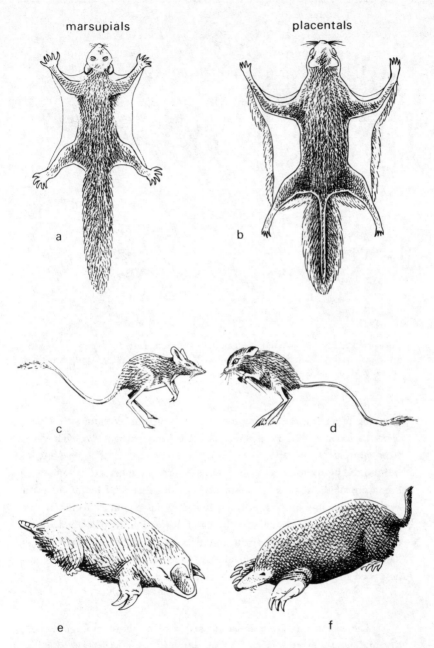

marsupials   placentals

a

b

c

d

e

f

Figure 16.6 Examples of parallel evolution. A and B, a marsupial flying phalanger and a placental flying squirrel; C and D, marsupial and placental jerboas; E and F, marsupial and placental moles. (After Hardy, 1965)

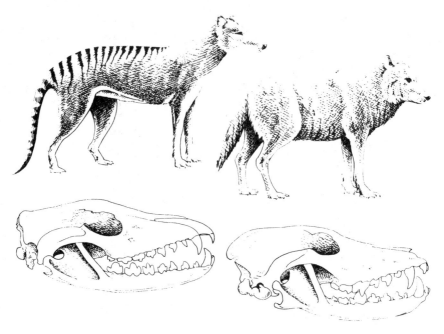

Figure 16.7 Another example of parallel evolution. Left, the Tasmanian wolf, a marsupial; right, the familiar placental wolf. The corresponding skulls are shown below. (After Hardy, 1965)

of living organisms are not imposed upon them from without—by God or natural selection—but are inherent in the organisms themselves. They arise from morphic fields; and these fields are not coded in the genes but are transmitted by morphic resonance. Usually they are inherited by subsequent organisms of the same species; but occasionally they may be picked up by organisms of quite different species, and show up as mutant forms. If these plagiaristic mutants are favoured by natural selection, their forms will tend to be repeated again and again and thus by morphic resonance become habitual characteristics of the plagiarizing species. They may then of course be passed on to other species descended from it.

The idea of formative causation sheds a new light on biological evolution and greatly extends Darwin's conception of natural selection to include the natural selection of morphic fields. It emphasizes the role of habit, as Darwin himself did, and allows for a transfer of habits by morphic resonance, not only *within* a species but also to *other* species. It thus provides a new understanding of the phenomena of atavism and of parallel and convergent evolution. But the discussion in this chapter has largely been confined to a few aspects of the evolution of the forms of plants and animals. It should be possible to extend this reinterpretation to cellular and molecular evolution on the one hand, and on the other hand to the evolution of instincts, animal societies, symbiotic associations, ecosystems, and human societies and cultures. The hypothesis of formative causation is still at an early stage of its development, and its evolutionary implications in these areas have as yet scarcely begun to be explored.

CHAPTER 17

Formative Causation in Cosmic Evolution

The neo-Darwinian theory is an attempt to understand the evolution of life in the context of a mechanistic universe—a theoretical universe of eternal matter and energy, governed by eternal laws of nature (chapter 3). This mechanistic world view has been like a Procrustean bed into which biological evolution has been forced to fit. Over the last few decades, many biologists have believed with deep conviction that neo-Darwinism is the only way in which evolution can be conceived of scientifically, without recourse to mystical explanations or to the creative power of God. This theory, with its central emphasis on the natural selection of genes, was elaborated in the 1930s and 1940s, and its subsequent development has been built on the theoretical foundations established in those decades.

However, the mechanistic world view presupposed by the neo-Darwinian theory has in the meantime been superseded by a great revolution in cosmology. The cosmos now seems more like a developing organism than an eternal machine. Currently, in the late 1980s, theoretical physicists are in the process of developing entirely new, evolutionary conceptions of matter and of the fundamental fields of nature.

The hypothesis of formative causation, unlike the mechanistic theory of nature, is based on the idea that all nature is evolutionary.

In this chapter, we consider the idea of formative causation in the context of the evolutionary theories of contemporary physics.

The Evolution of the Known Fields of Physics

Einstein's unfulfilled vision was a unified field theory: a theory that would enable the fundamental fields recognized by physicists, the gravitational, electro-magnetic and quantum matter fields, to be understood in terms of a single fundamental field. The ultimate goal was to find a single set of equations that could be used to predict all of the characteristics of these different kinds of fields. This vision continues to attract many theoretical physicists and is the goal towards which much contemporary theorizing is directed. If this goal were achieved—if physicists understood "the basic laws of the creation and subsequent evolution of the universe"[1]—then theoretical physics would reach its end. Some theoretical physicists, notably Stephen Hawking,[2] think that this end is already in sight.

A step in the direction of a unified field theory has already been taken by the unification of the electro-magnetic field and the "weak force" field associated with particles such as electrons and neutrons. Over the last few years, several approaches to further unifications have grown out of high-energy particle physics. One type of conceptual scheme goes under the name of grand unified theories, or GUTs; another is called supersymmetry. In the words of Paul Davies:

> Together these investigations point towards a compelling idea, that all nature is ultimately controlled by the activities of a single *superforce*. The superforce would have the power to bring the universe into being and furnish it with light, energy, matter, and structure. But the super-force would amount to more than just a creative agency. It would represent an amalgamation of matter, spacetime, and force into an integrated and harmonious framework that bestows upon the universe a hitherto unsuspected unity.[3]

One conjecture is that the various fields and forces arise from eleven dimensions, ten of space and one of time.

> Although the extra seven dimensions are invisible to us, they still manifest their existence as *forces*. What we think of as, say, an electro-magnetic force is really an unseen space dimension at work. The geometry of the seven extra dimensions reflects the symmetries inherent in the forces. It follows from this work that there really are no force fields at all, only empty eleven-dimensional spacetime curled into patterns. The world, it seems, can be built more or less out of structured nothingness.[4]

Until about the end of 1984, a theory of "supergravity" cast within this eleven-dimensional framework was favoured by many of the leading theoreticians as the most promising approach to an ultimate Theory of Everything (TOE). But since late 1984 a new approach using only ten dimensions, the superstring theory, has rapidly become fashionable. In this theory, particles are no longer treated as points, but as vibrating and rotating "strings." Some superstring theories treat these as open strings with free ends; others postulate closed strings joined into loops. Such theories involve "a profound generalization of the conventional field theory framework."[5]

These new field theories embody a conception of an original unified field, whose unified nature is manifested at ultra-high energies such as occurred very briefly at the beginning the universe. As the universe expanded, one by one the known fields of physics separated in identity from the unified field, which nevertheless continues to exist even though its unified nature is no longer manifested (Fig. 17.1). As the fields separated out, energy gave rise to matter: "Step by step the particles which go to build all the matter in the universe acquired their present identities. It was also at this stage that the beginnings of galaxies were generated."[6]

A study of thousands of clusters of galaxies has shown that their pattern of distribution in space cannot be accounted for simply in terms of gravitational processes. A recent hypothesis accounts for this in terms of "cosmic

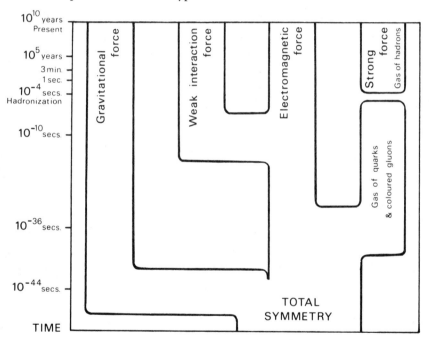

Figure 17.1 The evolutionary tree of the fields of nature which underlie the four known kinds of force. According to modern field theories, at the very earliest times, the fields of nature that are now seen as distinct were unified; the known fields of physics evolved through successive processes of symmetry-breaking. (After Pagels, 1985)

strings," which correspond roughly to "superfluid vortices" in the cosmic vacuum. The cosmic strings are supposed originally to have formed a dense network.

> But [they] quickly evolve as the Universe expands into a thin net of infinite (open) strings and a debris of closed oscillating string loops that have broken off the net. It is the loops which are thought to form galaxies by gravitationally accumulating matter around themselves. Because the locations of the loops are not determined by gravity, but rather by how and where they broke off from the net of open strings, unusual correlations are expected to be found in the locations of galaxies and clusters. . . . To the extent that the statistical properties of a string system evolve self-similarly in time, a scale-free autocorrelation is expected to develop among string loops.[7]

This whole subject is in a state of intense ferment and seems to be leading towards an entirely new conception of physical fields. One ingredient in this conception is that fields have evolved in such a way that intelligent life emerged, at least on Earth. This is the "Anthropic Cosmological Principle" (p. 9), which some physicists take to mean that there is an underlying direction and purpose in the cosmic evolutionary process.[8]

The Evolution of Morphic Fields

Physicists have, naturally enough, focussed their attention on the known fields of physics. These are fields at the opposite ends of the scale of magnitude: on the one hand the fields of universal extent, gravitational and electro-magnetic, and on the other hand the submicroscopic fields of sub-atomic particles, giving rise to the "strong force" and the "weak force." So far, physicists have hardly begun to consider the idea of fields of systems that lie between these extremes, partly for the simple reason that the natural sciences are divided into departments: the study of molecules and crystals is the province not of physicists, but of chemists, crystallographers, biochemists, and molecular biologists; living organisms lie in the realm of biology, and minds in psychology.

In these areas the old atomistic philosophy is usually implicit: minds should be reducible to brains, brains and all other living systems to physics and chemistry, and chemistry itself should ultimately be reducible to the properties of atoms and subatomic particles and be fully explicable in terms of quantum theory. These reductions have not in fact proved possible, but the hope is that they will be achieved at some time in the future. This hope depends on several implicit assumptions. One is that physicists know about all the fundamental fields of nature. Consequently, chemists, crystallographers, biologists, and psychologists do not usually feel free to postulate new kinds of fundamental fields, because fundamental field theory is the province of physics. But physicists themselves have felt no such restraint; *below* the level of the atom, fields have proliferated prodigiously. Each kind of particle—and now scores have been recognized—has its own matter field.

In spite of the profound revolutions it has passed through, physics is still pervaded by the reductionist spirit that it inherited from the atomistic philosophy. Most physicists still believe that there are fundamental entities in the atomistic sense, even though these are now thought of as quarks or as superstrings rather than as atoms themselves. Hence the fields of these ultimate entities are regarded as fundamental fields, and together with gravitational and electro-magnetic fields are the fundamental fields of nature. The fields of systems at higher levels of complexity are not fundamental in the same sense, but derivative.

By contrast, the hypothesis of formative causation, in the spirit of the philosophy of organism, regards the morphic fields of systems at all levels of complexity as fundamental rather than as derivative from the known fields of physics. From this point of view, the matter fields of quantum physics could be regarded as the morphic fields of particles, nuclei and atoms. But just as the different kinds of subatomic field cannot be reduced to each other, neither can the morphic fields of holons such as cells, plants, and societies be reduced to each other and still less to the fields of subatomic particles. Rather, there are nested hierarchies of fields: the fields of molecules contain and embrace those of atoms, nuclei, and subatomic particles; the fields of cells contain and embrace those of molecules; and so on.

All these fields are stabilized by morphic resonance from previously existing similar systems. In the case of subatomic particles, atoms, molecules, and crystals, which have existed in countless numbers for billions of years, the fields are so strongly stabilized that for all intents and purposes they appear to be changeless; the conventional idea that they are governed by changeless laws is approximately correct. But although these laws may appear to resemble Platonic Ideas, on the present hypothesis they refer to habits, deep-seated, but habits just the same. The same is true of all systems, physical and biological, that have been repeated very frequently. It is only in the case of new structures of activity that the cumulative effects of morphic resonance are likely to be detectable experimentally as new habits are built up.

A natural extension of the morphic field approach would be to regard living ecosystems as complex organisms with morphic fields that embrace the communities of organisms within them, and indeed to regard entire planets as organisms with characteristic morphic fields, and likewise planetary systems, stars, galaxies, and clusters of galaxies. Certainly, stars, galaxies, and galactic clusters can be classified into types or species. From the present point of view, the individual examples of each type are organized by characteristic morphic fields stabilized by morphic resonance from previous similar systems. Their development can be thought of as following chreodes: the various kinds of stars, for example, are thought to undergo a more or less predictable progressive development, some ending up as exploding supernovae, others collapsing into themselves and becoming black holes. And perhaps the string theory of galaxy formation from closed, vortical, vibratory loops of cosmic strings is already pointing towards a conception of the morphic fields of galaxies.

Nothing is known about planetary systems in other parts of the universe because they are undetectable by today's astronomical instruments. Nevertheless, it is generally assumed that there must be many billions of planetary systems. Perhaps they too fall into distinct types, associated with characteristic fields. Our own planetary system may not be unique; and if there are others like it, then the field of ours may be influenced by morphic resonance from them and may in turn influence them. The same could be true of the various planets: these too may represent "species" that occur elsewhere, a Mercury species, Venus species, Earth species, and so on. And of course in other solar systems there may be many other kinds of planets that do not occur in our own.

The possibility that there may be other planets very similar to ours immediately opens another vista of speculation. For if such planets exist, Earth may be following a developmental pathway that is already established and stabilized by morphic resonance; and perhaps the entire process of biological evolution is organized by a well-worn chreode.

On the other hand, it is possible that Earth is the first planet on which life of our own kind has evolved; in this case there would be no established evolutionary chreode; rather, a new one is developing. If similar forms of life have arisen subsequently on other, similar planets, or arise in the future, their general course of evolution may be shaped by evolution here and may tag along behind it, as it were.

When a new pattern of organization appears on Earth—say, a new kind of molecule or a new pattern of animal behaviour—if the same pattern has already existed billions of times elsewhere, the morphic fields should be well stabilized, assuming, as postulated, that morphic resonance does not fall off at astronomical distances. Consequently this background resonance would swamp any incremental effect of morphic resonance from the new patterns of organization here on Earth. None of the experimental tests of the hypothesis of formative causation would work, because they depend on detecting *changes* in the strength of morphic fields.

Alternatively, the failure of these tests for the effects of morphic resonance could, of course, simply mean that the hypothesis is wrong. The laws and fields of nature may indeed be changeless, as conventionally assumed.

On the other hand, if the experiments, or even only some of them, give positive results, this might mean that morphic resonance does fall off and become negligible at astronomical distances, or else it may mean that these new patterns of activity really have originated here, hence that they were unique when they first appeared. A truly creative evolutionary process may be happening here on Earth that is not, at least in detail, merely repeating what has happened elsewhere.

What we know of the rest of the universe indicates that similar patterns of organization appear again and again throughout the entire extent of space. These patterns are apparent at the highest level of organization, in galaxies and stars, and also at the lowest: the spectra of light emitted by the stars indicate that they arose by processes in atoms which behave in the same way in our own sun and on Earth. We could, of course, suppose that this simply shows that they all obey changeless universal laws. But we could also suppose that these similarities are maintained by morphic resonance over astronomical distances. There may be a universal network of morphic resonance among galaxies, stars, and atoms. And if this is so, then it is reasonable to suppose that there is also a universal network of resonance between molecules, crystals, and forms of life.

In considering the possible effects of morphic resonance over astronomical distances the question of how fast its influence can travel inevitably arises. There are at least three possibilities. Either this influence propagates at its own characteristic speed, which might be greater or less than that of light. Or it propagates at the speed of light. Or its effects may be in some way analogous to the non-local correlations in quantum theory, which are in a sense instantaneous. There seems at present no basis for deciding between these possibilities.

Universal Self-Resonance

In classical physics the endurance of matter and motion was taken for granted. Atoms were considered to be changeless and eternal, and the conservation principles asserted that mass and energy, momentum, electric charge, et cetera, are conserved. They are eternal, just as the laws of nature are eternal (chapter 2).

However, evolutionary cosmology raises not only the question of the evolution of the fields of nature, but also the question of why anything endures, persists, or continues.

The concept of morphic fields helps to explain why patterns of organization at all levels of complexity are repeated again and again. It also suggests an explanation for the persistence of any particular system in space and time: its morphic fields are stabilized by a cumulative resonance from its own past states. In general, a system resembles itself most closely in the immediate past, hence the most specific self-resonance will be from its own most recent states.

Perhaps the persistence of moving photons of light-energy may likewise depend on self-resonance with their own past vibratory movement. This can apparently continue indefinitely: the light reaching us from distant galaxies embodies a memory of them as they were millions of light years ago, and light is still reaching us from stars that have long since died. And the cosmic background radiation is thought to have originated in the Big Bang itself and to have continued in its movement ever since.

The universe contains structures of activity at all levels of complexity and magnitude, from subatomic particles to galactic clusters. *Ex hypothesi,* all of these are associated with characteristic morphic fields. Perhaps it makes sense to think of the entire universe as an all-inclusive organism. If so, then by analogy with all the kinds of organism within it, the entire universe would have a morphic field which would include, influence, and interconnect the morphic fields of all the organisms it contains.

If such a universal field exists, its properties and structure will be shaped by morphic resonance. But the universe is by definition unique. It could, perhaps, in principle be influenced by morphic resonance from other previous universes. But we have no way of knowing if there were any, and so this question seems unanswerable. In any case, the universal field will be subject to morphic resonance from its own past states, most specifically from the immediate past, but going right back to the beginning. This self-resonance would help to explain the continuity of the universe, as well as the continuity of material systems within it. Their own persistence, as we have seen, may depend on self-resonance with their own past states; but the self-resonance of the universal field within which they lie and through which they are interconnected may help to sustain their positions, movements, and interactions. This may suggest, among other things, a deeper understanding of the phenomenon of inertia.

The Implicate Order

Radical though it is, the new evolutionary physics has for the most part confined itself to the traditional subject matter of physics. Life and consciousness are recognized as a precondition for physics, both through the role of human theorists and observers and through the anthropic cosmological principle. Nevertheless, the nature of life and consciousness are not in practice taken into account in the actual theories of physics. These are the concern of other departments. But if a truly unified theory is ever to emerge, living organisms and conscious minds must be included within it along with the particles and fields of physics. There is a need for a new natural philosophy that goes further than physics alone can go but remains in harmony with it.

Perhaps the most profound of the new natural philosophies is the theory of the implicate order proposed by the physicist David Bohm. According to this theory, there are three major realms of existence: the explicate order, the implicate order, and a source or ground beyond both. The explicate order is the world of seemingly separate and isolated "thing-events" in space and time. The implicate order is a realm in which all things and events are enfolded in a total wholeness and unity, which as it were underlies the explicate order of the world we experience through our senses.

The implicate order is not somehow inserted into material systems in space and time; rather, material systems *and* space and time themselves all "unfold" from this underlying order. Any describable event, object, or entity in the ordinary, explicate world is "an abstraction from an unknown and undefinable totality of flowing movement." This universal flux Bohm calls the holomovement. "The holomovement, which is 'life implicit,' is the ground both of 'life explicit' and of 'inanimate matter,' and this ground is what is primary, self-existent and universal."[9] The holomovement "carries" the implicate order, and is an "unbroken and undivided totality."[10]

Bohm argues that a conception of undivided wholeness is implicit in both relativity and quantum physics. Einstein proposed that reality be regarded from the very beginning as constituted of fields. Particles are regions of intense field that can move through space. The idea that they are separate and independently existent is, at best, "an abstraction furnishing a valid approximation only in a certain limited domain."[11]

Quantum theory implies unbroken wholeness for three reasons. First, action is composed of indivisible quanta, and hence interactions between different entities (e.g., electrons) constitute "a single structure of indivisible links." Second, entities such as electrons can show differing properties (particle-like, wave-like or something in between) depending on their environmental context. Third, entities that have originally been combined show a peculiar non-local relationship "which can best be described as a non-causal connection of things that are far apart."[12] (This is the Einstein-Podolsky-Rosen paradox.[13])

One analogy for the implicate order is provided by holograms, in which the interference pattern in each region of the photographic plate is relevant to the whole structure, and each region of the structure is relevant to the whole of the interference pattern on the plate.[14] However, this analogy has the obvious limitation of being static and does not capture the idea of holomovement.

Bohm emphasizes the importance for physics, biology, and psychology of the notion of formative causation as "an ordered and structured inner movement that is essential to what things are." Any formative cause must evidently have an end or goal which is at least implicit—what Aristotle called a final cause. Thus, for example, it is not possible to refer to the inner movement from the acorn giving rise to the oak tree without simultaneously referring to the oak tree that is going to result from this movement. Bohm points out that in the ancient view, "the notion of formative cause was considered to be of essentially the same nature for the mind as it was for life and for the cosmos as a whole."[15]

Bohm relates the idea of formative cause to the holomovement, and sees the organization of physical particles, living organisms, and minds in terms of the hierarchy of implicate orders in this undivided process of flux. We experience this forming activity of the mind in the flowing movement of our own awareness. Each moment of consciousness has a certain explicit content and an implicit context which is a corresponding background. The actual structure, function, and activity of thought are in the implicate order. "The distinction between implicit and explicit in thought is thus being taken . . . to be essentially equivalent to the distinction between implicate and explicate in matter in general."[16]

Bohm's theory of the implicate order is more fundamental than the hypothesis of formative causation, but the two approaches appear to be quite compatible. Bohm and I have discussed their possible relationship,[17] and he has summarized his interpretation of morphic fields as follows:

> The implicate order can be thought of as a ground beyond time, a totality, out of which each moment is projected into the explicate order. For every moment that is projected out into the explicate there would be another movement in which that moment would be injected or "introjected" back into the implicate order. If you have a large number of repetitions of this process, you'll start to build up a fairly constant component to this series of projection and injection. That is, a fixed disposition would become established. The point is that, via this process, past forms would tend to be repeated or replicated in the present, and that is very similar to what Sheldrake calls a morphogenetic field and morphic resonance. Moreover, such a field would not be located anywhere. When it projects back into the totality (the

implicate order), since no space and time are relevant there, all things of a similar nature might get connected together or resonate in totality. When the explicate order enfolds into the implicate order, which does not have any space, all places and all times are, we might say, merged, so that what happens in one place will interpenetrate what happens in another place.[18]

What if Morphic Resonance Is Not Detectable?

What if experiments designed to test for the effects of morphic resonance fail again and again to reveal the predicted effects? There would be at least three possible interpretations, two of which we have already considered.

First, there is the possibility that most if not all new patterns of activity that appear on Earth have already appeared frequently elsewhere in the universe or in other or previous universes. Morphic resonance from these systems may swamp the predicted effects.

Second, the assumption of the hypothesis of formative causation that morphic resonance takes place only from the past may be wrong. It may emanate from the future as well, or even instead. If so, the resonance from innumerable future systems could make changes in the strength of morphic fields undetectable.

Third, the hypothesis may simply be wrong. This obvious conclusion would probably necessitate a return to the conventional idea that the laws of nature are changeless. In the light of evolutionary cosmology, such changeless laws would somehow have to exist prior to the universe and thus transcend space and time. They would indeed appear to be very like Platonic Ideas, and this is of course how physicists have traditionally thought of them. But what was previously a pure assumption, metaphysical in nature, would now be supported by experimental evidence; attempts to challenge and refute the assumption would have failed. Hence there would be stronger reasons for accepting it than if it had not been challenged or tested.

On the other hand, if experimental tests for morphic resonance support the hypothesis of formative causation, giving results that agree with its predictions, there could again be several possible interpretations.

First and most obvious is the possibility that the hypothesis is true, in the sense that it agrees with the facts.

Second, the results obtained in these experiments may simply be accepted as factual, without the need to accept the theoretical superstructure of the hypothesis of formative causation and concepts such as morphic fields and morphic resonance. The effects predicted by this hypothesis could be described by means of some general law that makes no attempt to offer an explanatory structure. For example, such a law could be expressed as follows: "Structures of activity tend to occur more readily the more often they have occurred before." This would be a law with predictive value, and its development and elaboration would involve attempts to define terms such as "structures of activity" more precisely.

Third, the theoretical terminology of the hypothesis of formative causation could be translated into other terminologies that seem more suitable for linking the hypothesis of formative causation to other areas of enquiry. For example, at one extreme there is the esoteric terminology of "subtle bodies" and "akashic records";[19] at the other extreme, the terminology of quantum physics with its non-local connections and correlations.

Fourth, the essential features of the hypothesis could be incorporated into other theoretical frameworks, such as Bohm's theory of the implicate order.

CHAPTER 18

Creativity Within a Living World

The Mystery of Creativity

The hypothesis of formative causation accounts for the regularities of nature in terms of habits. It explains the way in which patterns of organization are repeated again and again—for example, in the formation of haemoglobin molecules, in the growth of wheat plants, in the nest-buiding instincts of birds—in terms of morphic fields maintained by morphic resonance from previous similar systems. But it does not explain how any *new* pattern of organization, such as a new kind of crystal or a new instinct or scientific theory, comes into being in the first place. These new patterns are organized by new morphic fields. Where do these new fields come from? How are they created?

Creativity is a profound mystery precisely because it involves the appearance of patterns that have never existed before. Our usual way of explaining things is in terms of pre-existing causes: the cause somehow contains the effect; the effect follows from the cause. If we apply this way of thinking to the creation of a new form of life, a new work of art, or a new scientific theory, we are led to the conclusion that in some sense the new pattern of organization was already present: it was a latent possibility. Given the appropriate circumstances, this latent pattern becomes actual. Creativity thus consists in the manifestation or discovery of this pre-existing possibility. In other words, the new pattern has not been created at all; it has only been manifested in the physical world, whereas previously it was unmanifest.

This is in essence the Platonic theory of creativity. All possible forms have always existed as timeless Forms, or as mathematical potentialities implicit in the eternal laws of nature: "The possible would have been there from all time, a phantom awaiting its hour; it would therefore have become reality by the addition of something, by some transfusion of blood or life," as Henri Bergson expressed it.[1] He went on to point out that this is the conception inherent in the traditional European philosophies:

> The ancients, Platonists to a greater or lesser degree, . . . imagined that Being was given once and for all, complete and perfect, in the immutable system of Ideas; the world which unfolds before our eyes could therefore add nothing to it; it was, on the contrary, only diminution or degradation; its successive states measured as it were the increasing or decreasing distance between what is, a shadow projected in time, and what ought to be, Idea set in eternity. The moderns, it is true, take a quite different point of view. They no longer treat Time as an intruder, a disturber of eternity; but they would very much like to reduce it to a simple appearance. The temporal is, then, only the confused form of the rational. What we perceive as being a succession of states is conceived by our intellect, once the fog has settled, as a system of relations. The real becomes once more the eternal, with this simple difference, that it is the eternity of the Laws in which the phenomena are resolved instead of being the eternity of the Ideas which serve them as models.[2]

Both the Platonic philosophy and the theories of mechanistic physics were conceived in the context of a world that did not evolve. Eternal Forms or laws seemed appropriate enough in an eternal universe. But they are inevitably thrown into question by the idea of evolution as a process of creative development. We can now no longer ignore the possibility that creativity is real; everything may not be given in advance; new patterns of organization may be made up as the world goes on. Everything new that happens is possible in the tautological sense that only the possible can happen. But we need not attribute to these possibilities, which are unknowable until they actually happen, a pre-existent reality transcending time and space.

In this chapter we consider a variety of ways in which the creativity of the evolutionary process can be envisioned; but it is important to recognize at the outset that none of them can ultimately succeed in dispelling the mystery. If we choose to adopt a Platonic approach, we are left with the mystery of a transcendent realm of latent possibilities. If, on the other hand, we accept that there is a genuine creativity in the evolutionary process, how can we explain it? We can attribute it to God, or to intelligent spirits such as angels, to goddesses, Nature herself, chance, life, or fields. But then we cannot explain why any of these should have the capacity to create new patterns of organization: we sooner or later reach the limits of our understanding. If we attribute creativity to divine powers or superhuman intelligences on the one hand, or to chance on the other, we reach these limits sooner; if we recognize that creative capacities are inherent in morphic fields themselves, we reach these limits later, but we reach them just the same.

We begin by considering the conception of creativity inherent in the mechanistic world view in its original, seventeenth-century form and the radical change that has been brought about by the theory of evolution.

How Evolution Brings Nature Back to Life

In the mechanistic philosophy of nature as it was originally conceived in the seventeenth century, all creativity was ascribed to God. He was the sole source of all matter and motion, of all the laws of nature, and of all the designs of plants and animals. Nature itself was inanimate—blind, unconscious, and mechanical, with no freedom or spontaneity at all. Nature was not *creative;* it was *created.*

Before the advent of the mechanistic philosophy, nature had been thought of as alive; the world itself was animate, as were all beings within it. They had a life of their own, and their own internal purposes. When nature was personified, she was the Great Mother. When she was depersonified, the Mother became matter in motion, still the source and substance of all things, but no longer with any life or spontaneity; she was governed in every respect by the eternal laws of the Heavenly Father. In effect, the mechanistic philosophy treated the entire material world as if it were dead;[3] it had no life of its own (chapters 2 and 3). In so far as the structures of flowers, the structures of organs such as the eye, or the nest-building instincts of birds seemed to have purposive designs, these, like all other aspects of the natural world, reflected the supreme designing intelligence of the God of the world machine.

But this mechanistic world of Newtonian physics did not evolve: everything had been designed and created by God at the outset—or else, for those who rejected this idea of God, the universe and the laws that governed it were eternal and self-subsistent; there was no need for any creativity at all, because everything proceeded as it did with an inexorable, mechanical necessity and was in principle entirely predictable.

As the evolutionary vision developed in the nineteenth century, it began to bring nature back to life again. A creative spontaneity re-emerged in the natural world.

Darwin made this very clear. The source of evolutionary creativity is not *beyond* nature, in the eternal designs and plans of a machine-making God, the God of Paley's natural theology (chapter 3). The evolution of life has taken place spontaneously, *within* the material world. Nature itself, or herself, has given rise to all the myriad forms of life.

Darwin could not help personifying nature (p. 272). And in personified terms, what his theory tells us is that Mother Nature, rather than the Heavenly Father, is the source of all life. The Great Mother is prodigiously fertile; but she is also cruel and terrible, the devourer of her own offspring. It was this destructive aspect of Nature that impressed Darwin so deeply, and in the form of natural selection he made it the primary creative power, "a power incessantly ready for action."[4]

Thus in the light of the Darwinian theory of evolution, nature becomes creative, and takes on at least some of the attributes of the archaic Mother Goddess, from whose womb all life comes forth and to whom all life returns. When depersonified, she can simply be called nature, or matter, or life, or emergent evolution. And thus evolutionary creativity can be attributed to either the Great Mother herself or the depersonified abstractions that replace her and from which new forms of life come forth.

In dialectical materialism, for example, the creative source of everything is called matter and undergoes a continual, spontaneous, dialectical development, resolving conflicts and contradictions in successive syntheses. But, clearly, "matter" in this sense has prodigious creative properties that the matter of Newtonian physics did not have; the permanent billiard-ball atoms had no such power to create cells or giraffes or the philosophical theories of Marx and Engels. Even the dynamic, self-organizing atoms of modern quantum physics have no such creative power. And if we extend the meaning of the word *matter* to include not merely matter as physicists conceive of it but also fields and energy, and indeed all physical reality, then we might as well call it nature; but not of course the inanimate, uncreative nature of Newtonian physics, but the creative nature of an evolutionary world.

Henri Bergson attributed this creativity to the *élan vital* or vital impetus. Like Darwinians, Marxists, and other believers in emergent evolution, he denied that the evolutionary process was designed and planned in advance in the eternal mind of a transcendent God; rather, it is spontaneous and creative:

> Nature is more and better than a plan in course of realization. A plan is a term assigned to a labour: it closes the future whose form it indicates. Before the evolution of life, on the contrary, the portals of the future remain wide open. It is a creation that goes on for ever in virtue of an initial movement. This movement constitutes the unity of the organized world—a prolific unity, of an infinite richness, superior

to any that the intellect could dream of, for the intellect is only one of its aspects or products.[5]

The neo-Darwinian theory of evolution shares this vision of evolution as a vast, spontaneous, creative process. As the molecular biologist Jacques Monod put it in his book on the neo-Darwinian world view, *Chance and Necessity,* "evolutionary emergence, owing to the fact that it arises from the essentially unforseeable, is the creator of *absolute* newness."[6] What Bergson attributed to the *élan vital,* Monod ascribed to "the inexhaustible resources of the well of chance,"[7] expressed through random mutations in DNA.

In this conception, the creative role of chance, of that which is indeterminate, is expressed in its interplay with necessity, that which is determinate. Here again, it is illuminating to see what happens when these abstract principles are personified. Just as nature becomes the Great Mother, they too come to life in the form of goddesses. In pre-Christian Europe, Necessity was one of the names for Fate or Destiny, often represented by the Three Fates, the stern spinning-women who spin, allot, and cut the thread of life, dispensing to mortals their destiny at birth. This ancient image is paralleled in neo-Darwinian thinking in a curiously literal manner. The "thread of life," which determines an organism's genetic destiny, consists of the helical DNA molecules arranged in threadlike chromosomes.

On the other hand, Chance is one of the names of the goddess Fortune. The turnings of her wheel, the Wheel of Fortune, confer both prosperity and misfortune. She is the patroness of gamblers; another of her traditional names is Lady Luck.[8] The goddess Fortuna is blind. And so is chance:

> Chance *alone* is at the source of every innovation, of all creation in the biosphere. Pure chance, absolutely free but blind, at the very root of the stupendous edifice of evolution: this central concept of modern biology is no longer one among other possible or conceivable hypotheses. It is today the *sole* conceivable hypothesis, the only one compatible with observed or tested fact. And nothing warrants the supposition (or the hope) that conceptions about this should, or ever could, be revised.[9]

However, the material world, the realm in which chance and necessity hold sway, is only one aspect of the mechanistic world view. The other is the Platonic realm of eternal Forms, laws, or mathematical formulae. Some biologists prefer to see in this realm, rather than in the workings of blind chance, the source of all new forms of life. The evolution of the dinosaurs or of starfish or of palm trees represents the manifestation of pre-existing non-material archetypes (p. 107). These archetypes themselves do not evolve, being beyond time and space. Either they are Ideas in the mind of God, or, if we dispense with God, they have an independent existence which is inexplicable in terms of anything else.

Thus neo-Darwinism leads to an impasse. In so far as evolutionary creativity depends on the manifestation of eternal Forms or principles of order, it is not true creativity at all, but only the manifestation of patterns that have always existed in a non-material realm. And in so far as creativity depends on blind chance, it is essentially unintelligible, and we just have to leave it at that.

Traditionally, in Europe, the transcendent realm has been regarded as the province of the Heavenly Father and the material realm the province of the Great Mother. In these personified terms, the Platonic approach stresses the rational, male creative principle; while the materialist approach stresses the non-rational, female aspects of creativity. Do these personified archetypes represent a deeper way of understanding the mystery of creativity than the depersonified abstractions of modern thought? Or do these impersonal abstractions represent a higher form of understanding that has outgrown the primitive, personified modes of thought found in the realms of myth and religion? This is obviously a matter of opinion; but whichever way we prefer to see it, the archaic and modern ways of explaining creativity show striking parallels.

This is about as far as we can go in the context of neo-Darwinism. The evolutionary philosophy of organism allows us to go further. The organizing principles of nature are not beyond it, in a transcendent realm, but within it. Not only does the world evolve in space and time, but these immanent organizing principles themselves evolve. According to the hypothesis of formative causation, these organizing principles are morphic fields, which contain an inherent memory.

In general terms, fields have inherited many of the properties traditionally ascribed to souls in the pre-mechanistic philosophies of nature, and the growth of field theories can be regarded as another of the ways in which nature has been coming back to life, as I now consider. I go on to discuss how creativity is expressed within the context of *existing* morphic fields, and then consider how *new* morphic fields might arise.

Fields, Souls, and Magic

What the mechanistic philosophy of the seventeenth century rejected, and what mechanists still reject, is the idea that the world and all living beings within it are animate, in other words that they are organized by non-material souls or animas or psyches. This ancient idea was worked out in systematic detail by Aristotle and by the neo-Platonists. It pervaded medieval and Renaissance philosophy. It persisted into the present century in biology within the vitalist tradition. And over the last sixty years it has been revived in a modern evolutionary form in the holistic philosophy of organism. In this process the idea of souls as purposive organizing principles has been replaced by concepts of organizing fields, organizing relations, principles of self-organization, mind in nature, patterns that connect, the implicate order, information, and organizing principles under yet other names.

Of course, from a mechanistic point of view, both traditional animism and modern organicism involve an invalid projection of the qualities and purposes of human life onto the inanimate world around us. This is the "pathetic fallacy." In Monod's words, animist belief, in which he included both organicism and dialectical materialism, "consists essentially in a projection into inanimate nature of man's awareness of the intensely teleonomic functioning of his own nervous system."[10]

This kind of projection would inevitably result in delusion if nature is in fact both inanimate and mechanistic. But this begs the question, for it leaves unexplained the intensely teleonomic functioning of our nervous systems themselves, the teleonomy of all living organisms, and the self-organizing properties of natural systems at all levels of complexity.

Ironically, the mechanistic approach itself seems to be *more* anthropomorphic than the animistic. It projects one particular kind of human activity, the construction and use of machines, onto the whole of nature. The mechanistic theory derives its plausibility precisely from the fact that machines *do* have purposive designs whose source is in living minds.

As a matter of fact, classical physics is replete with terms that imply all sorts of correspondences between human life and the realm of nature, words whose animistic associations have long been more or less unconscious: for example *law, force, work, energy,* and *attraction.* Quantum physics has mischievously added more, such as *charm.* And in orthodox biology too we find essential explanatory terms that do not properly belong in an inanimate world: *function, adaptation, selection, information, programs,* and so on.

Mechanistic science developed against an animistic background, in a world where magic was still taken seriously. A number of ancient magical conceptions are surprisingly similar to essential elements of classical and modern physics. In this connection, the following discussion by the anthropologist James Frazer is particularly interesting:

> If we analyse the principles of thought on which magic is based, they will probably be found to resolve themselves into two: first, that like produces like, or that an effect resembles its cause; and second, that

things which have once been in contact with each other continue to act on each other at a distance after the physical contact has been severed. The former principle may be called the Law of Similarity, the latter principle the Law of Contact or Contagion. From the first of these principles, namely the Law of Similarity, the magician infers that he can produce any effect he desires merely by imitating it: from the second he infers that whatever he does to a material object will affect equally the person with whom the object was once in contact, whether it formed part of his body or not. . . . The same principles which the magician applies in the practice of his art are implicitly believed by him to regulate the operations of inanimate nature; in other words, he tacitly assumes that the Laws of Similarity and Contact are of universal application and not limited to human actions.[11]

In various ways these same two principles play an essential role in classical physics; and in the light of the non-locality inherent in quantum theory the law of contact has taken on a new significance.

Physicists, like Frazer's magicians, gain their power by imitating nature: mathematics has proved a very effective means of doing so. Physicists make mathematical *models* of natural processes, mental constructions in imaginary mathematical space. Not all the models they make are successful. But the successful ones seem to correspond in some mysterious way to various aspects of the physical world. It is by virtue of these models that aspects of reality can be predicted, controlled, and manipulated. Such models lie at the heart of all modern technologies.

Like the world of the magician, the world of the physicist is full of unseen connections traversing apparently empty space. As Frazer put it, the laws of magic

assume that things act on each other at a distance through a secret sympathy, the impulse being transmitted from one to the other by means of what we may conceive as a kind of invisible ether, not unlike that which is postulated by modern science for a precisely similar purpose, namely, to explain how things can physically affect each other through a space which appears to be empty.[12]

Now, of course, fields themselves rather than fields of ether are thought to be the medium of the secret sympathies of nature.

According to the old animistic philosophies, the *anima mundi*, the soul of the world, and the souls of all beings within it were immutable. They influenced the matter with which they were associated, but their nature was not changed by it: they did not evolve; they stayed the same. Until very recently the fields of physics were thought of in a similar way. They stayed the same: their nature was not changed by the energy they contained and organized or by what happened within them. But now they are thought to have evolved: they have histories.

Contemporary theories of the evolution of physical fields uneasily straddle two very different paradigms: the traditional conception of eternal mathematical laws and the idea of the universe as a vast evolving organism. Are the mathematical structures of a grand unified theory or a theory of everything more real than the fields through which they are manifested in space and time? Or are the fields more real than the mathematics by which they are described and modelled?

If mathematical laws are more real than the fields, then the ultimate reality is still in the transcendent realm of eternal Ideas or laws. This is what most physicists have traditionally assumed.

If, on the other hand, the fields are more real than the mathematics we use to model them, then we find ourselves in an evolving universe whose organizing principles are evolving with it.

Creative Morphic Fields

The *evolution* of organizing fields is a very unfamiliar idea. It is alien to the traditional animisms, alien to the traditions of physics, and alien to the mechanistic philosophy. For if fields evolve, it is no longer appropriate to explain them in terms of immutable essences or changeless laws; nor does the concept of blind chance seem sufficient to explain the appearance of such integrated structures of ordering.

Before we explore in more detail the possible roles of morphic fields in evolutionary creativity, let us recall the hypothetical properties of these fields at all levels of complexity:

1. They are self-organizing wholes.
2. They have both a spatial and a temporal aspect, and organize spatio-temporal patterns of vibratory or rhythmic activity.
3. They attract the systems under their influence towards characteristic forms and patterns of activity, whose coming-into-being they organize and whose integrity they maintain. The ends or goals towards which morphic fields attract the systems under their influence are called attractors.
4. They interrelate and co-ordinate the morphic units or holons that lie within them, which in turn are wholes organized by morphic

fields. Morphic fields contain other morphic fields within them in a nested hierarchy or holarchy.

5. They are structures of probability, and their organizing activity is probabilistic.

6. They contain a built-in memory given by self-resonance with a morphic unit's own past and by morphic resonance with all previous similar systems. This memory is cumulative. The more often particular patterns of activity are repeated, the more habitual they tend to become.

In the course of this book, we have considered the expression of these properties at the molecular and crystalline levels, in the morphogenesis of plants and animals, in animal and human behaviour, in human learning and memory, in social organization and culture, and in the evolutionary process. So far the question of creativity has been left open. In an attempt to address it, we will first examine how creativity is expressed within existing morphic fields and then consider how entirely new fields could conceivably originate.

The kind of creativity expressed within the context of already existing morphic fields is creativity in a weak sense of the word. The end-points or goals or attractors given by the fields remain the same; what are new are the ways of reaching them. This kind of creativity is commonly expressed by words such as *adaptability, flexibility, ingeniousness,* and *resourcefulness.* The appearance of entirely new fields with their new goals or attractors involves a higher order of creativity or originality. Some people would probably prefer to restrict the use of *creativity* for this latter, stronger sense. But this word is often used in contemporary discourse in a wide and general way that includes both strong and weak senses, and the following discussion adopts this general usage.

The main reason that developmental biologists proposed the idea of morphogenetic fields in the first place was because organisms can retain their wholeness and recover their form even if parts of them are damaged or removed (Fig. 5.3). The field in some sense contains the form or pattern of the entire morphic unit, and it attracts the developing or regenerating system towards it. If the process of development is displaced from its normal pathway, it can return to it—just as a ball pushed up the side of a hill can roll back into the valley and rejoin the usual canalized pathway of change in Waddington's model of a chreode (Fig. 6.2).

In all processes of regulation and regeneration, the developmental process adjusts in such a way that a more or less normal structure of activity is regained by a more or less new route. In other words, there is an element of novelty or creativity in the developmental process. A striking example is the way in which a newt's eye regenerates after the surgical removal of its lens. In normal embryonic development the lens develops from an infolding of the embryonic skin tissue that overlies the developing eye; but in response to the removal of the lens from an eye that is already mature, a new lens arises from the edge of the iris (Fig. 18.1).

Many further examples of regulation or adjustment are provided by the way in which developing organisms respond to genetic mutations. Mutant organisms are not merely the product of mutant genes: they are the result of developmental processes that have adjusted to the new internal conditions in such a way that entire, integrated organisms are still produced, even if they are abnormal in various ways. In so far as random mutations are a source of evolutionary creativity, the creativity is inherent not so much in the chromosomal and genetic changes as in the ways that organisms respond and adjust to them: it is an expression of the organizing activity of the morphic fields.

Environmental changes, like genetic mutations, impose new necessities on organisms. Necessity is the mother of invention: but the inventions themselves are made by organisms. The adjustment of the form and function of plants and animals to the conditions of life, their *adaptation* to the environment, occurs in countless ways; and such purposive adaptations, which so impressed both Lamarck and Darwin, tend to become increasingly hereditary and habitual the more often they are repeated, and are a major source of creativity in the evolutionary process.

Likewise, in the realm of behaviour, we find comparable abilities to adjust creatively to genetic mutations, to damage, and to change in the

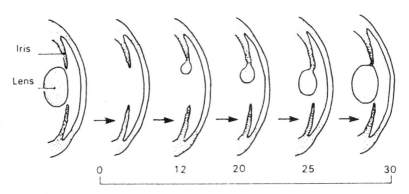

Days after surgical removal of lens

Figure 18.1 Regeneration of a lens from the margin of the iris in a newt's eye after the surgical removal of the lens. (Cf. Needham, 1942)

environment. Animals born with abnormal bodies sometimes manage to survive in spite of them through appropriate adjustments of their movements and behaviour. Creatures that have lost a limb or some other structure often adjust to the damage more or less effectively: for example a dog can learn to run on three legs, and blind people develop ways of finding their way around by relying on other senses. Often damage to termites' nests is repaired. If obstacles are put in the path of animals or of people who want to get somewhere, they may find a way round the obstacle and get there by a different route. Usually animals and people transferred to new and unfamiliar environments can adjust to them more or less appropriately.

Of course, not all kinds of mutation, damage, and environmental change elicit successful responses. Many are immediately lethal. Others are too extreme for adaptation to occur successfully. But within certain limits, innovative responses occur at all levels of organization. Morphic fields appear to have an inherent creativity, which is recognizable precisely because the new pathways of development or behaviour often seem so adaptive and purposeful.

To some extent, *every* individual organism and *every* element of its structure and behaviour represents a creative response to its internal and external conditions. No two organisms of the same kind are exactly identical; they are in different places, in different microenvironments, made up of different atoms and molecules, and subject to chance fluctuations from the quantum level upwards. Morphic fields are not rigid; they are probability structures and bring about their ordering effects through a probabilistic influence; they have an inherent flexibility. They encompass the uniqueness of individual morphic units within a field of probability that defines the structure and the limits of the type; they are, in the language of dynamics, attractors or basins of attraction (pp. 101–2).

"Where there's a will, there's a way." The will is given by the goal or the morphic attractor, which, from the present point of view of the organism, lies in the future. The progress of a system towards its morphic attractor involves adjustments, great and small, of its component parts and their interrelationships; it finds a way. In so far as it is prevented from following the usual, habitual path, it may find a more or less novel means of reaching the same goal.

Much human creativity is of this general type: it involves finding new ways to achieve habitual goals or ends—inventive ways of saying or doing or making things; ingenious ways of repairing things; the solving of puzzles and problems; the making of better mousetraps.

The finding of new ways is different in degree from the process by which we learn our usual ways of behaving, speaking, and thinking, but it does not seem to be different in kind. When we learn anything at all, success in achieving the given goals depends on doing it in a manner that is adapted to our own bodies, skills, and circumstances. And every time we do something or speak or think, our habits adjust themselves more or less well to the conditions that obtain on that particular occasion.

A great deal of this adjustment is unconscious. But even when we use our conscious minds to adjust, adapt, find a new way, or solve a problem, we generally find it hard to say how we did it. The answer just comes, it happens, we stumble upon it, or the penny drops. It is as if the new ways come *through* consciousness; but the creative processes themselves are unconscious, beneath or behind our awareness.

The inherent tendency of systems to find a way to their morphic attractors, or to find a way back to them, is also expressed in the context of social and cultural fields. The co-ordinated behaviour of social insects such as bees, for example, is organized by the morphic fields of the society; and if the hive is damaged and members of the colony are killed, the behaviour of the surviving insects is often regulated in such a way that the damage is made good and the harmonious functioning of the colony is restored. The adjustment of human families, communities, and larger societies to accidents, loss of life, external or internal threats, disturbances, and calamities seems comparable: individuals respond as the collective field, the group mind or *conscience collective,* adjusts to the new conditions and progressively restores the society to a co-ordinated integrity.

These fields work through their influence on the people within them. Some people may have more awareness than others of what needs to be done, and leaders of various kinds generally have the ability to communicate it. Both this awareness and people's responses to it are influenced by the collective field and are not just the product of separate, individual minds. Nor do rulers, patriarchs, matriarchs, shamans, prophets, priests, leaders, or other persons of authority claim that they are speaking merely as individuals: they do so under the aegis of the gods or guardian spirits or ancestors, the values or traditions of the group. Nor do they claim to be speaking and acting for themselves alone, but for the sake of the whole group's life and survival.

Habit and Creativity

The idea underlying the preceding discussion is that morphic fields have an inherent creativity. The idea emphasized in earlier chapters of this book is that they are habitual in nature. These two aspects of fields are complementary, rather than contradictory. Morphic fields contain goals or attractors that are indeed habitual and conservative; the creativity that occurs within them involves finding new ways of reaching these goals. Moreover, the expression of *any* habitual pattern of development or activity requires a certain flexibility and adaptation to circumstances; habits could hardly be viable without some degree of creative adaptability.

Nevertheless, as fields evolve and as habitual chreodes become established within them, there is a sense in which their inherent creativity is reduced. The evolutionary radiations or explosive phases that seem to occur fairly early in the history of a new phylum, order, family, genus, or species involve various differentiations or adaptations of the ancestral form (pp. 284–85). Comparable explosive phases may have occurred in the evolution of patterns of instinctive behaviour, as well as in the evolution of human languages and social, political, and cultural forms. A similar process occurs in the evolution of religions, arts, and sciences as distinct sects, schools, and traditions arise within them. In the realm of technology, there is often a comparable proliferation of versions and models following the invention of a new kind of machine: think, for example, of the variety of cars on the market, or the variety of microcomputers.

There is an obvious reason why the appearance of new variations on basic themes tends to become less frequent as time goes on: the number of possible variant forms may be finite. As new versions appear and either die out or become increasingly habitual, there are progressively fewer remaining potentialities that have not already been explored.

However, no amount of creativity expressed within the context of any morphic field at any level of complexity can explain the appearance of that field itself for the very first time.

The Origin of New Fields

The appearance of a new kind of field involves a creative jump or synthesis. A new morphic attractor comes into being, and with it a new pattern of relationships and connections. Consider a new kind of molecule, for example, or a new kind of instinct or a new theory.

One way of thinking about these creative syntheses involves looking from below, from the bottom up: we then see the "emergence" of ever more complex forms at higher levels of organization. The progressive appearance of new syntheses is elevated to a general principle in dialectical materialism and in other philosophies of emergent evolution. Evolution then becomes more than a word describing a process; it involves a creative principle inherent in matter, or energy, nature, life, or process itself. New patterns of organization, new morphic fields, come into being as a result of this intrinsic creativity. But why should matter, energy, nature, life, or process be creative? This is inevitably mysterious. Not much more can be said than that it is their nature to be so.

Another approach is to start from above, from the top down, and to consider how new fields may have originated from pre-existing fields at a higher and more inclusive level of organization. Fields arise within fields. For example, a new habit of behaviour, such as the opening of milk bottles by tits (Fig. 9.5), involves the appearance of a new morphic field. From the "bottom up" point of view, this must have emerged by the synthesis of pre-existing behavioural patterns, such as the tearing of strips of bark from twigs (pp. 178–80), in a new, higher-level whole. From the "top down" point of view, this new field arose in the higher-level, more inclusive morphic field that organizes the searching for food and all the activities involved in feeding. This higher-level field may somehow have formed within itself a new lower-level field, that of milk-bottle opening.

This creative process is interactive in the sense that the higher-level fields within which new fields come into being are modified by these new patterns of organization within them. They have a greater internal complexity, which is the context in which the further creation of new fields within them is expressed.

These principles may well apply at all levels of organization, from new kinds of protein molecules that have arisen within the fields of cells to galaxies within the field of the growing universe. In every case, the higher-level fields are influenced by what has happened in the past and what is happening within them now; their creativity is evolutionary.

Ultimately this way of thinking leads us back to the primal morphic field of the universe as the ultimate source and ground of all the fields within it. In the context of modern evolutionary cosmology, this is the original unified field from which all the fields of nature were derived as the universe grew and developed (Fig. 17.1).

In summary, we can either think of the creation of new fields as an ascending process, with new syntheses emerging at progressively higher levels of organization, or we can think of it as a descending process, with new fields arising within higher-level fields, which are their creative source. Or, of course, we can think of evolutionary creativity as somehow involving a combination of these processes.

The Primal Field of Nature

What could the idea of a primal, unified, universal field possibly mean?

The sceptic in all of us is inclined to think that it doesn't mean much. It is just another speculative theory that takes us beyond anything that we can directly observe. We are leaving empirical science behind us and entering the realm of metaphysics. There is no point in going further, for we will only enmesh ourselves in tangled webs of speculation.

If we do want to go further, we have to recognize that we are indeed in the field of metaphysics. For well over two thousand years, philosophers have discussed the source of pattern and order in the world, the nature of flux and change, the nature of space and time, and the relation of the changing world of our experience to eternity and changelessness. In one major tradition, rooted in the cosmology of Plato, these questions have been answered in terms of the *anima mundi,* the world soul, a conception not unlike the world field of modern cosmology. The world is contained within the world soul, which in turn is contained within the mind of God, the realm of Ideas beyond both time and space. The world soul differs from the realm of Ideas in that it has within it time, space, and becoming. It is the creative source of all of the souls within it, just as the world field is the source of all of the fields of nature.

Just as the notion of the world field raises the problem of its relationship to eternal laws, so the notion of the world soul raised the problem of its relationship to the eternal realm of Ideas. For the neo-Platonic philosopher Plotinus, these Ideas dwelt within what he called the Intelligence. The Intelligence differed from the Soul in possessing perfect self-awareness, and in contemplating the Forms themselves rather than images of the Forms. Just as the Intelligence "like some huge organism contains potentially all other intelligences," so the Soul contains potentially all other souls.

> The Intelligence is not merely one: it is one and many. In the same way, there is both Soul and many souls. From the one Soul proceed a multiplicity of different souls. . . . The function of the Soul as intellective is intellection. But it is not limited to intellection. If it were, there would be no distinction between it and the Intelligence. It has functions besides the intellectual, and these, by which it is not simply intelligence, determine its distinctive existence. In directing itself to what is above, it thinks. In directing itself to itself, it preserves itself. In directing itself to what is lower, it orders, administers, and governs.[13]

Below the influence of the Soul "we can find nothing but the indeterminateness of Matter."[14] But at all levels of existence the contents of the world are organized by souls; none are entirely indeterminate or inanimate.

> The whole constitutes a harmony, in which each inferior grade is "in" the next above. . . . The bond of unity between the higher and lower products of Soul is the aspiration, the activity, the life, which is the reality of the world of becoming.[15]

However we interpret the similarities and differences between the old idea of the world soul and the new idea of the world field, both inevitably raise the further question of their own origin and the source of the activity within them. The world soul was traditionally believed to arise from and to be contained within the Being of God. Some contemporary physicists believe that the world field is in some sense contained within or arises from eternal, transcendent laws. But then what is the source of these laws? How could such transcendent, non-physical laws have given rise to the physical reality of the universe? And in an evolutionary universe, why should we assume that all these laws were already fixed in advance?

We may, of course, simply regard the origin of the universe and the creativity within it as an impenetrable mystery and leave it at that. If we choose to look further, we find ourselves in the presence of several long-established traditions of thought about the ultimate creative source, whether this is conceived of as the One, Brahma, the Void, the Tao, the eternal embrace of Shiva and Shakti, or the Holy Trinity.

In all these traditions, we sooner or later arrive at the limits of conceptual thought, and also at a recognition of these limits. Only faith, love, mystical insight, contemplation, enlightenment, or the grace of God can take us beyond them.

EPILOGUE

We live in a world that was born some fifteen billion years ago, a world that has always been growing and still grows, a world of developing galaxies and stars and planetary systems and planets. On this particular planet, life has been developing for over three billion years in an evolutionary process that continues in ourselves. The development of science is part of this very process—a process that science itself has discovered, first in the realm of life on Earth and now in the whole of nature. In brief, we now have an evolutionary cosmology.

But many of our habits of thought grew up within the image of an eternal, machine-like universe. There was no need for memory in the mechanistic universe, because it was permeated at all times and in all places by timeless principles of order, the eternal laws of nature.

But do these old ideas still make sense in an evolutionary universe? Were the laws for everything in the world—from protozoans to galaxies, from orchestras to planetary systems, from molecules to flocks of geese—all present in advance, awaiting the time when their harmonious, ordering properties could be manifested in the evolutionary process? Or is memory inherent in nature? Do habits build up as evolution goes on?

These are the questions that we have been asking in this book. We have explored the implications of both the view of eternal law and the view of evolving habit. We have looked at a wide range of phenomena, in the chemical, biological, social, cultural, and mental realms, from both points of view, comparing the interpretations they offer; and we have considered various ways in which experimental tests could reveal to us which of these alternatives is in better accordance with the way things are.

At present the question is open. It is possible that we do, after all, live in an amnesic world that is governed by eternal laws. But it is also possible that memory is inherent in nature; and if we find that we are indeed living in such a world, we shall have to change our way of thinking entirely. We shall sooner or later have to give up many of our old habits of thought and adopt new ones: habits that are better adapted to life in a world that is living in the presence of the past—and is also living in the presence of the future, and open to continuing creation.

NOTES

Chapter 1: Eternity and Evolution

1. For a non-technical account, see Capra (1974).
2. Davies (1984), p. 105.
3. Barrow and Tipler (1986), pp. 412–3.
4. Teilhard de Chardin (1959).
5. Quoted in Burtt (1932), p. 9.
6. Monod (1972), p. 160.
7. See Laszlo (1987).
8. Guth and Steinhardt (1984).
9. Weinberg (1977); Pagels (1985).
10. Davies (1984), p. 8.
11. Press (1986).
12. Barrow and Tipler (1986), p. 16.
13. Ibid., p. 21.
14. Ibid., p. 23.
15. Pagels (1985), p. 347.
16. For a recent discussion, see Cartwright (1983).
17. Quoted in Potters (1967), p. 190.
18. Ibid.
19. Nietzsche (1911).
20. In Murphy and Ballou (1961).
21. E.g., Bergson (1911a,b); Fawcett (1916).
22. Butler (1880), pp. 175–6.
23. Ibid. (1878), p. 297.

24. For reviews of this discussion, see Russell (1916), pp. 335–44 and Gould (1977), pp. 96–100.
25. Hyatt (1893), p. 4.
26. E.g., Semon (1921); Rignano (1926).

Chapter 2: Changeless Laws, Permanent Energy

1. Burkert (1972), p. 40.
2. Philip (1966); Gorman (1979).
3. Philip (1966).
4. Burkert (1972).
5. Ibid., ch. 6, p. 482.
6. Jaki (1978); Wilber (1984), pp. 101–111.
7. Gilson (1984).
8. Burtt (1932), p. 42.
9. Dijksterhuis (1961), pp. 225–33.
10. Burtt (1932), p. 44.
11. Ibid., p. 45.
12. Ibid., p. 48.
13. Ibid., p. 54.
14. Ibid., p. 64.
15. Ibid., p. 75.
16. Translation in Wallace (1910), pp. 79–80.
17. Dijksterhuis (1961).
18. Koestler (1967).
19. Translation in Wallace (1910), pp. 79–80.
20. Ibid., p. 85.
21. Burnet (1930).
22. Merchant (1982).
23. Leclerc (1972); Dobbs (1975); Westfall (1980); Castillejo (1981).
24. Quoted in Burtt (1932), p. 257.
25. Einstein (1954), pp. 103–4.
26. In Wilber, ed. (1984), p. 185.
27. Ibid., p. 137.
28. Ibid., p. 116.
29. Ibid., p. 51.
30. However, in general relativity, complications are introduced by the breakdown of normal definitions of energy and momentum at distances comparable with the size of the entire universe.
31. Feynman (1965), p. 59.

32. Prigogine (1980), pp. 3–4.
33. Quoted in Pagels (1983), p. 336.
34. Davis and Hersch (1983).
35. Pagels (1983), p. 331.
36. Popper (1983), p. 134.
37. Ibid., pp. 138–9.

Chapter 3: From Human Progress to Universal Evolution

1. Eliade (1954).
2. Ibid.
3. *Revelation* 21:5.
4. Russell (1968); Griffiths (1982).
5. Eliade (1954), p. 104.
6. Nisbet (1980).
7. *Epistle to the Hebrews* 11:1,7,8,13–16.
8. Cohn (1957).
9. Bacon (1627).
10. For a discussion of Bacon's views on nature from a feminist perspective, see Griffin (1978); Merchant (1982).
11. See Gilson (1984), ch. 3.
12. Ibid.
13. Bowler (1984), ch. 8; Mayr (1982), ch. 11.
14. Bowler (1984), ch. 2.
15. Gillespie (1960), ch. 7.
16. Bowler (1984), ch. 2.
17. Gillespie (1960), ch. 7.
18. Gillespie (1979).
19. Ibid.
20. Darwin (1872), ch. 7.
21. Ibid.
22. Gould and Eldredge (1977).
23. Alvarez, et al. (1980); Maddox (1984a).
24. Maddox (1984b).
25. Hallam (1984).
26. Maddox (1985b).
27. Genesis 2:8–9.
28. Quoted in Dawkins (1986), pp. 4–5.
29. Darwin (1872), ch. 4.
30. Dawkins (1986), pp. ix–x.

31. Wallace (1911), pp. 394–5.
32. Bergson (1911b), pp. 27, 29.
33. Whitehead (1925), ch. 6.
34. Ibid.
35. Whyte (1974), p. 43.
36. Ibid., p. 40.

Chapter 4: The Nature of Material Forms

1. Quoted in Bynum, et al. (1981), p. 300.
2. E.g., Capra (1974).
3. For an illuminating discussion of this process, see Popper (1983).
4. Bowler (1984), pp. 59–63.
5. Ibid., pp. 120–1.
6. Driesch (1914), p. 141.
7. Thompson (1942), pp. 869–870.
8. See also Riedl (1978).
9. Webster and Goodwin (1982).
10. Webster, in Ho and Saunders (1984), pp. 193–217.
11. Goodwin (1982), p. 51.

Chapter 5: The Mystery of Morphogenesis

1. Needham (1959), p. 205.
2. Ibid., p. 238.
3. Holder (1981).
4. Needham (1959), p. 210.
5. Holder (1981).
6. Reprinted in Moore (1972).
7. Driesch (1908).
8. Weismann (1893).
9. Darwin (1875), ch. 27.
10. Wolpert and Lewis (1975).
11. E.g., Crick (1966).
12. Monod (1972), p. 37.
13. Quoted in Driesch (1914), p. 119.
14. Ibid. (1908).
15. Ibid.
16. Weismann (1893).

17. Dawkins (1976), p. 21.
18. Ibid., p. 22.
19. Ibid. (1982), p. 113.
20. For a lucid critique of the genetic program and kindred concepts, see Oyama (1985).
21. Gerhardt, et al. (1982), p. 111.
22. Ibid., p. 112.
23. For accounts of the vitalist–mechanist controversies, see Nordenskiöld (1928); Coleman (1977).
24. Dawkins (1982), p. 294.
25. Varela (1979), p. 9.
26. Wiener (1961), p. 132.
27. Dawkins (1982), p. 21.
28. Only one putative morphogen has been identified so far in animal embryos, retinoic acid in the chick limb bud (Slack, 1987; Thaller and Eichele, 1987).
29. Prigogine (1980); Prigogine and Stengers (1984).
30. Meinhardt (1982), p. 13.
31. Wolpert (1978), p. 154.
32. E.g., Gerhardt, et al. (1982), pp. 87–114.
33. Quoted in Lewin (1984), p. 1327.
34. Ibid.
35. E.g., von Bertalanffy (1971).
36. Koestler (1967), p. 385.
37. Sheldrake (1981), p. 73.
38. Whyte (1974), p. 43.
39. von Bertalanffy, in Koestler and Smythies (1969); Capra (1982).
40. Eigen and Winkler (1982).
41. Miller (1978).

Chapter 6: Morphogenetic Fields

1. Davies (1984), p. 7.
2. Gurwitsch (1922); translation by Spemann (1938).
3. Weiss (1939), p. 291.
4. Waddington (1957). Waddington at first spelt this word *chreode* but later changed the spelling to *chreod*. I have adopted his original spelling because of its priority and its greater euphony.
5. Quoted in Haraway (1976), p. 58.
6. Ibid., p. 61.

7. Abraham and Shaw (1984).
8. Thom (1975), p. 320.
9. Ibid., p. 159.
10. Haraway (1976), p. 58.
11. Thom (1975), p. 320.
12. Ibid. (1983), p. 141.
13. Goodwin, in Ho and Saunders (1984), p. 229.
14. Ibid., p. 239.
15. Weiss (1939), p. 292.
16. In Waddington (1972), p. 138.
17. Gierer (1981), p. 4.
18. Ibid., p. 5.
19. Goodwin and Cohen (1969).
20. Goodwin (1980).
21. Goodwin, in Ho and Saunders (1984).
22. Gierer (1981), p. 44.
23. Dawkins (1986), ch. 3.
24. Rapp (1979; 1987).
25. Williams (1979).
26. Rapp (1979).
27. E.g., Bunning (1973).
28. E.g., Changeux (1986).
29. Sheldrake (1981), p. 57.
30. Oyama (1985), pp. 1–2.

Chapter 7: Fields, Matter, and Morphic Resonance

1. Hesse (1961).
2. Nersessian (1984).
3. Berkson (1974).
4. Quoted in Hesse (1961), p. 210.
5. Nersessian (1984).
6. Quoted in Hesse (1961), p. 211.
7. Nersessian (1984), p. 199.
8. Ibid., p. 207.
9. D'Espagnat (1976).
10. Davies (1979).
11. Pagels (1983), ch. 8.
12. Murrell, et al. (1978).
13. E.g., Pecher (1939); Verveen and de Felice (1974).

14. Morphic fields can be regarded as propensity fields in the sense of Popper (1982).
15. Maddox (1986).
16. Sheldrake (1981), pp. 64–71.
17. Alberts, et al. (1983), pp. 111–3.
18. Creighton (1978).
19. Creighton (1983).
20. Janin and Wodak (1983).
21. Anfinsen and Scheraga (1975).
22. Creighton (1978).
23. Alberts, et al. (1983), p. 118.
24. Creighton (1978), p. 235.
25. Alberts, et al. (1983), p. 119.
26. Vainshtein, et al. (1975).
27. The idea of experiments on protein folding was suggested to me in the course of a discussion at Harvard with Dr. Stephen Jay Gould and some of his students. Unfortunately, neither Dr. Gould nor I can remember the name of the student who suggested it.
28. It would be best to select relatively complex enzymes for this experiment, since their refolding is relatively slow, taking an hour or more. The rate of refolding can be monitored by the recovery of enzyme activity (Teipel and Koshland, 1971).
29. The main problem with this experimental design is that it may not be possible to make proteins unfold into states different from those that occur in nature. In this case, from the normal starting points refolding would follow the normal chreodes, which would already be stabilized by morphic resonance from innumerable past molecules. This background resonance would swamp any influence from the experimentally refolded proteins. But very little is known about the way in which proteins fold up inside cells, and it is not known whether or not the refolding process under laboratory conditions follows exactly the same pathway as under natural conditions (Baldwin and Creighton, 1980). Consequently, a negative result in this particular experiment would be inconclusive. Nevertheless, in spite of this methodological disadvantage the experiment would still be worth performing, because a positive result would be of such startling significance.
30. Sheldrake (1981), pp. 64–71.
31. Maddox (1985a).
32. McLachlan (1957); see also Schrack (1985).
33. Maddox (1985a).
34. Danckwerts (1982).

35. Sheldrake (1981), pp. 103–7.
36. Jantz (1979).
37. Quoted in Griffin (1982).
38. Griffin (1982).

Chapter 8: Biological Inheritance

1. Alberts, et al. (1983).
2. King and Wilson (1975).
3. Ibid.
4. This could, for example, be due to an effect on the chemical or physical properties of the primordia, or through an effect on their size. For a mathematical model of the development of these mutant leaves, see Young (1983).
5. Struhl (1981).
6. Lewis (1978).
7. Lawrence and Morata (1983); Sanchez-Herrero, et al. (1985).
8. North (1983); McGinnis, et al. (1984).
9. E.g., Beachy, et al. (1985).
10. Goodwin, in Ho and Saunders (1984).
11. Mayr (1982), p. 356.
12. Darwin (1859, 1875).
13. Medvedev (1969); Joravsky (1970).
14. Rignano (1911); Semon (1912); Kammerer (1924).
15. Hudson and Richens (1946); see also "The problem of Lysenkoism" in Levins and Lewontin (1985).
16. Waddington (1975).
17. Ibid. (1952).
18. Ibid. (1975), p. 59.
19. In Koestler and Smythies (1969), p. 383.
20. Ho, et al. (1983).
21. It might be expected that the Ho group's flies were influenced by morphic resonance from Waddington's. However, they were using a quite different strain, so any such effect may have been small. If flies of the *same* strain are used in subsequent similar experiments, morphic resonance would indeed be expected to have significant effects.
22. Waddington (1956a).
23. Ho, et al. (1983), table 2.
24. Lewis and John (1972), p. 137.
25. Sheldrake (1981), pp. 128–131.

26. Ibid., ch. 9.
27. In *A New Science of Life*, behavioural fields were referred to as *motor fields*, in order to emphasize that they are concerned with the organization of movements rather than with the coming into being of form.
28. Thorpe (1963).
29. Hinde (1982).
30. Smith (1978).
31. Parsons (1967).
32. Manning (1975), p. 80.
33. Rothenbuhler (1964).
34. Dilger (1962).
35. Brockelman and Schilling (1984).

Chapter 9: Animal Memory

1. Rose (1986), p. 40.
2. Boakes (1984).
3. Quoted by Lashley (1950), p. 454.
4. Boakes (1984).
5. Lashley (1950).
6. Ibid. (1929), p. 14.
7. Ibid. (1950), p. 472.
8. Ibid., p. 479.
9. Pribram (1971); Wilber, ed. (1982).
10. Boycott (1965), p. 48.
11. Rosenzweig, et al. (1972).
12. Weisel (1982).
13. Rose (1981; 1984); Rose and Harding (1984); Rose and Csillag (1985); Horn (1986).
14. In similar experiments with chicks, detailed studies have shown that there are changes in the number of vesicles in the synapses following learning (Rose, 1986).
15. Cipolla-Neto, Horn, and McCabe (1982).
16. Kolata (1984).
17. Fox (1984).
18. Ibid.
19. Crick (1984).
20. Kandel (1970).
21. Ibid. (1979).
22. Manning (1979).

23. Tinbergen (1951), p. 147.
24. Manning (1975).
25. Boakes (1984).
26. For discussions of animal thought, see Walker (1983) and Griffin (1984).
27. Kammerer (1924), p. 189.
28. Ibid., p. 190.
29. Darwin (1873).
30. Kammerer (1924); Munn (1950).
31. Pavlov (1923).
32. Razran (1958), p. 759.
33. Ibid., p. 760.
34. McDougall (1938).
35. Crew (1936).
36. Agar, et al. (1954).
37. Tryon (1929).
38. Drew (1939).
39. Robert Boakes and Michael Morgan, personal communications, 1981.
40. Rosenthal (1976).
41. Fisher and Hinde (1949).
42. Hinde and Fisher (1951), p. 396.
43. Ibid., p. 395.
44. Diamond (1986), pp. 107–8.

Chapter 10: Morphic Resonance in Human Learning

1. Thom (1975).
2. Lyons (1970).
3. The Listener (6 April 1978) pp. 434–5.
4. Recent studies on language acquisition in children support this proposal: see Eimas (1985).
5. Chomsky (1976).
6. Lynn (1982).
7. Flynn (1983; 1987).
8. Ibid. (1984).
9. Tuddenham (1948).
10. Anderson (1982).
11. Flynn (1984).
12. Jensen (1980).
13. Flynn (1984), p. 29.

14. *New Scientist* (28 October 1982), p. 766.
15. *New Scientist* (28 April 1983), p. 218.
16. See Sheldrake (1985). In fact, two first prizes of $10,000 each were awarded. Robert L. Schwartz of the Tarrytown Group generously provided the extra $5,000 prize money to upgrade the $5,000 second prize to make an extra first prize.
17. The subjects were also asked to take a standard personality test which enabled them to be ranked on an "extraverted-introverted feeling" scale. Mahlberg found that there was a statistically significant correlation between more rapid learning of the real Morse code and "introverted feeling", raising the possibility that "introverted feeling" may be associated with a greater receptivity to morphic resonance, at least under the conditions in this experiment (Mahlberg, 1987).
18. Mabee (1943).
19. Salthouse (1984), p. 94.
20. Norman and Fisher (1982).
21. Hirsch (1970).
22. In experiments in which children with no experience of typing were asked to key in single letters to computer keyboards, the ABCDE layout proved easier than the QWERTY. However, there may well be an inherent difference between learning to type running text and keying in single capital letters (Nicolson and Gardner, 1985).
23. Michaels (1971), p. 424.

Chapter 11: Remembering and Forgetting

1. Koffka (1935), p. 43.
2. H. G. Hartgenbusch, quoted in Koffka (1935), p. 44.
3. Koffka (1935), p. 510.
4. Bartlett (1932).
5. Koestler (1967).
6. Bower (1970).
7. This idea has something in common with the "neural Darwinism" of G. M. Edelman (conveniently summarized by Rosenfield, 1986). Edelman postulates the selection of "neural cell groups" with characteristic patterns of activity as the basis of categorization. These patterns, like morphic fields, are strengthened by repetition. But Edelman takes for granted the conventional assumption that memories are stored in the brain and adopts a version of the usual theory of synaptic modification.
8. Baddeley (1976), p. 285.

9. Ibid., p. 211.
10. Ibid., p. 212.
11. Yates (1969).
12. E.g., Lorayne (1950).
13. In Neisser (1982), pp. 386–7.
14. E.g., Wood (1936); Neisser (1982).
15. Baddeley (1976).
16. Ibid.

Chapter 12: Minds, Brains, and Memories

1. For a historical survey and lucid summary of materialist arguments, see Popper, in Popper and Eccles (1977).
2. Popper and Eccles (1977).
3. Koestler (1978), p. 235.
4. Penfield (1975).
5. Eccles (1953).
6. Taylor (1979), p. 300.
7. Young (1978), p. 7.
8. Crick (1979), p. 137.
9. Capra (1982), p. 318.
10. Jantsch (1980), ch. 4.
11. Ibid., p. 161.
12. Johnson-Laird (1985), p. 115.
13. E.g., Hofstadter (1980); Marr (1982); Sutherland (1982); Poggio, Torre, and Koch (1985).
14. *Nature* (1981), pp. 517, 531.
15. A number of modern models of neural activity have much in common with the idea of morphic fields: in one, for example, problem-solving is modelled in terms of "trajectories of neural dynamics" in "flow maps," in which there are "valleys" in state space, with properties very like Waddington's concept of chreodes (Hopfield and Tank, 1986).
16. Young (1978).
17. Plotinus (1956); see the sections on "Problems of the Soul."
18. Malcolm (1977); Bursen (1978); Russell (1984).
19. Bursen (1978).
20. Hunter (1964).
21. E.g., Squire (1986). For vivid descriptions of some clinical cases see Sacks (1985).
22. Luria (1970; 1973); Gardner (1974).

23. John (1982), p. 251.
24. Teuber (1975).
25. Penfield and Roberts (1959).
26. Rose (1976).
27. Quoted in Wolf (1984), p. 175.
28. E.g., Stevenson (1970).
29. E.g., Palmer (1979).
30. E.g., Wolman (1977).
31. E.g., Stevenson (1974).
32. Jung (1959), p. 48.

Chapter 13: The Morphic Fields of Animal Societies

1. Wilson (1971), p. 317.
2. Ibid., p. 317.
3. Wilson (1980).
4. Ibid. (1971).
5. Ibid.
6. Ibid.
7. von Frisch (1975).
8. Wilson (1971), p. 228.
9. von Frisch (1975).
10. Wilson (1971), p. 229.
11. Ibid.
12. von Frisch (1975).
13. Marais (1973), pp. 119–120.
14. Ibid., p. 121.
15. Wilson (1980), pp. 207–8.
16. Partridge (1981), p. 492.
17. Ibid., pp. 493–4.
18. Selous (1931), p. 9.
19. Ibid., p. 83.
20. Ibid., p. 10.
21. Potts (1984).
22. E.g., Nollman (1985), pp. 106–108.
23. Wilson (1980).
24. McFarland (1981).

Chapter 14: The Fields of Human Societies and Cultures

1. For recent attempts to apply mathematical models to the study of cultural inheritance, see Boyd and Richerson (1985).
2. Wilson (1980), p. 284.
3. Dawkins (1982), p. 290.
4. Ibid. (1976), p. 207.
5. Quoted in Kidd (1911).
6. Jones (1980).
7. Kidd (1911).
8. Abercrombie, et al. (1984), pp. 215–6
9. Lévi-Strauss (1972), p. 328.
10. Piaget (1971).
11. Leach (1970).
12. Field approaches to the social sciences are already being explored; see for example de Green (1978).
13. Quoted in Lukes (1975), p. 4.
14. Freud (1985), p. 221.
15. McDougall (1920), p. 9.
16. Bateson (1973; 1979).
17. Turner (1985), p. 842.
18. Ibid.
19. Ibid.
20. Canetti (1973), p. 16.
21. Ibid., p. 17.
22. Ibid., p. 32.
23. Novak (1976), pp. 135–6.
24. Murphy and White (1978), p. 146.
25. Lukes (1975), p. 231.
26. Jung (1959), p. 42.
27. Ibid. (1953), p. 188.
28. Ibid., p. 149.
29. von Franz (1985).

Chapter 15: Myths, Rituals, and the Influence of Tradition

1. Eliade (1960), p. 18.
2. T. G. H. Strehlow, quoted in Lévi-Strauss (1972), p. 235.
3. Lévi-Strauss, (1972), p. 236.

4. Ringwood (1986).
5. Abercrombie, et al. (1984).
6. Lévi-Strauss (1966), p. 236.
7. Ibid., pp. 236–7.
8. La Fontaine (1985).
9. Abercrombie, et al. (1984).
10. Govinda (1960), p. 28.
11. The Book of Common Prayer of the Church of England.
12. Lang (1942), p. 540.
13. E.g., Lovejoy (1936).
14. E.g., Gablik (1977); Durand (1984).
15. Kuhn (1970), p. 176.
16. In fact, Kuhn uses the word *paradigm* in a wide variety of ways, as Masterman (1970) has shown; but in the Postscript to the second edition of *The Structure of Scientific Revolutions* he points out that these fall into two main categories, as described in the passage quoted.
17. Kuhn (1970), p. 182.
18. Ibid.
19. Ibid., p. 5.
20. Ibid., p. 138.
21. Ibid.
22. Ibid., p. 189.
23. Gablik (1977), p. 159.
24. Kuhn (1970), p. 208.
25. Ibid., p. 209.
26. Quoted in Koestler (1970), p. 495.
27. Ibid., p. 115.
28. Ibid., p. 117.
29. Mayr (1982), p. 495.
30. Kuhn (1970), p. 123.

Chapter 16: The Evolution of Life

1. E.g., Dawkins (1986).
2. Halstead (1985).
3. Rose, Kamin, and Lewontin (1984).
4. Monod (1972).
5. Bergson (1911b).
6. Teilhard de Chardin (1959).
7. Gillespie (1979).

8. Darwin (1875), pp. 7–8.

9. Darwin (1875).

10. E.g., Mivart (1871); Bateson (1894); de Vries (1906); Goldschmidt (1940); Gould (1980).

11. Darwin (1875), v. 2, p. 354.

12. E.g., Bateson (1894); de Vries (1906); Willis (1940).

13. Goldschmidt (1940), p. 397.

14. Darwin (1859; 1875).

15. Mayr (1982), p. 356.

16. Lamarck (1914), p. 122.

17. Darwin (1875), v. 2, p. 359.

18. Ibid. (1872), ch. 5.

19. Quoted in Huxley (1959), p. 18.

20. Huxley (1959), p. 8.

21. Taylor (1983).

22. Darwin (1875), v. 2, p. 356.

23. Ibid., p. 489.

24. Waddington (1953), p. 91.

25. Ibid., p. 96.

26. Darwin (1872), ch. 4.

27. Stanley (1981), p. 3.

28. Dawkins (1985), p. 683.

29. Darwin (1888), ch. 2.

30. Gillespie (1979), ch. 5.

31. Quoted by Mayr (1982), p. 544.

32. Darwin (1872), ch. 15.

33. Quoted in Mayr (1982), p. 409.

34. Ibid., pp. 477–8.

35. Margulis and Sagan (1986).

36. Rensch (1959).

37. Stanley (1981)

38. Such mutations could lead to the "unmasking" of "redundant" ancestral DNA and hence to a tuning in to ancestral morphic fields. For a discussion of the unmasking idea see Britten, in Duncan and Weston-Smith (1977).

39. Darwin (1875), v. 2, p. 5.

40. Ibid., p. 27.

41. Ibid., p. 44.

42. Riedl (1978).

43. Gould (1983).

44. Hall (1984).

45. Gould (1983).
46. Ibid., p. 184.
47. Dawkins (1986), p. 94.
48. Went (1971), p. 198.
49. Ibid., p. 201.
50. Ibid., p. 221.
51. Rensch (1959).
52. Stanley (1981).
53. Taylor (1983).
54. Dawkins (1986), p. 95.

Chapter 17: Formative Causation in Cosmic Evolution

1. Pagels (1985), p. 355.
2. Hawking (1980).
3. Davies (1984), pp. 5–6.
4. Ibid., p. 7.
5. Green (1985).
6. Davies (1984), p. 8.
7. Hogan (1986), p. 572.
8. Barrow and Tipler (1986).
9. Bohm (1980), p. 195.
10. Ibid., p. 151.
11. Ibid., p. 174.
12. Ibid., p. 175.
13. For a non-technical discussion, see for example Pagels (1983).
14. Bohm (1980), p. 146.
15. Ibid., p. 13.
16. Ibid., p. 204.
17. Sheldrake and Bohm (1982); Weber (1986).
18. Bohm and Weber (1982), pp. 35–6.
19. E.g., Blavatsky (1897).

Chapter 18: Creativity Within a Living World

1. Bergson (1946), p. 101.
2. Ibid., pp. 104–5.
3. Green (1978); Merchant (1982).
4. Darwin (1859), ch. 3.

5. Bergson (1911), p. 110.
6. Monod (1972), p. 113.
7. Ibid., p. 110.
8. B. G. Walker (1983).
9. Monod (1972), p. 110.
10. Ibid., p. 38.
11. Frazer (1911), p. 52.
12. Ibid., p. 15.
13. Plotinus (1964), p. 65.
14. Inge (1929), p. 221.
15. Ibid., p. 221.

REFERENCES

Abercrombie, N., S. Hill, and B. S. Turner. 1984. *Dictionary of Sociology.* Harmondsworth: Penguin.

Abraham, R. H., and C. D. Shaw. 1984. *Dynamics: the Geometry of Behavior.* Santa Cruz: Aerial Press.

Agar, W. E., F. H. Drummond, O. W. Tiegs, and M. M. Gunson. 1954. Fourth (final) report on a test of McDougall's Lamarckian experiment on the training of rats. *Journal of Experimental Biology* 31:307–321.

Alberts, B., B. Bray, J. Lewis, et al. 1983. *Molecular Biology of the Cell.* New York: Garland.

Alvarez, L. W., W. Alvarez, F. Asaro, and H. V. Michel. 1980. Extraterrestrial cause for Cretaceous-Tertiary extinction. *Science* 208:1095–1108.

Anderson, A. M. 1982. The great Japanese IQ increase. *Nature* 297:180–181.

Anfinsen, C. B., and H. A. Scheraga. 1975. Experimental and theoretical aspects of protein folding. *Advances in Protein Chemistry* 29:205–300.

Bacon, F. 1627. *New Atlantis.* London: Rawley.

Baddeley, A. D. 1976. *The Psychology of Memory.* New York: Basic Books.

Baldwin, R. L., and T. E. Creighton. 1980. Recent experimental work on the pathway and mechanism of protein folding. In *Protein Folding,* ed. R. Jaenicke. Amsterdam: Elsevier.

Barnett, S. A. 1981. *Modern Ethology.* Oxford: Oxford University Press.

Barrow, J. D., and F. J. Tipler. 1986. *The Anthropic Cosmological Principle.* Oxford: Clarendon Press.

Bartlett, F. C. 1932. *Remembering.* Cambridge: Cambridge University Press.

Bateson, G. 1973. *Steps to an Ecology of Mind.* London: Paladin.

———— 1979. *Mind and Nature.* London: Wildwood House.

Bateson, W. 1894. *Materials for the Study of Variation.* London: Macmillan.

Beachy, P. A., S. A. Helfard, and D. S. Hogness. 1985. Segmental distribution of bithorax complex proteins during Drosophila development. *Nature* 313: 545–551.

Bentley, W. A., and W. J. Humphreys. 1962. *Snow Crystals.* New York: Dover.

Bergson, H. 1911a. *Matter and Memory.* London: Allen and Unwin.

———— 1911b. *Creative Evolution.* London: Macmillan.

———— 1946. *The Creative Mind.* New York: Philosophical Library.

Berkson, W. 1974. *Fields of Force.* London: Routledge and Kegan Paul.

Blavatsky, H. P. 1897. *The Secret Doctrine.* London: Theosophical Publishing House.

Boakes, R. 1984. *From Darwin to Behaviourism.* Cambridge: Cambridge University Press.

Bohm, D. 1980. *Wholeness and the Implicate Order.* London: Routledge and Kegan Paul.

———— and R. Weber. 1982. Nature as creativity. *ReVision* 5(2):35–40.

Bower, G. H. 1970. Organizational factors in memory. *Cognitive Psychology* 1:18–46.

Bowler, P. J. 1984. *Evolution: The History of an Idea.* Berkeley: University of California Press.

Boycott, B. B. 1965. Learning in the octopus. *Scientific American* 212(3):42–50.

Boyd, R., and P. J. Richerson. 1985. *Culture and the Evolutionary Process.* Chicago: University of Chicago Press.

Brockelman, W. Y., and D. Schilling. 1984. Inheritance of stereotyped gibbon calls. *Nature* 312:634–636.

Bunning, E. 1973. *The Physiological Clock.* London: English Universities Press.

Burkert, W. 1972. *Lore and Science in Ancient Pythagoreanism.* Cambridge, Mass.: Harvard University Press.

Burnet, J. 1930. *Early Greek Philosophy.* London: Black.

Bursen, H. A. 1978. *Dismantling the Memory Machine.* Dordrecht: Reidel.

Burtt, E. A. 1932. *The Metaphysical Foundations of Modern Physical Science.* London: Kegan Paul, Trench and Trubner.

Butler, S. 1878. *Life and Habit.* London: Cape.

———— 1880. *Unconscious Memory.* London: Cape.

Bynum, W. F., E. J. Browne, and R. Porter. 1981. *Dictionary of the History of Science.* London: Macmillan.

Canetti, E. 1973. *Crowds and Power.* Harmondsworth: Penguin.

Capra, F. 1974. *The Tao of Physics.* London: Wildwood House.

———— 1982. *The Turning Point.* London: Wildwood House.

Cartwright, N. 1983. *How the Laws of Physics Lie.* Oxford: Clarendon Press.

Castillejo, D. 1981. *The Expanding Force in Newton's Cosmos.* Madrid: Ediciones de Arte y Bibliofilia.

Changeux, J. 1986. *Neuronal Man.* Oxford: Oxford University Press.

Chomsky, N. 1976. *Reflections on Language.* London: Temple Smith.

Cipolla-Neto, J., G. Horn, and B. J. McCabe. 1982. Hemispheric asymmetry and imprinting: the effect of sequential lesions to the Hyperstriatum ventrale. *Experimental Brain Research* 48:22–27.

Cohn, N. 1957. *The Pursuit of the Millenium.* London: Secker and Warburg.

Cole, F. J. 1930. *Early Theories of Sexual Generation.* Oxford: Clarendon Press.

Coleman, W. 1977. *Biology in the Nineteenth Century.* Cambridge: Cambridge University Press.

Creighton, T. E. 1978. Experimental studies of protein folding and unfolding. *Progress in Biophysics and Molecular Biology* 33:231–297.

———— 1983. *Proteins.* San Francisco: Freeman.

Crew, F.A.E. 1936. A repetition of McDougall's Lamarckian experiment. *Journal of Genetics* 33:61–101.

Crick, F.H.C. 1966. *Of Molecules and Men.* Seattle: University of Washington Press.

———— 1979. Thinking about the brain. *Scientific American* 241(3):181–188.

———— 1984. Memory and molecular turnover. *Nature* 312:101.

Danckwerts, P. V. 1982. Letter. *New Scientist* 96 (11 November):380–381.

Darwin, C. 1859. *The Origin of Species,* 1st ed. London: Murray.

———— 1872. *The Origin of Species,* 6th ed. London: Murray.

———— 1873. Inherited instinct. *Nature* 7:281.

———— 1875. *The Variation of Animals and Plants Under Domestication.* London: Murray.

———— 1888. *The Descent of Man,* 2d ed. London: Murray.

Davies, P.C.W. 1979. *The Forces of Nature.* Cambridge: Cambridge University Press.

———— 1984. *Superforce.* London: Heinemann.

Davis, P. J., and R. Hersch. 1983. *The Mathematical Experience.* London: Pelican.

Dawkins, R. 1976. *The Selfish Gene.* Oxford: Oxford University Press.

———— 1982. *The Extended Phenotype.* Oxford: Oxford University Press.

———— 1985. What was all the fuss about? *Nature* 316:683–684.

———— 1986. *The Blind Watchmaker.* London: Longmans.

de Green, K. B. 1978. Force fields and emergent phenomena in sociotechnical macrosystems: theories and models. *Behavioural Science* 23:1–14.

D'Espagnat, B. 1976. *The Conceptual Foundations of Quantum Mechanics.* Reading, Mass.: Benjamin.

de Vries, H. 1906. *Species and Varieties: Their Origin by Mutation.* London: Kegan Paul.

Diamond, J. M. 1986. Rapid evolution of urban birds. *Nature* 324:107–108.

Dijksterhuis, E. J. 1961. *The Mechanization of the World Picture.* Oxford: Oxford University Press.

Dilger, W. C. 1962. The behavior of lovebirds. *Scientific American* 206(1):88–98.

Dobbs, B.J.T. 1975. *The Foundations of Newton's Alchemy.* Cambridge: Cambridge University Press.

Drew, G. C. 1939. McDougall's experiments with the inheritance of acquired habits. *Nature* 143:188–191.

Driesch, H. 1908. *Science and Philosophy of the Organism.* London: Black.

———— 1914. *The History and Theory of Vitalism.* London: Macmillan.

Duerden, J. E. 1920. Inheritance of callosities in the ostrich. *American Naturalist* 54:289.

Duncan, R., and M. Weston-Smith. 1977. *Encyclopedia of Ignorance.* Oxford: Pergamon Press.

Durand, G. 1984. La beauté comme présence paraclétique: essai sur les résurgences d'un bassin sémantique. *Eranos Yearbook* 53:127–173.

Eccles, J. C. 1953. *The Neurophysiological Basis of Mind.* Oxford: Oxford University Press.

Eigen, M., and R. Winkler. 1982. *Laws of the Game.* London: Allen Lane.

Eimas, P. D. 1985. The perception of speech in early infancy. *Scientific American* 252(1):34–40.

Einstein, A. 1954. *Ideas and Opinions.* New York: Crown.

Eliade, M. 1954. *The Myth of the Eternal Return.* Princeton: Bollingen.

———— 1960. *Myths, Dreams and Mysteries.* New York: Harper and Row.

Fawcett, E. D. 1916. *The World as Imagination.* London: Macmillan.

Feynman, R. 1965. *The Character of Physical Law.* London: BBC.

Fisher, J., and R. A. Hinde. 1949. The opening of milk bottles by birds. *British Birds* 42:347–357.

Flynn, J. R. 1983. Now the great augmentation of the American IQ. *Nature* 301:655.

———— 1984. The mean IQ of Americans: massive gains 1932 to 1978. *Psychological Bulletin* 95:29–51.

———— 1987. Massive IQ gains in 14 nations: what IQ tests really measure. *Psychological Bulletin* 101:171–191.

Fox, J. L. 1984. The brain's dynamic way of keeping in touch. *Science* 225:820–821.

Franke, H. W. 1966. *Sinnbild der Chemie.* Basel: Basilius Press.

Frazer, J. G. 1911. *The Golden Bough: The Magic Art and the Evolution of Kings.* London: Macmillan.

Freud, S. 1952. *Totem and Taboo.* In *The Origins of Religion.* A. Dickson, ed. 1985. Harmondworth: Penguin.

Gablik, S. 1977. *Progress in Art.* New York: Rizzoli.

Gardner, H. 1974. *The Shattered Mind.* New York: Vintage Books.

Gerhardt, J. C., et al. 1982. The cellular basis of morphogenetic change. In *Evolution and Development,* Dahlem Conferences, ed. J. T. Bonner. Berlin: Springer Verlag.

Gierer, A. 1981. Generation of biological patterns and form. *Progress in Biophysics and Molecular Biology* 37:1–47.

Gillespie, C. C. 1960. *The Edge of Objectivity.* Princeton: Princeton University Press.

Gillespie, N. C. 1979. *Charles Darwin and the Problem of Creation.* Chicago: University of Chicago Press.

Gilson, E. 1984. *From Aristotle to Darwin and Back Again.* Indiana: University of Notre Dame Press.

Goldschmidt, R. 1940. *The Material Basis of Evolution.* New Haven: Yale University Press.

Goodwin, B. C. 1980. Pattern formation and regeneration in the protozoa. *Society for General Microbiology Symposium* 30:377–404.

———— 1982. Development and evolution. *Journal of Theoretical Biology* 97:43–55.

———— and M. H. Cohen. 1969. A phase-shift model for the spatial and temporal organization of developing systems. *Journal of Theoretical Biology* 25:49–107.

Gorman, P. 1979. *Pythagoras: A Life.* London: Routledge and Kegan Paul.

Gould, S. J. 1977. *Ontogeny and Phylogeny.* Cambridge, Mass.: Harvard University Press.

———— 1980. Return of the hopeful monster. In *The Panda's Thumb.* New York: Norton.

———— 1983. *Hen's Teeth and Horse's Toes.* New York: Norton.

———— and N. Eldredge. 1977. Punctuated equilibria: the tempo and mode of evolution reconsidered. *Paleobiology* 3:115–151.

Govinda, L. A. 1960. *Foundations of Tibetan Mysticism.* London: Rider.

Green, M. B. 1985. Unification of forces and particles in superstring theories. *Nature* 314:409–414.

Griffin, D. R. 1982. Review of *A New Science of Life. Process Studies* 12:34–40.

———— 1984. *Animal Thinking.* Cambridge, Mass.: Harvard University Press.

Griffin, S. 1978. *Woman and Nature.* New York: Harper and Row.

Griffiths, B. 1982. *The Marriage of East and West.* London: Collins.

Gurwitsch, A. 1922. Ueber den Begriff des embryonalen Feldes. *Archiv für Entwicklungsmechanik* 51:383–415.

Guth, A. H., and P. J. Steinhardt. 1984. The inflationary universe. *Scientific American* 250(5):90–102.

Haeckel, E. 1892. *The History of Creation.* London: Kegan Paul.

———— 1910. *The Evolution of Man.* London: Watts.

Hall, B. K. 1984. Developmental mechanisms underlying the formation of atavisms. *Biological Reviews* 59:89–124.

Hallam, A. 1984. The causes of mass extinctions. *Nature* 308:686.

Halstead, B. 1985. Anti-Darwinian theory in Japan. *Nature* 317:587–589.

Haraway, D. J. 1976. *Crystals, Fabrics and Fields: Metaphors of Organicism in Twentieth-Century Developmental Biology.* New Haven: Yale University Press.

Hardy, A. 1965. *The Living Stream.* London: Collins.

Hawking, S. 1980. *Is the End in Sight for Theoretical Physics?* Cambridge: Cambridge University Press.

Hesse, M. 1961. *Forces and Fields.* London: Nelson.

Hinde, R. A. 1982. *Ethology.* London: Fontana.

———— and J. Fisher. 1951. Further observations on the opening of milk bottles by birds. *British Birds* 44:393–396.

Hirsch, R. S. 1970. Effect of standard versus alphabetical keyboard formats on typing performance. *Journal of Applied Psychology* 54:484–490.

Ho, M. W., and P. T. Saunders, eds. 1984. *Beyond Neo-Darwinism.* London: Academic Press.

Ho, M. W., C. Tucker, D. Keeley, and P. T. Saunders. 1983. Effects of successive generations of ether treatment on penetrance and expression of the bithorax phenocopy in Drosophila melanogaster. *Journal of Experimental Zoology* 225:357–368.

Hofstadter, D. 1979. *Godel, Escher, Bach.* Brighton: Harvester Press.

Hogan, C. 1986. Galaxy superclusters and cosmic strings. *Nature* 320:572.

Holder, N. 1981. Regeneration and compensatory growth. *British Medical Bulletin* 37:227–232.

Hopfield, J. J., and D. W. Tank. 1986. Computing with neural circuits: a model. *Science* 233:625–633.

Horn, G. 1986. *Memory, Imprinting and the Brain: An Inquiry into Mechanisms.* Oxford: Clarendon Press.

Hudson, P. S., and R. H. Richens. 1946. *The New Genetics in the Soviet Union.* Cambridge: Imperial Bureau of Plant Breeding and Genetics.

Hunter, I.M.L. 1964. *Memory.* Harmondsworth: Penguin.

Huxley, F. 1959. Charles Darwin: life and habit. *The American Scholar* (Fall/Winter):1–19.

Hyatt, A. 1893. Phylogeny of an acquired characteristic. *Proceedings of the American Philosophical Society* 32:349–647.

Inge, W. R. 1929. *The Philosophy of Plotinus.* London: Longmans, Green and Co.

Jaki, S. L. 1978. *The Road of Science and the Ways to God.* Edinburgh: Scottish Academic Press.

Janin, J., and S. J. Wodak. 1983. Structural domains in proteins and their role in protein folding. *Progress in Biophysics and Molecular Biology* 42:21–78.

Jantsch, E. 1980. *The Self-Organizing Universe.* Oxford: Pergamon.

Jantz, R. L. 1979. On the levels of dermatoglyphic variation. *Birth Defects: Original Article Series* 15:53–61.

Jennings, H. S. 1906. *Behavior of the Lower Organisms.* New York: Columbia University Press.

Jensen, A. R. 1980. *Bias in Mental Testing.* London: Methuen.

John, E. R. 1982. Multipotentiality: a theory of recovery of function after brain injury. In *Neurophysiology After Lashley,* ed. J. Orbach. Hillsdale, N.J.: Lawrence Erlbaum.

Johnson-Laird, P. N. 1985. Modularity of brain and mind. *Nature* 318:115–116.

Jones, G. 1980. *Social Darwinism and English Thought.* Brighton: Harvester Press.

Joravsky, D. 1970. *The Lysenko Affair.* Cambridge, Mass.: Harvard University Press.

Jung, C. G. 1953. *Two Essays on Analytical Psychology.* London: Routledge and Kegan Paul.

————— 1959. *The Archetypes and the Collective Unconscious.* London: Routledge and Kegan Paul.

Kammerer, P. 1924. *The Inheritance of Acquired Characteristics.* New York: Boni and Liveright.

Kandel, E. R. 1970. Nerve cells and behavior. Reprinted in *Progress in Psychobiology,* ed. R. F. Thompson, 1976. San Francisco: Freeman.

————— 1979. Small systems of neurons. *Scientific American* 241(3):61–71.

Kidd, B. 1911. *Sociology.* In *Encyclopaedia Britannica,* 11th ed. Cambridge: Cambridge University Press.

King, M. C., and A. C. Wilson. 1975. Evolution at two levels in humans and chimpanzees. *Science* 188:107–116.

Koestler, A. 1967. *The Ghost in the Machine.* London: Hutchinson.

————— 1970. *The Act of Creation.* London: Pan Books.

————— 1971. *The Case of the Midwife Toad.* London: Hutchinson.

————— 1978. *Janus: A Summing Up.* London: Hutchinson.

————— and J. R. Smythies, eds. 1969. *Beyond Reductionism.* London: Hutchinson.

Koffka, K. 1935. *Principles of Gestalt Psychology.* London: Routledge and Kegan Paul.

Kolata, G. 1984. New neurons form in adulthood. *Science* 224:1325–1326.

Kruk, Z. L., and C. J. Pycock. 1983. *Neurotransmitters and Drugs.* London: Croom Helm.

Kuhn, T. S. 1970. *The Structure of Scientific Revolutions,* 2d ed. Chicago: University of Chicago Press.

La Fontaine, J. S. 1985. *Initiation.* Harmondworth: Penguin.

Lamarck, J. B. 1914. *Zoological Philosophy.* London: Macmillan.

Lang, P. H. 1942. *Music in Western Civilization.* London: Dent.

Lashley, K. S. 1929. *Brain Mechanisms and Intelligence.* Chicago: Chicago University Press.

————— 1950. In search of the engram. *Symposium of the Society for Experimental Biology* 4:454–483.

Laszlo, E. 1987. *Evolution: The Grand Synthesis.* Boston: Shambala.

Lawrence, P. A., and G. Morata. 1983. The elements of the bithorax complex. *Cell* 35:595–601.

Leach, E. 1970. *Lévi-Strauss.* London: Fontana.

Leclerc, I. 1972. *The Nature of Physical Existence.* London: Allen and Unwin.

Lévi-Strauss, C. 1966. *The Savage Mind.* London: Weidenfeld and Nicholson.

——— 1972. *Structural Anthropology.* Harmondsworth: Penguin.

Levins, R., and R. Lewontin. 1985. *The Dialectical Biologist.* Cambridge, Mass.: Harvard University Press.

Lewin, R. 1984. Why is development so illogical? *Science* 224:1327–1329.

Lewis, E. B. 1978. A gene complex controlling segmentation in Drosophila. *Nature* 276:565–570.

Lewis, K. R., and B. John. 1972. *The Matter of Mendelian Heredity.* London: Longman.

Lorayne, H. 1950. *How to Develop a Super-Power Memory.* Preston: Thomas.

Lorenz, K. 1958. The evolution of behavior. *Scientific American* 199:67–78.

Lovejoy, A. O. 1936. *The Great Chain of Being.* Cambridge, Mass.: Harvard University Press.

Lukes, S. 1975. *Emile Durkheim: His Life and Work.* Harmondsworth: Penguin.

Luria, A. R. 1968. *The Mind of a Mnemonist.* London: Cape.

——— 1970. The functional organization of the brain. *Scientific American* 222(3):66–78.

——— 1973. *The Working Brain.* Harmondsworth: Penguin.

Lynn, R. 1982. IQ in Japan and the United States shows a growing disparity. *Nature* 297:222–223.

Lyons, J. 1970. *Chomsky.* London: Fontana.

Mabee, C. 1943. *The American Leonardo: A Life of Samuel F. B. Morse.* New York: Knopf.

Mackie, G. O. 1964. Analysis of locomotion in a siphonophore colony. *Proceedings of the Royal Society* B 159:366–391.

McDougall, W. 1920. *The Group Mind.* Cambridge: Cambridge University Press.

——— 1938. Fourth report on a Lamarckian experiment. *British Journal of Psychology* 28:321–345.

McFarland, D. 1981. Cultural Behaviour. In *The Oxford Companion to Animal Behaviour,* ed. D. McFarland. Oxford: Oxford University Press.

McGinnis, W., R. L. Garber, J. Wirz, et al. 1984. A homologous protein coding sequence in Drosophila homeotic genes and its conservation in other metazoans. *Cell* 37:403–408.

McLachlan, D. 1957. The symmetry of dendritic snow crystals. *Proceedings of the National Academy of Sciences* 43:143–151.

Maddox, J. 1984a. Extinctions by catastrophe? *Nature* 308:685.

——— 1984b. Nuclear winter and carbon dioxide. *Nature* 312:593.

——— 1985a. No pattern yet for snowflakes. *Nature* 313:93.

——— 1985b. Periodic extinctions undermined. *Nature* 315:62.

——— 1986. Making molecules into atoms. *Nature* 323:391.

Mahlberg, A. 1987. Evidence of collective memory: a test of Sheldrake's theory. *Journal of Analytical Psychology* 32:23–34.

Malcolm, N. 1977. *Memory and Mind.* Ithaca: Cornell University Press.

Manning, A. 1975. Behaviour genetics and the study of behavioural evolution. In *Function and Evolution in Behaviour* eds. G. P. Baerends, C. Beer, and A. Manning. Oxford: Oxford University Press.

——— 1979. *An Introduction to Animal Behaviour,* 3d ed. London: Arnold.

Marais, E. N. 1973. *The Soul of the White Ant.* Harmondsworth: Penguin.

Margulis, L. and D. Sagan. 1986. *Microcosmos.* New York: Summit Books.

Marr, D. 1982. *Vision: A Computational Investigation.* San Francisco: Freeman.

Marsh, O. C. 1892. Recent polydactyle horses. *American Journal of Science* 43:339–355.

Masterman, M. 1970. The nature of a paradigm. In *Criticism and the Growth of Knowledge,* eds. I. Lakatos and A. Musgrave. Cambridge: Cambridge University Press.

Mayr, E. 1982. *The Growth of Biological Thought.* Cambridge, Mass.: Harvard University Press.

Medvedev, Z. A. 1969. *The Rise and Fall of T. D. Lysenko.* New York: Columbia University Press.

Meinhardt, H. 1982. *Models of Biological Pattern Formation.* London: Academic Press.

Merchant, C. 1982. *The Death of Nature.* London: Wildwood House.

Michaels, S. E. 1971. QWERTY versus alphabetic keyboards as a function of typing skill. *Human Factors* 13:419–426.

Miller, J. G. 1978. *Living Systems.* New York: McGraw-Hill.

Mivart, St. G. J. 1871. *Genesis of Species.* London: Macmillan.

Monod, J. 1972. *Chance and Necessity.* London: Collins.

Moore, J. A., ed. 1972. *Readings in Heredity and Development.* Oxford: Oxford University Press.

Morgan, T. H. 1901. *Regeneration.* New York: Macmillan.

Munn, N. L. 1950. *Handbook of Psychological Research on the Rat.* Boston: Houghton Mifflin.

Murrell, J. N., S.F.A. Kettle, and J. M. Tedder. 1978. *The Chemical Bond.* Chichester: Wiley.

Murphy, G., and R. O. Ballou, eds. 1961. *William James on Psychical Research.* London: Chatto and Windus.

Murphy, M., and R. A. White. 1978. *The Psychic Side of Sports.* Reading, Mass.: Addison-Wesley.

Nature survey of the neurosciences. 1981. *Nature* 293:515–534.

Needham, J. 1942. *Biochemistry and Morphogenesis.* Cambridge: Cambridge University Press.

——— 1959. *A History of Embryology.* Cambridge: Cambridge University Press.

Neisser, U., ed. 1982. *Memory Observed: Remembering in Natural Contexts.* San Francisco: Freeman.

Nersessian, N. J. 1984. Aether/or: the creation of scientific concepts. *Studies in the History and Philosophy of Science* 15:175–212.

Nicolson, R. I., and P. H. Gardner. 1985. The QWERTY keyboard hampers schoolchildren. *British Journal of Psychology* 76:525–531.

Nietzsche, F. W. 1911. Eternal recurrence: the doctrine expounded and substantiated. In *The Complete Works of Friedrich Nietzsche,* Vol. 16, ed. O. Levy. Edinburgh: Foulis.

Nisbet, R. 1980. *History of the Idea of Progress.* New York: Basic Books.

Nollman, J. 1985. *Dolphin Dreamtime.* London: Blond.

Nordenskiöld, E. 1928. *The History of Biology.* New York: Tudor.

Norman, D. A., and D. Fisher. 1982. Why alphabetic keyboards are not easy to use. *Human Factors* 24:509–519.

North, G. 1986. Descartes and the fruit fly. *Nature* 322:404–405.

Novak, M. 1976. *The Joy of Sports.* New York: Basic Books.

Oyama, S. 1985. *The Ontogeny of Information.* Cambridge: Cambridge University Press.

Pagels, H. R. 1983. *The Cosmic Code.* London: Joseph.

——— 1985. *Perfect Symmetry.* London: Joseph.

Palmer, J. 1979. A community mail survey of psychic experiences. *Journal of the American Society for Psychical Research* 73:221–251.

Parsons, P. A. 1967. *The Genetic Analysis of Behaviour.* London: Methuen.

Partridge, B. 1981. Schooling. In *The Oxford Companion to Animal Behaviour,* ed. D. McFarland. Oxford: Oxford University Press.

Pavlov, I. P. 1923. New researches on conditioned reflexes. *Science* 58:359–361.

Pecher, C. 1939. La fluctuation d'éxcitabilité de la fibre nerveuse. *Archives Internationales de Physiologie* 49:129–152.

Penfield, W. 1975. *The Mystery of the Mind.* Princeton: Princeton University Press.

——— and L. Roberts. 1959. *Speech and Brain Mechanisms.* Princeton: Princeton University Press.

Philip, J. A. 1966. *Pythagoras and Early Pythagoreanism.* Toronto: University of Toronto Press.

Piaget, J. 1971. *Structuralism.* London: Routledge and Kegan Paul.

Plotinus. *The Enneads,* trans. S. Mackenna, 1956. London: Faber.

——— *The Essential Plotinus,* trans. E. O'Brien, 1964. New York: Mentor.

Poggio, T., V. Torre, and C. Koch. 1985. Computational vision and regularization theory. *Nature* 317:313–319.

Popper, K. R. 1982. *Quantum Theory and the Schism in Physics.* London: Hutchinson.

——— 1983. *Realism and the Aim of Science.* London: Hutchinson.

——— and J. C. Eccles. 1977. *The Self and Its Brain*. Berlin: Springer International.

Potters, V. G. 1967. *C. S. Peirce on Norms and Ideals*. Worcester: University of Massachusetts Press.

Potts, W. K. 1984. The chorus line hypothesis of manoeuvre co-ordination in avian flocks. *Nature* 309:344–345.

Press, W. H. 1986. A place for teleology? *Nature* 320:315.

Pribram, K. H. 1971. *Languages of the Brain*. Englewood Cliffs: Prentice Hall.

Prigogine, I. 1980. *From Being to Becoming*. San Francisco: Freeman.

——— and I. Stengers. 1984. *Order Out of Chaos*. London: Heinemann.

Rapp, P. E. 1979. An atlas of cellular oscillations. *Journal of Experimental Biology* 81:281–306.

——— 1987. Why are so many biological systems periodic? *Progress in Neurobiology* (in the press).

Razran, G. 1958. Pavlov and Lamarck. *Science* 128:758–760.

Rensch, B. 1959. *Evolution Above the Species Level*. London: Methuen.

Richardson, J. S. 1981. The anatomy and taxonomy of protein structure. *Advances in Protein Chemistry* 34:167–339.

Riedl, R. 1978. *Order in Living Organisms*. Chichester: Wiley.

Rignano, E. 1911. *Inheritance of Acquired Characters*. Chicago: Open Court.

——— 1926. *Biological Memory*. New York: Harcourt, Brace and Co.

Ringwood, A. E. 1986. Terrestrial origin of the moon. *Nature* 322:323–328.

Rose, S.P.R. 1976. *The Conscious Brain*. Harmondsworth: Penguin.

——— 1981. What should a biochemistry of learning and memory be about? *Neuroscience* 6:811–821.

——— 1984. Strategies in studying the cell biology of learning and memory. In *Neuropsychology of Memory*, eds. L. R. Squire and N. Butters. New York: Guilford Press.

——— 1986. Memories and molecules. *New Scientist* 112 (27 Nov):40–44.

——— and A. Csillag. 1985. Passive avoidance training results in lasting changes in deoxyglucose metabolism in left hemisphere regions of chick brain. *Behavioural and Neural Biology* 44:315–324.

Rose, S.P.R., and S. Harding. 1984. Training increases ³H fucose incorporation in chick brain only if followed by memory storage. *Neuroscience* 12:663–667.

Rose, S.P.R., L. J. Kamin, and R. Lewontin. 1984. *Not in Our Genes*. Hardmondsworth: Pelican.

Rosenthal, R. 1976. *Experimenter Effects in Behavioral Research*. New York: Irving.

Rosenfield, I. 1986. Neural Darwinism: a new approach to memory and perception. *New York Review of Books* (October 9):21–27.

Rosenzweig, M. R., E. L. Bennett, and M. C. Diamond. 1972. Brain changes in response to experience. Reprinted in *Progress in Psychobiology*, ed. R. F. Thompson (1976). San Francisco: Freeman.

Rothenbuhler, W.C. 1964. Behavior genetics of nest cleaning in honeybees. *American Zoologist* 4:111–123.

Russell, E. S. 1916. *Form and Function*. London: Murray.

Russell, J. 1984. *Explaining Mental Life*. London: Macmillan.

Russell, J. L. 1968. Time in Christian Thought. In *The Voices of Time*, ed. J. T. Fraser. London: Allen Lane.

Sacks, O. 1985. *The Man Who Mistook His Wife for a Hat*. London: Duckworth.

Salthouse, I. 1984. The skill of typing. *Scientific American* 250(2):94–99.

Sanchez-Herrero, E., I., Vernos, R. Mareo, and G. Morata. 1985. Genetic organization of the Drosophila bithorax complex. *Nature* 313:108–113.

Schrack, R. A. 1985. Electrical aspects of the snowflake crystal. *Nature* 314:324.

Selous, E. 1931. *Thought Transference, or What? in Birds*. London: Constable.

Semon, R. 1912. *Das Problem der Vererbung Erworbener Eigenschaften*. Leipzig: Engelmann.

———— 1921. *The Mneme*. London: Allen and Unwin.

Sheldrake, A. R. 1981. *A New Science of Life: The Hypothesis of Formative Causation*. London: Blond and Briggs.

———— 1985. *A New Science of Life: The Hypothesis of Formative Causation*, new ed. London: Blond.

———— and D. Bohm. 1982. Morphogenetic fields and the implicate order. *ReVision* 5:41–48.

Slack, J.M.W. 1987. We have a morphogen! *Nature* 327:553–540.

Smith, A. P. 1978. An investigation of the mechanisms underlying nest construction in the mud wasp Paralastor sp. *Animal Behaviour* 26:232–240.

Spemann, H. C. 1938. *Embryonic Development and Induction.* New Haven: Yale University Press.

Squire, L. R. 1986. Mechanisms of memory. *Science* 232:1612–1619.

Stanley, S. M. 1979. *Macroevolution.* San Francisco: Freeman.

———— 1981. *The New Evolutionary Timetable.* New York: Basic Books.

Stevenson, I. 1970. *Telepathic Impressions.* Charlottesville: University of Virginia Press.

———— 1974. *Twenty Cases Suggestive of Reincarnation.* Charlottesville: University of Virginia Press.

Struhl, G. 1981. A homoeotic mutation transforming leg to antenna in Drosophila. *Nature* 292:635–638.

Sutherland, S. 1982. The vision of David Marr. *Nature* 298:691–692.

Taylor, G. R. 1979. *The Natural History of the Mind.* New York: Dutton.

———— 1983. *The Great Evolution Mystery.* London: Secker and Warburg.

Teilhard de Chardin, P. T. 1959. *The Phenomenon of Man.* London: Collins.

Teipel, J. W., and D. E. Koshland. 1971. Kinetic aspects of conformational change in proteins I. Rate of regain of enzyme activity from denatured proteins. *Biochemistry* 10:792–798.

Teuber, H. 1975. Recovery of function after brain injury in man. In *Outcome of Severe Damage to the CNS.* Ciba Foundation Symposia 34. Amsterdam: Elsevier.

Thaller, C., and G. Eichele. 1987. Identification and spatial distribution of retinoids in the developing chick limb bud. *Nature* 327:625–628.

Thom, R. 1975. *Structural Stability and Morphogenesis.* Reading, Mass.: Benjamin.

———— 1983. *Mathematical Models of Morphogenesis.* Chichester: Horwood.

Thompson, D. W. 1942. *On Growth and Form,* 2d ed. Cambridge: Cambridge University Press.

Thorpe, W. H. 1963. *Learning and Instinct in Animals.* London: Methuen.

Tinbergen, N. 1951. *The Study of Instinct.* Oxford: Oxford University Press.

Tryon, R. C. 1929. The genetics of learning ability in rats. *University of California Publications in Psychology* 4:71–89.

Tuddenham, R. D. 1948. Soldier intelligence in World Wars I and II. *American Psychologist* 3:54–56.

Turner, R. H. 1985. Collective behaviour. In *Encyclopaedia Britannica,* 15th ed., Chicago.

Vainshtein, B. K., et al. 1975. Structure of leghaemoglobin from lupin root nodules. *Nature* 254:163–164.

van Spronsen, J. W. 1969. *The Periodic System of Chemical Elements.* Amsterdam: Elsevier.

Varela, F. J. 1979. *Principles of Biological Autonomy.* New York: North Holland.

Verveen, A. A., and L. J. de Felice. 1974. Membrane noise. *Progress in Biophysics and Molecular Biology* 28:189–265.

von Bertalanffy, L. 1971. *General Systems Theory.* London: Allen Lane.

von Franz, M. L. 1985. The transformed beserk. *ReVision* 8(1):20.

von Frisch, K. 1975. *Animal Architecture.* London: Hutchinson.

Waddington, C. M. 1952. Selection of the genetic basis for an acquired character. *Nature* 169:278–279.

——— 1953. Experiments in acquired characteristics. *Scientific American* 189:92–97.

——— 1956a. Genetic assimilation of the bithorax phenotype. *Evolution* 10:1–13.

——— 1956b. *Principles of Embryology.* New York: Macmillan.

——— 1957. *The Strategy of the Genes.* London: Allen and Unwin.

——— 1975. *The Evolution of an Evolutionist.* Edinburgh: Edinburgh University Press.

——— ed. 1972. *Towards a Theoretical Biology. 4: Essays.* Edinburgh: Edinburgh University Press.

Walker, B. G. 1983. *The Woman's Encyclopedia of Myths and Secrets.* San Francisco: Harper and Row.

Walker, S. 1983. *Animal Thought.* London: Routledge and Kegan Paul.

Wallace, A. R. 1911. *The World of Life: A Manifestation of Creative Power, Directive Mind and Ultimate Purpose.* London: Chapman and Hall.

Wallace, W. 1910. Descartes. *Encyclopaedia Britannica,* 11th ed., New York.

Weber, R. 1986. *Dialogues with Scientists and Sages: the Search for Unity.* London: Routledge and Kegan Paul.

Webster, G., and B. C. Goodwin. 1982. The Origin of Species: a structuralist approach. *Journal of Social and Biological Structure* 5:15–47.

Weinberg, S. 1977. *The First Three Minutes.* London: Deutsch.

Weisel, T. N. 1982. Postnatal development of the visual cortex and the influence of environment. *Nature* 299:583–591.

Weismann, A. 1893. *The Germ-Plasm: A Theory of Heredity.* London: Scott.

Weiss, P. 1939. *Principles of Development.* New York: Holt.

Went, F. W. 1971. Parallel evolution. *Taxon* 20:197–226.

Westfall, R. S. 1980. *Never at Rest: A Biography of Isaac Newton.* Cambridge: Cambridge University Press.

Whitehead, A. N. 1925. *Science and the Modern World.* New York: Macmillan.

Whyte, L. L. 1955. *Accent on Form.* London: Routledge and Kegan Paul.

——— 1974. *The Universe of Experience.* New York: Harper and Row.

Wiener, N. 1961. *Cybernetics,* 2d ed. Cambridge, Mass.: MIT Press.

Wilber, K., ed. 1982. *The Holographic Paradigm and Other Paradoxes.* Boulder: Shambala.

——— ed. 1984. *Quantum Questions.* Boulder: Shambala.

Williams, R.J.P. 1979. The conformational properties of proteins in solution. *Biological Reviews* 54:389–437.

Willis, J. C. 1940. *The Course of Evolution by Differentiation or Divergent Mutation.* Cambridge: Cambridge University Press.

Wilson, E. O. 1971. *The Social Insects.* Cambridge, Mass.: Harvard University Press.

——— 1980. *Sociobiology* (abridged edition). Cambridge, Mass.: Harvard University Press.

Wolf, F. A. 1984. *Star Wave.* New York: Macmillan.

Wolman, B. B., ed. 1977. *Handbook of Parapsychology.* New York: Van Nostrand Reinhold.

Wolpert, L. 1978. Pattern formation in biological development. *Scientific American* 239:154–164.

———— and J. Lewis. 1975. Towards a theory of development. *Federation Proceedings* 34:14–20.

Wood, E. E. 1936. *Mind and Memory Training.* London: Theosophical Publishing House.

Yates, F. A. 1969. *The Art of Memory.* Harmondsworth: Penguin.

Young, J.P.W. 1983. Pea leaf morphogenesis: a simple model. *Annals of Botany* 52:311–316.

Young, J. Z. 1978. *Programs of the Brain.* Oxford: Oxford University Press.

GLOSSARY

adaptation: An attribute of an organism that appears to be of value for something, generally its survival or reproduction. The purposive, or seemingly purposive, nature of adaptations can be thought of in terms of teleology or teleonomy (q.v.).

allele: Each gene (q.v.) occupies a particular region of a chromosome, its locus. At any given locus, there may exist alternative forms of the gene. These are called alleles of each other.

atavism: The reappearance of characteristics of more or less remote ancestors. Also called reversion or throwing back.

atom: In the philosophy of atomism (q.v.), the eternal, invariant, impenetrably hard, homogeneous, ultimate unit of matter. In chemistry, the smallest unit or part of an element that can take part in a chemical reaction. In modern physics, a complex structure of activity, with a central nucleus orbited by electrons. Nuclei and their constituent particles are in turn complex structures of activity.

atomism: The doctrine that all things are composed of ultimate, indivisible atoms of matter endowed with motion. These ultimate particles are the enduring basis of all reality. In the modern form of this philosophy, atoms have been superseded by fundamental subatomic particles.

attractor: A term used in modern dynamics to denote a limit towards which trajectories of change within a dynamical system move. Attractors generally lie within basins of attraction. Attractors and basins of attraction are essential features of the mathematical models of morphogenetic fields due to René Thom.

chreode: A canalized pathway of change within a morphic field.

chromosomes: Microscopic, threadlike structures found in the nuclei of living cells, and also in cells without nuclei such as bacteria. They are made up of DNA and protein and contain chains of genes.

cybernetics: The theory of communication and control mechanisms in living systems and machines.

dialectical materialism: A form of materialism that sees matter not as something static, on which change and development have to be imposed, but as containing within its own nature those tensions or "contradictions" that provide the motive force for change.

DNA: Deoxyribonucleic acid, a molecule consisting of a large number of chemical units called nucleotides attached together in single file to form a long strand. Usually two such strands are linked together parallel to each other and coiled into a helix. DNA is the material of genetic inheritance, but in higher organisms only a small proportion of the DNA appears to be in genes. DNA contains four kinds of nucleotide, and the sequence of the nucleotides is the basis of the genetic code. DNA strands pass on their structure to copies of themselves in the process of replication, and the genetic code of genes can be "translated" into the sequences of amino acids which are joined together in chains to form proteins. Protein synthesis takes place on the basis of strands of RNA (ribonucleic acid), which serve as templates. These are "transcribed" from the DNA of genes.

dominance: In genetics, a dominant gene is one that brings about the same phenotypic (q.v.) effects whether it is present in a single dose along with a specified allele (q.v.), or in a double dose. The allele that is ineffective in the presence of the dominant gene is said to be *recessive.*

dualism: The philosophical doctrine that mind and matter exist as independent entities, neither being reducible to the other (cf. materialism).

energy: In general, the capacity or power to produce an effect. In the technical sense of physics, energy is the property of a system that is a measure of its capacity for doing work. Work is technically defined as what is done when a force moves its point of application. Energy can be potential or kinetic, and it comes in a variety of forms: electrical, thermal, chemical, nuclear, radiant, and mechanical.

entelechy: In Aristotelian philosophy, the principle of life, identified with the soul or psyche. The entelechy is both the formal or formative cause and the final cause, or end, of a living body; thus there is always an internalized purpose in life. In the vitalism (q.v.) of Hans Driesch, entelechy is the non-material vital principle, a directive, teleological causal factor which brings about harmonious developmental, behavioural, and mental processes (cf. genetic program and morphic field).

epigenesis: The origin of new structures during embryonic development (cf. preformation).

evolution: Literally, a process of unrolling or opening out. In biology, originally applied to the development of individual plants and animals, which according to the doctrine of preformation depended on the unrolling or unfolding of pre-existing parts. Only in the 1830s was this word first applied to the historical transmutation of organisms; by the 1860s and 1870s it had come to refer to a general process of transmutation, which was generally assumed to be directional or progressive. Darwin's theory of evolution by natural selection enabled this process to be thought of as blind and purposeless, and this interpretation is central to neo-Darwinism (q.v.), the dominant orthodoxy in modern biology. A variety of other evolutionary philosophies postulate an inherently creative principle in matter or in life; and some see in the evolutionary process the manifestation of a directional or purposive principle. According to modern cosmology, the entire universe is an evolutionary system.

field: A region of physical influence. Fields interrelate and interconnect matter and energy within their realm of influence. Fields are not a form of matter; rather, matter is energy bound within fields. In current physics, several kinds of fundamental field are recognized: the gravitational and electro-magnetic fields and the matter fields of quantum physics. The hypothesis of formative causation broadens the concept of physical fields to include morphic fields as well as the known fields of physics.

force: In general, active power; strength or energy brought to bear. In physics, an external agency capable of altering the state of rest or motion of a body.

form: The shape, configuration, or structure of something as distinguished from its material. In the Platonic tradition, the term *Form* is used to translate the Greek term *eidos* and is interchangeable with the term *Idea*. Particular things we experience in the world participate in their eternal Forms, which transcend space and time. By contrast, in the Aristotelian tradition, the forms of things are immanent in the things themselves. From the nominalist point of view, forms have no objective reality independent of our own minds.

formative causation, hypothesis of: The hypothesis that organisms or morphic units (q.v.) at all levels of complexity are organized by morphic fields, which are themselves influenced and stabilized by morphic resonance (q.v.) from all previous similar morphic units.

gene: A unit of the material of inheritance. Genes consist of DNA and are situated in chromosomes; an individual gene is a short length of chromosome that influences a particular character or set of characters of an organism in a particular way. Alternative forms of the same gene are called alleles. The unit of the gene is defined in different ways for different purposes: for molecular biologists it is usually regarded as a *cistron,* a length of DNA that codes for a chain of amino acids in a protein. For some schools of neo-Darwinism, the gene is the unit of selection, and evolution is the change of gene frequencies in populations.

genetic program: A program is a plan of intended proceedings, as in a concert or computer program. The concept of the genetic program implies that organisms inherit plans of intended proceedings; these plans are assumed to be carried in the genes. The genetic program is the principal metaphor through which conceptions of purposive activity and of formative causes are introduced into modern biology (cf. entelechy).

genotype: The genetic constitution of an organism (cf. phenotype).

gestalt: A German term roughly meaning form, configuration, shape, or essence. The term is used to refer to unified wholes, complete structures or totalities which cannot be reduced to the sum of their parts.

habit: A bodily or mental disposition; a settled tendency to appear or behave in a certain way, generally acquired by frequent repetition; a settled practice, custom, or usage. The word *habit* also means dress or attire, as in a monk's habit. In biology, it is used to refer to the characteristic mode of growth or appearance of a plant or animal; and crystallographers refer to the habits of crystals, meaning the characteristic forms they assume. On the hypothesis of formative causation, the nature of morphic units at all levels of complexity tends to become increasingly habitual through repetition, owing to morphic resonance.

heredity: The transmission of characters from ancestors to their descendents. Originally understood in a broad sense which included the inheritance of acquired characteristics and habits of life; restricted in modern biology to mean the inheritance of genes (see Mendelian inheritance, neo-Darwinism). According to the hypothesis of formative causation, heredity includes both genetic inheritance and the inheritance of morphic fields by morphic resonance.

holism: The doctrine that wholes are more than the sum of their parts (cf. reductionism).

holon: A whole that can also be part of a larger whole. Holons are organized in multi-levelled nested hierarchies or holarchies. This term, due to Arthur Koestler, is equivalent in meaning to morphic unit (q.v.).

homoeotic mutation: A mutation causing one part of the body to develop in a manner appropriate to another part: for example, a leg growing where an antenna normally does in a fruit fly.

information: To inform literally means to put into form or shape. Information is now generally taken to be the source of form or order in the world; information is informative and plays the role of a formative cause, as for example in the concept of "genetic information."

information theory: A branch of cybernetics (q.v.) that attempts to define the amount of information required to control a process of given complexity. Information in this narrow technical sense is measured in bits. A bit is the amount of information required to specify one of two alternatives, for example to distinguish between 1 and 0 in the binary notation used in computers.

interactionism: A form of dualism (q.v.) according to which mental events can cause physical events, and vice versa.

Lamarckian inheritance: The inheritance of acquired characteristics. Until the late nineteenth century, it was generally believed that characteristics acquired by organisms in response to the conditions of life or as a result of their own habits could be inherited by their descendents, and both Lamarck and Darwin shared this general opinion. The possibility of this type of inheritance is denied on theoretical grounds by the current orthodoxy of genetics (cf. Mendelian inheritance).

materialism: The doctrine that whatever exists is either matter or entirely dependent on matter for its existence.

matter: That which has traditionally been contrasted with form or with mind. In the philosophy of materialism, matter is the substance and basis of all reality, and is usually conceived of in the spirit of atomism. In Newtonian physics, matter, distinguished by mass and extension, was contrasted with energy. According to relativity theory, mass and energy are mutually transformable, and material systems are now regarded as forms of energy.

mechanics: In its broad, traditional sense, the body of practical and theoretical knowledge concerned with the invention and construction of machines, the explanation of their operation, and the calculation of their efficiency. In physics, the study of the behaviour of matter under the action of force. In the present century, Newtonian mechanics has been substantially modified by relativity theory and has been replaced by quantum mechanics as a method of interpreting physical phenomena occurring on a very small scale.

mechanistic theory: The theory that all physical phenomena can be explained mechanically (see mechanics), without reference to goals or purposive designs (cf. teleology). The central metaphor is the machine. In the seventeenth century, the universe was conceived of as a vast machine, designed, made, and set running by God and governed by his eternal laws. By the late nineteenth century, it was commonly regarded as an eternal machine which was slowly running down. In biology, the mechanistic theory states that living organisms are nothing but inanimate machines or mechanical systems: all the phenomena of life can in principle be understood in terms of mechanical models and can ultimately be explained in terms of physics and chemistry.

meme: A term coined by Richard Dawkins, who defines it as "a unit of cultural inheritance, hypothesized as analogous to the particulate gene and as naturally selected by virtue of its 'phenotypic' consequences on its own survival and replication in the cultural environment."

memory: The capacity for remembering, recalling, recollecting, or recognizing. From the mechanistic point of view, animal and human memory depend on material memory traces within the nervous system. From the point of view of the hypothesis of formative causation, memory in its various forms, both conscious and unconscious, is due to morphic resonance.

Mendelian inheritance: Inheritance by means of pairs of discrete hereditary factors, now identified with genes. One member of each pair comes from each parent. The genes may blend in their effects on the body, but they do not themselves blend and are passed on intact to future generations.

mind: In Cartesian dualism, the conscious thinking mind is distinct from the material body; the mind is non-material. Materialists derive the mind from the physical activity of the brain. Depth psychologists point out that the conscious mind is associated with a much broader or deeper mental system, the unconscious mind. In the view of Jung, the unconscious mind is not merely individual but collective. On the hypothesis of formative causation, mental activity, conscious and unconscious, takes place within and through mental fields, which like other kinds of morphic fields contain a kind of in-built memory.

molecule: A chemical unit. The smallest amount of a chemical substance that is capable of independent existence. Each kind of molecule has a characteristic atomic composition, a specific structure, and specific physical and chemical properties.

morphic field: A field within and around a morphic unit which organizes its characteristic structure and pattern of activity. Morphic fields underlie the form and behaviour of holons or morphic units at all levels of complexity. The term *morphic field* includes morphogenetic, behavioural, social, cultural, and mental fields. Morphic fields are shaped and stabilized by morphic resonance from previous similar morphic units, which were under the influence of fields of the same kind. They consequently contain a kind of cumulative memory and tend to become increasingly habitual.

morphic resonance: The influence of previous structures of activity on subsequent similar structures of activity organized by morphic fields. Through morphic resonance, formative causal influences pass through or across both space and time, and these influences are assumed not to fall off with distance in space or time, but they come only from the past. The greater the degree of similarity, the greater the influence of morphic resonance. In general, morphic units closely resemble themselves in the past and are subject to self-resonance from their own past states.

morphic unit: A unit of form or organization, such as an atom, molecule, crystal, cell, plant, animal, pattern of instinctive behaviour, social group, element of culture, ecosystem, planet, planetary system, or galaxy. Morphic units are organized in nested hierarchies of units within units: a crystal, for example, contains molecules, which contain atoms, which contain electrons and nuclei, which contain nuclear particles, which contain quarks.

morphogenesis: The coming into being of form.

morphogenetic fields: Fields that play a causal role in morphogenesis. This term, first proposed in the 1920s, is now widely used by developmental biologists, but the nature of morphogenetic fields has remained obscure. On the hypothesis of formative causation, they are regarded as morphic fields stabilized by morphic resonance.

mutation: A sudden change. Mutations are observed in the phenotypes of organisms, and can generally be traced to changes in the genetic material. The term *mutation* is now generally taken to mean a random change in a gene.

nature: Traditionally personified as Mother Nature. The creative and controlling power operating in the physical world, and the immediate cause of all phenomena within it. Or the inherent and inseparable combination of qualities essentially pertaining to anything and giving it its fundamental character. Or the inherent power or impulse by which the activity of living organisms is directed or controlled. From the conventional point of view of science, nature is made up of matter, fields, and energy and is governed by the laws of nature, usually thought to be eternal.

neo-Darwinism: The modern version of the Darwinian theory of evolution by natural selection. It differs from Darwin's theory in that it denies the possibility of Lamarckian inheritance (q.v.); heredity is explained in terms of genes passed on by Mendelian inheritance (q.v.). Genes mutate at random, and the proportions of alternative versions of genes, or alleles, within a population are influenced by natural selection. In its most extreme form, neo-Darwinism reduces evolution to changes of gene frequencies in populations.

organicism: A form of holism according to which the world consists of organisms (or holons or morphic units, q.v.) at all levels of complexity. Organisms are wholes made up of parts, which are themselves organisms, and so on; they are organized in nested hierarchies. The parts of organisms can be understood only in relation to their activities and functions in the ongoing whole. Organisms in this sense include atoms, molecules, crystals, cells, tissues, organs, plants and animals, societies, cultures, ecosystems, planets, planetary systems, and galaxies. In this spirit, the entire cosmos can be regarded as an organism rather than a machine (cf. mechanistic theory).

paradigm: An example or pattern. In the sense of T. S. Kuhn (1970), scientific paradigms are general ways of seeing the world shared by members of a scientific community, and they provide models of acceptable ways in which problems can be solved.

phenotype: The actual appearance of an organism; its manifested attributes. Contrasted with the genotype, which is the particular genetic material the organism has inherited from its parents.

physicalism: A modern form of materialism. The doctrine that all scientific propositions can in principle be expressed in the terminology of the physical sciences, including propositions about mental activity.

Platonism: The philosophical tradition that, following Plato, postulates the existence of an autonomous realm of Ideas or Forms or essences existing outside space and time and independently of manifestations of them in the phenomenal world.

protein: A complex organic molecule composed of many amino acids linked together in chains, called polypeptide chains. The sequence of amino acids is specified by the sequence of nucleotides in the DNA of genes. There may be one or more such chains in a protein, and the chains are folded up into characteristic three-dimensional configurations. Proteins are found in all living organisms, and there are many different kinds of protein molecule. Many proteins are enzymes, the catalysts of biochemical reactions; others play a variety of structural and other roles.

preformation: The theory (now known to be false) that the entire diversity of structure of adult organisms pre-exists in the fertilized egg. Embryonic development supposedly consisted merely of the manifestation of this pre-formed structure as it enlarged and unfolded, or "evolved" (cf. epigenesis).

Pythagoreanism: The belief that the universe is somehow essentially mathe-matical. Its fundamental mathematical reality transcends space and time. Closely akin to Platonism.

reductionism: The doctrine that more complex phenomena can be reduced to less complex ones (cf. holism). In philosophy, the theory that human behaviour can ultimately be reduced to the behaviour of inanimate matter governed by the laws of nature. In biology, the belief that all the phenomena of life can ultimately be understood in terms of chemistry and physics. Closely associated with the mechanistic theory, materialism, and atomism (q.v.).

regulation: In embryology, the normal development of an embryo, or part of an embryo, in spite of the disturbance of its structure in some way, as by removing some of it, adding to it, or rearranging it. For example, half of a young sea-urchin embryo will develop into a small but normally propor-tioned larva and eventually into a normal sea urchin.

synapse: An area of functional contact between nerve cells or between nerve cells and effectors such as muscle cells.

systems theory: A form of holism concerned with the organization and properties of "systems" at all levels of complexity. Much of the early inspira-tion for this approach came from an attempt to establish parallels between physiological systems in biology and social systems in the social sciences. The systems approach has been deeply influenced by cybernetics (q.v.). The central metaphor in much systems thinking is the self-regulating machine.

teleology: The study of ends or final causes; the explanation of phenomena by reference to goals or purposes.

teleonomy: The science of adaptation. "In effect, teleonomy is teleology made respectable by Darwin" (Dawkins, 1982). The apparently purposive structures, functions, and behaviour of organisms are regarded as evolutionary adaptations established by natural selection.

vitalism: The doctrine that living organisms are truly vital or alive, as opposed to the mechanistic theory that they are inanimate and mechanical. Living organization depends on purposive vital factors, such as entelechy (q.v.), which are not reducible to the ordinary laws of physics and chemistry. Vitalism is a less far-reaching form of holism than organicism (q.v.), in so far as it accepts the mechanistic assumption that the systems studied by physicists and chemists are inanimate and essentially mechanical.

RESEARCH ON MORPHIC RESONANCE

Awards for Student Research on Morphic Resonance

The Institute of Noetic Sciences is offering awards totalling $5,000 for the best experimental tests of the hypothesis of formative causation to be carried out by students. Awards are offered in each of the following categories:

- Pre-university students
- University undergraduate students
- Post-graduate students

Students' experimental results may either support or go against the hypothesis. Entries will be judged by an international panel of scientists. The closing date for entries is September 30, 1990. For information please write to:

> The Morphic Resonance Research Competition
> Institute of Noetic Sciences
> P.O. Box 97
> Sausalito, CA 94966
> U.S.A.

The Fund for Morphic Resonance Research

Research on morphic resonance at universities is being supported by grants from the Fund for Morphic Resonance Research. You are invited to help promote research on morphic resonance by contributing to this fund. Patrons receive regular updates on research through the *Morphic Resonance Research Newsletter.* For details, please write to:

The Fund for Morphic Resonance Research
BM RESONANCE
London WC1N 3XX
England

INDEX

lucy   common
turkana boy